WRITE TO THE POINT

WRITE TO THE POINT
Effective Communication
in the Workplace

MICHAEL B. GOODMAN

Northeastern University
Boston, Massachusetts

PRENTICE-HALL, INC.
Englewood Cliffs, New Jersey 07632

Library of Congress Cataloging in Publication Data

Goodman, Michael B.
    Write to the point.

Bibliography: p.
Includes index.
1. Commercial correspondence. 2. Business report
writing. 3. English language—Business English.
I. Title.
HF5721.G64 1984   808'.066651   83-19107
ISBN 0-13-971762-5

Editorial/production supervision and
    interior design: Margaret McAbee
Cover design: Judith Butler
Manufacturing buyer: Harry Baisley

Printed in the United States of America

10 9 8 7 6 5 4 3

ISBN 0-13-971762-5

Prentice-Hall International, Inc., *London*
Prentice-Hall of Australia Pty. Limited, *Sydney*
Editora Prentice-Hall do Brasil, Ltda., *Rio de Janeiro*
Prentice-Hall Canada Inc., *Toronto*
Prentice-Hall of India Private Limited, *New Delhi*
Prentice-Hall of Japan, Inc., *Tokyo*
Prentice-Hall of Southeast Asia Pte. Ltd., *Singapore*
Whitehall Books Limited, *Wellington, New Zealand*

For Dora Goodman (1917–1982)

Contents

Preface   **xiii**

**PART I WRITING FOR INSTITUTIONS
AND BUSINESSES**   **1**

1 Writing and the Workplace   **3**

Determining Objectives 4
Analyzing and Developing an Awareness
  of Your Audience 4
Limiting the Topic 6
Brainstorming 7
Writing and Brainstorming 8
Forced Association and Generating Ideas 10
Journal Writing as a Method of Invention 11
Summary 13

**2 The Force of the Right Word:
Style and Writing
for the Working World**   **15**

Recognizing the Qualities of Good Business
  and Professional Writing 15
Achieving Clarity through Intelligent Word
  Choice (Connotation-Denotation) 16

Definition as a Means to Clarity   19
Emphasizing Directness through Active
      Verbs   24
Preferring Active Voice over Passive Voice for
      Completeness and Clarity   26
Conciseness for Economy and Vitality   29
Effect of Language Inflation   31
Concreteness versus Abstraction   36
Levels of Abstraction   37
Dangers of Abstract Language   40
Summary   42

3   **Unwritten Messages in Writing**   **43**

Barriers Between Writer and Reader   44
Developing an Appropriate Tone   45
Positive Expression   49
Impact of the Message   51
Injecting Tact, Courtesy, and Naturalness   52
Summary   53

4   **Clarity and Directness
with Hard-Working Sentences
and Paragraphs**   **56**

Sentence Patterns   56
Structuring Paragraphs for Directness   61
Summary   63

PART II   **TYPES OF WRITING, MEMOS,
LETTERS, PROPOSALS,
AND REPORTS**   **67**

5   **Useful Memos of Substance
and Selectivity**   **69**

Deciding When to Use a Memo—Not Crying
      "Wolf" in the Office   70

Posterior Coverage as a Motive for
    Writing   72
Determining Who Should Receive a
    Memo   72
Evaluating Memos—What to Look For   76
Summary   83

**6   Business Letters:
    Strategy, Structure, and Style   91**

Advice from the Pros   91
Strategy for Letters   93
Categories of Letters   98
Evaluating Letters   106
Parts of a Business Letter   106
Summary   111

**7   Elements of a Report:
    The Theoretical Foundation   120**

Definitions   120
Describing Objects, Mechanisms, Concepts in
    Reports   126
Putting Instructions in Writing   132
Classifying Information   133
Interpreting Facts   134
Arranging Information   134
Analysis Used for Organization   137
Position: The Unwritten Message of
    Order   139
Summary   139

**8   The Short Report:
    How to Get It Written   143**

Putting Ideas on Paper   143
Determining the Subject of the Report   148
Goals of the Short Report   150
Meeting the Reader's Needs   151
Limiting the Scope   151
Planning and Organizing a Short Report   152
Recognizing Unity, Coherence,
    Emphasis   156

Ending a Short Report   158
Revision and Editing: Accepting Your Ugly
    Child   159
Introducing the Report: Beginnings, Introduc-
    tions, Summaries   160
Summary   163

9 **Proposals: Responding,**
   Analyzing, Writing   **170**

Responsiveness   171
Solicited and Unsolicited Proposals   176
The Storyboard—A Dynamic Management
    and Brainstorming Tool   176
Responding to an RFP   179
Content of Proposal Sections   184
Evaluation of Proposals   189
Audience Analysis   194
Expectations and Criticisms   194
Debriefings   195
The Losing Proposal   196
The Winner   197
Summary   198

10 **The Long Report:**
    Research and Documentation   **200**

Preliminary Considerations   200
Selecting a Manageable Subject   203
Developing Ideas: Logic and Argument   206
Accumulating Strong Information   210
Assessing the Material You Gather   217
Primary Sources of Information   217
Graphics   222
Revising and Editing the Draft   223
Documentation   226
Preparing the Final Copy   236
Proofreading   236
Evaluating the Report   238
Summary   240

PART III   PRACTICAL COMMUNICATIONS   245

11   Graphics and Oral Presentations

Graphic Material in Reports   247
Illustrating Your Own Reports   252
Electronic Printing and Graphics   254
Composition of the Text as Graphic Aid   255
Oral Presentations   258
Approaches to Oral Presentation   269
Preparing a Presentation   270
Evaluation Checklist   279(?)
Visual Aids for Presentations   281
Summary   284

12   Writing in the Automated Office   286

Office Automation and Writing   286
Basic Functions of Automated Equipment   290
Advanced Functions of Text Processing Equipment   291
Advantages for the Writer   293
Writers' Reactions   296
Drawbacks of Electronic Equipment   296
Technical Problems Beyond Your Control: Three Personal Experiences   298
The Illusion of Paperless Writing   299
Writing with a Computer   300
Summary   301

13   The Résumé, Letter of Application, and Job Interview   302

Cover Letters That Set You Apart   303
Writing a Résumé That Gets Read, Not Discarded   307

Writing a Résumé   308
Formats for Résumés   310
The Interview   312
Summary   322

**14   Writing Confidently and Clearly
on the Job   326**

Suggestions for Maintaining Good Writing
    Habits   326

**Appendix A Punctuation and
Spelling   329**

Appendix B Sample Report   341

Selected Readings   353

Index   359

Preface

Write to the Point addresses working professionals who must write in the routine performance of a job, as well as college students who are about to enter the job market. Hundreds of books on writing are available, making writing about writing somewhat of a national sport. This text approaches professional writing as a set of skills which, when presented to motivated readers, provides new insights into the key elements of effective writing. This book develops a way of thinking about writing that is appropriate for the professional world. Examples from actual work situations are discussed here to place the theory of writing in a practical context, and numerous exercises are included to strengthen these new skills. The methods discussed here can be applied directly to your writing work tomorrow.

This book grew out of writing seminars and workshops designed for institutions and businesses. The participants were managers, supervisors, and executives from large institutions, as well as small businesses. Their formal education ranged from high school to doctoral work. Most had extensive technical backgrounds—personnel, finance, accounting, nursing, computer science, administration, engineering, insurance, education. A shared need to improve the kind of writing done on the job drew these people together. Not that they could not write; they wanted to write more efficiently, directly, and clearly. They were not looking for formulas, but for ways to improve the skills they possessed to match growing requirements for more and better business communication.

Muddled writing in institutions and businesses is part of a larger problem. It is a symptom of many factors, including minimal emphasis on writing in school curricula for several years past; the absence of a writing portion on college entrance examinations; the audio-visual impact of radio, television, and other electronic media; and social factors related indirectly to the learning process.

To begin to remedy the problem requires a fresh way to think about writing work-related documents. The manager of a state government information agency, and a published author, observed that writing for the official world requires a redirection of skills similar to those used to create short stories. Such an ability to write commercial-quality documents can be cultivated. This book presents writing as a skill that can enhance both the effectiveness of business ideas and the professional image of individuals and their organizations. The materials included here assume a level of quality from today's business and professional writer that goes far beyond the rote skill previously considered adequate. This text presents the process of writing from brainstorming to the presentation of a completed document, including word choice, audience analysis, illustrations, and organization.

Business writers generally ignore the need to generate their raw material in a fresh way in favor of mechanical issues such as punctuation and spelling. You need not be a poet to write a quarterly report, but the steps you take are similar. Many books on business communications promise formulas, but as Brooks and Warren point out in their *Modern Rhetoric,* "formulas for better writing do not exist." Every written communication is different, depending on the information, the context, the writer, and the person who reads the written message. This uniqueness of interactive communication conflicts with a major characteristic of business and government, the need to duplicate effort consistently in much the same way that coffee pots come rolling off an assembly line.

This book discusses writing as an extremely high form of human interaction and emphasizes the process of writing. The importance of invention in writing is not new; it dates back to Aristotle's *Rhetoric.* Not only does this help to clarify the process and improve the quality of your writing, but the discussion of brainstorming methods places the elements of style, organization, and grammar into a context that makes you better aware of the steps required to construct even a routine, uncomplicated memo or letter.

To do this, this text uses letters, memos, proposals, and reports that students and participants in the workshops have freely given me, as well as examples from student papers. References to actual institutions and businesses have been changed to obvious ones to protect the identity of the writers. These examples can serve as case studies. They should provide you with an opportunity to learn from the successes and shortcomings in the writing of others and from your own work, and to apply that knowledge toward revision of incorrect or confusing passages.

This book serves three audiences. First, it can be used as a text in conventional college courses that treat business, technical, and professional writing. Second, it can provide an individual working alone with suggestions and practical exercises to improve writing. Such an individual might be aware of the personal need for improvement in writing, but unable to take time for a formal course. And third, it can be used for a seminar or workshop given within a business, institution, or agency by an existing training and personnel development department.

This text does cover some familiar ground too, although what is familiar is presented differently. For example, the discussion of style appears in the context of audience analysis. Diction is treated as an exercise in selecting the most effective word that accomplishes your purpose and considers the needs of your audience. Other discussions include the placement of information in a document and its effect, the impact that arrangement has on a reader, and some unwritten messages about you that come through to your audience from a piece of writing. The book also presents the research methods required by longer reports and a practical strategy for writing them. This book does not dictate formulas for letters, memos, personnel evaluations, or progress reports. Most institutions have clearly defined formats already, if they have a policy governing written communication. Instead this book offers a way to think about selecting the best and clearest method of organizing information into a report, a proposal, a letter, or a memo.

Business generates new business with proposals, and chapter 9 covers this extensively. In this discussion of proposal writing, the chapter presents storyboarding—a dynamic technique which a group of professionals can use to write a complicated document effectively, yet meet severe time limits. Storyboarding provides a powerful management and brainstorming tool for writing. This chapter includes the unique preparation of proposals, an analysis of a request for proposal (RFP), the expectations of proposal evaluators, and the key elements of a winning proposal.

Chapter 12, Writing in the Automated Office, is new to treatments of writing on the job. It discusses the future of writing in the workplace, providing a perspective on what computers and text processing can do for you as a writer. It gives you a way to think about writing using electronic equipment, and acquaints you with the power of this new technology. It discusses the composition and editorial processes, and considerations of format. In effect, text processing makes you the writer, the typist, the typesetter, and the editor. To perform those four functions requires detailed knowledge of the full process of writing and some familiarity with printing. The chapter does not analyze particular features of individual machines because they differ widely.

Other chapters deal with the oral presentation of reports, graphic aids, audio-visual material, and the use of the telephone.

This book has no theoretical description of English grammar, but it does discuss the importance of correctness as a part of any document. Therefore, effective word choice, clear sentence and paragraph structure, practical letter and memo writing, readable reports, and efficient editing are covered in explanatory discussions and are illustrated by examples. Writing is, after all, something you do, not something you talk about.

Chapter 13, which you may wish to read first, concerns the writing needed to get a job. It discusses the résumé, the cover letter, and advice about the job interview.

The appendix includes a sample research paper, a brief guide to punctuation and spelling, and a list of selected readings for reference to other information.

You can use the practical exercises and sensible suggestions in *Write to the Point* to strengthen your writing effectiveness. Follow the sequence presented, or explore individual chapters independently—although some information presented early in the book could make reading a later chapter easier.

By all means, accept the suggestions presented in the chapters that follow and apply them right away. Do not wait until you cover all elements. Good writing habits begin now.

ACKNOWLEDGMENTS

I do thank the many participants in my workshops and seminars for their comments and suggestions, as well as the many students in my university writing courses for their questions, observations, and examples.

I also thank the following people for their support, criticism and advice concerning the material for this book:

George Spicer, New York University Medical Center;

Lemuel Coley, Dick Levine, Jack Ludwig, Tom Rogers, Pat Silber, John Thompson, Lester Paldy, and Edna Zemanian, State University of New York at Stony Brook;

Kinley Roby, Earl Harbert, Helen Loeb, Northeastern University;

Carl Lebowitz, New York University;

Maresh Gupta, Computer Sciences Dept., SUNY College at Old Westbury;

Kenneth Friedenreich, Saddleback College;

David Lebowitz, University Hospital, SUNY at Stony Brook;

Sylvia Sorrentino, Joan Herman, and George Schneider, United Hospital Fund of New York;

Lenore Beaky, LaGuardia Community College;

Rich Brandes, *Photo Review*;

Carey Horowitz, Hill and Knowlton;

Joe Wetherell, Liberty Mutual;

Tony Pecararo, Oppenheimer;

Jay Mitchell, Jay Mitchell and Co;

Charles Milazzo, Shearson;

Tom Gatten, Human Rights Commission, State of Connecticut;

Cherie Piebes, IBM;

Daisy Maresca, Wang;

Joel Cartun and Thomas Kane, Comvestrix;

Hugh Marsh, Santa Barbara Research Center;

Bill Tebo, Andy Kaufmann, Jack Saxe, and Omar Dreyer, Lee North,

Marilyn Moldovan, Bruce Yberg, Craig Langwost, Glen Kingsley, Mary Ann Thompson, Grumman Corp;

Phil Miller, Prentice-Hall;

Mary Bruno, Joanna Kalinowski, and Judy McCusker, who typed the manuscript; and Cecil Yorke and Margaret McAbee who edited it.

Special thanks to Andy Kaufmann, Phil Miller, Margaret McAbee, and George Spicer.

Of course I wish to acknowledge the welcome support and patience of my wife, Karen, and our sons, Craig and David.

MICHAEL B. GOODMAN

WRITE TO THE POINT

Part I

Writing for Institutions and Businesses

GETTING TO THE POINT: ESSENTIAL STEPS

- BRAINSTORM—IDEAS & GRAPHICS
- DETERMINE **OBJECTIVES**
- IDENTIFY THE **AUDIENCE**
- ANALYZE THE INFORMATION
- ARRANGE DATA FOR MAXIMUM EFFECT
- WRITE A DRAFT OF THE WHOLE DOCUMENT
- ALLOW TIME TO COOL OFF
- PUT IMPORTANT IDEAS "UP FRONT"
- REVISE, EDIT, REARRANGE
- PREPARE THE FINAL DRAFT, INCLUDING GRAPHICS

These ten steps reflect the dynamic process of writing, not a formula for it. Presenting ideas clearly on paper often requires you to perform several steps simultaneously, repeat some steps, or take some out of order. The creative and critical processes necessary to writing are elaborated throughout this book.

Brainstorming, of course, provides you with a written message to shape later. Initially, many beginning writers see only the final steps of the process. They start to write as if they were creating the final product the first time out, rather than thinking of the preliminary efforts required to shape written expression. Brainstorming techniques discussed in Chapters 1, 6, 8, 9, and 10—free writing, questions for generating metaphors, deadlining, forced associations, absurd revisions, storyboarding, dialoging—each allows you to think freshly about an idea and provides you with a tool to use your writing time efficiently. In professional writing the preliminary sketching of graphics that accompany the text should accompany the original

1

thoughts, if only for the simple reason that drawings, charts, and graphs take time to create.

The first written expression of an idea often includes the determination of why you are writing and what you want the finished document to achieve, in addition to clear identification of audience needs and expectations. These issues often occur during brainstorming, and some writers manage them intuitively. The creation, analysis, and subsequent arrangement of ideas depend on the circumstances and the context, which often changes according to the purpose and audience. The following chapters discuss these issues as part of the process of writing.

The initial steps define and clarify the message you wish to communicate. They also demonstrate the complexity of thought required to prepare a draft, and provide a way to identify the true subject. Simultaneously addressing these issues in writing provides you with raw material.

Once you have gathered most of the information that you need, write a draft of the entire document from beginning to end at one sitting, or a detailed outline if the document is to be extensive. This practice allows you to inject coherence into the document, and helps to reveal gaps in the information that must be filled. It also ensures a complete draft of the document. Writers often sketch a few pages, and then run out of steam. They look at what they have written, and then begin to revise, change, and edit the material just created. That is a symptom of writing the first draft as if it were the final one. An awareness and understanding that revision and editing follow a draft as separate activities free you from the debilitating burden of attempting to get the ideas right the first time.

Allow time to cool off between the creation of the draft and the revision, arrangement, and editing of it. Though this seems obvious, it is essential to separate with time the critical and creative functions that apply to writing. The amount of time can be as little as that required to walk to the water cooler, or much longer. No matter how short it is, taking time to cool off ensures higher quality and productivity in writing.

In reworking a whole draft, you can see more clearly the main points that need emphasis, and since professional writing requires early presentation of main points, you can identify and put them "up front."

The following chapters expand, discuss, illustrate, and demonstrate the dynamics of these steps essential to writing. Chapters 1, 2, 3, 6, 8, 9, and 10 address purpose and audience. Chapters 5, 7, 9, 10, and 11 discuss textual and graphic analysis, and the arrangement of information. Chapters 8 and 10 also provide a strategy for managing time, a necessity for putting your thoughts on paper. Revision, editing, and the preparation of the final draft are presented through the discussion of style in the early chapters, as well as in Chapters 8, 9, and 10.

No formula for writing can replace the effort necessary to write well. However, understanding the process can make the time spent writing more efficient and the final document more professional.

1

Writing and the Workplace

Fear presents the greatest barrier to clear and direct writing. But what do you have to be afraid of? Like the Emperor with his new clothes, the mask of confidence and the appearance of professionalism you project in the office might be exposed through clear writing.

By writing clearly you run the risk of knowing that someone will be able to understand the limits of your knowledge and ability. But through clear writing you also begin to understand your own limits. Once you identify your limits, you can choose between two courses of action:

1. seek the additional information and training needed to expand your limits, or
2. cover your tracks by putting up a verbal fog between the material and the reader.

Most poor writing is an unconscious effort to create such a barrier. Very often the muddled language gets between us and our own material. We lose sight of the meaning we sought to communicate and end with obvious babble. For example:

> A detailed knowledge relative to problem areas and approaches to their solutions has been accumulated, with the result that advanced activity programming by individuals in middle- and top-management positions, and the latter in particular, prior to a promotion situation or their initial assumption of their new responsibilities, is indicated and in fact recommended.

Reading this example, who could figure out quickly what this company vice-president had to say? How could people in his organization be promoted if they tried to follow such directions? Unwittingly he has engaged in language inflation, a growing national disease. As far as he is concerned, his memos appear to be authoritative, profound, and knowledgeable. Instead, his draft defies understanding. Who could act on this without calling the VP, or dropping into his office for clarification? Inflated writing from administrators makes subordinates think they are stupid for not understanding the message. In this case communication is not achieved, it is merely imitated.

Often an individual who reaches a responsible position mistakenly uses a bloated vocabulary and an ornate style to impress the audience with the power of the position, rather than to express important information, directions, or ideas. Stated as a general rule:

WRITE TO *EXPRESS* RATHER THAN TO *IMPRESS*.

When you write expressively, you impress your reader with your confidence and control over the material.

DETERMINING OBJECTIVES

Any piece of writing that is effective must have an identifiable purpose, even if that purpose is to cover mistakes. In order to do this, before you begin, write down what the letter, memo, or report should accomplish. Writing a purpose helps you focus your attention more sharply. After all, if you are unsure of what you want your writing to do, then it is a good bet that your reader will be unsure too.

Any piece of writing can entertain, instruct, direct, or influence. Decide which one, or which combination, best suits your purpose. However, most writing in the work place is not intended to entertain the way a novel, play, poem, or short story does. Generally, it reflects management styles: it instructs—tells or consults; directs - tells or joins; requests action, or commands it; or influences the reader's attitude or opinion—sells. The chapters on letters, memos, and reports apply these approaches, and demonstrate the importance of a clearly identified purpose to effective communication.

ANALYZING AND DEVELOPING AN AWARENESS OF YOUR AUDIENCE

After deciding what needs to be said and determining the purpose, identify the reader or audience as clearly as possible. Think of the person the letter or report goes to, and think of other possible readers as well. Then ask youself: What is the best way to present this information to the reader? "You have to be very smart," was the way T. S. Eliot responded to a similar question about what it took to be a great poet. Though he was a bit caustic, he was right. A capable writer should be able to present the same information in different ways. Chapter 2, on style, presents practical methods for developing that versatility.

So your reader can understand easily, present information in an insightful way. To do that effectively, identify and list certain characteristics of the readers' needs, whether or not the reader is real or imagined.

Ask the following questions about the reader in order to identify the important qualities that must be either considered or avoided. It is a useful habit to list information in order to isolate what you already know about your audience, and pinpoint what you need to find out.

- Who is the audience?
- What are the reader's expectations?
- What level of expertise does the reader have in this field, or with this subject?
- What preconceptions do they have concerning the project?
- If the audience includes several readers with different levels of expertise and with different needs, what is the common ground?
- What misconceptions does the reader have on this topic?
- Is the reader technically oriented; business oriented?
- What information does the reader need?
- Would this communication likely find its way to anyone else, a newspaper, for example?
- What is your attitude toward the audience?
- What tone will work most effectively with this audience?
- What do you want your writing to make your reader think, feel, do, or know?
- What does the reader think about you and the organization you represent?
- What are your reader's hobby horses—that is, pet projects, ideas, or prejudices?
- How much convincing will it take for you to get your reader to think, feel, do, or know what you intend?

Answers to all these questions, of course, require some enlightened guesswork on your part, especially if you have no knowledge of your readers and have received no phone calls or letters from them.

In that case, assume that the audience can generally be described as an individual who is intelligent, but uninformed. An intelligent reader can grasp complicated material that is presented in a clear, organized way. An uninformed reader does not know exactly what you wish to say, or how you wish to apply a principle to a specific instance. Our intelligent, but uninformed reader will always serve as a model when you are unable to find sufficient information about an audience.

Whenever you write, always have some reader in mind. These individuals may be in at least four general areas: 1) in your own group; 2) near your group in the company; 3) in another part of the company; 4) outside your organization. Single out a representative individual and write as if directing the message solely to that person. When you think of another human being holding your writing and reading it for information, then you have developed the right awareness. When you have an individual in mind, it is much easier to incorporate an effective consideration for your audience in the document.

Every piece of writing has a beginning and an end, and, for your reader, nothing precedes the beginning. And nothing follows the end. What you write must present your reader with all the information necessary to clearly understand your message, idea, or explanation.

How do you gather information about your reader? You can check published sources, if the individual or group is well known. You can observe the reader from first hand experience. You can gather information from others who have written to the same audience, avoiding their mistakes and embracing their successes. You can analyze their previous correspondence, as well as telephone and personal conversations with a particular group. The last way to find out about your audience is old-fashioned intuition and guesswork. After all, writing is a skill, not an exact science, and the intuitive aspects of writing make it an art beyond a quantifiable skill.

Every piece of writing has some artistry. Even the communication of technical or business information can be shaped in a readable and artistic manner.

LIMITING THE TOPIC

Identifying the purpose of the message and the type of audience who will receive it helps to determine the limits of the topic, and sharpens the focus. But several mechanical limits can, and should, be considered to limit the scope of any message. A properly focused discussion, rather than one that rambles, makes your writing more powerful by riveting your reader's attention to the information you have chosen. Focus, then, requires that you make many choices after you have gathered as much information as possible. Focus is another element of the writer's craft, that of shaping material. Implicitly, the act of shaping material requires that you be comfortable with what you know and be able to throw out a prized idea or well-turned sentence because it does not fit exactly and would cause confusion rather than clarification.

If the purpose is to inform, the focus should include as much information as possible at the expense of other types of comment. If the purpose is to interpret, then the focus should not be on just the facts, but on a discussion of their meaning. This gives you the opportunity to use facts in such a way that the presentation performs interpretation. All dramas and fiction work this way, drawing the audience toward an inescapable conclusion by the ordering of the material. (The ordering and organization of this type are discussed in the chapters on short and long reports.)

The audience also affects the kind of limits placed on a document. For example, your organization may be considering the purchase of new word-processing equipment for your office. The president, who is a marketing whiz, wants to know about the uses of the machine to generate mass mailings that appear personal and avoid the feel of a form letter. The auditor wants to know the figures on the cost, training, maintenance, and useful life of the system. The office manager might be concerned with the reaction to the new system and with the format of the training. The computer systems analyst will want to know some facts about the machine itself and how it will be made comparable with the existing data-processing equipment. Each of these readers who make up the audience would also like to know something about their particular area.

The document can gain focus in several mechanical ways:

- time
- space
- geography
- demography
- economics

To limit a topic, you can arbitrarily put a date in your title, for example, 1977–1979, making the subject more specific and limited. You can discuss an organization and use space, the top ten floors of the main office, for instance, as the principle that gives automatic focus and limitation. You can talk about the Northeast, the West, or even Maine and New Hampshire or Oregon and Washington, in an effort to limit the topic on geographical principles. Or you can talk about Americans between 25 and 40 to limit your discussion demographically, or households with $20–30,000 incomes for an economic focus. Functions of departments—sales or accounting—can also be a basis for the limitations placed on a report.

Whatever principle is indicated by the analysis of purpose and audience, limiting the topic is essential to powerful, clear, and direct writing.

BRAINSTORMING

Years ago, teachers of rhetoric called this element of writing *invention*. Now many people call it prewriting, or the more familiar term, brainstorming. It is almost in the literal sense a storm in your brain, the rapid discharge of electrical impulses associated with thought. Since the writing process is twofold—creative and critical—brainstorming is essential in creating something to say in a novel fashion, or in a direct, clear manner. Much of what is written in business and in institutions is routine. The creative aspect of the process is slighted in many treatments of writing for the professions and for the real world. The assumption, and I think a false one, is that the writer has already thought about the topic in a new way and can express the idea, but needs editing to get out the deadwood. Every writer has this problem, and this text offers several practical methods of editing your writing to make it more powerful. However, the creative part must also be stressed, even in the most routine of communications. In fact, the more routine the communication, the greater the need for a fresh presentation, and that calls for a reevaluation of perceptions.

Now, you ask, what does this have to do with writing about debits and credits, or the hospital's public relations campaign? Reevaluating perceptions sets you up for shifting your thinking a few degrees so that you can look at the same information from a different, fresh point of view.

Brainstorming, then, is an essential first step in the creation of the final product, separate from the critical steps of revision and editing. The tendency to create and be critical at the same time has been called the Dangerous Method. That method seeks to write it right the first time. But as Professor & writer John Thompson advises, "Don't get it right, get it written. Then get it right." The meaning is clear. The creative and the critical aspects of writing should and must be separated.

Brainstorming sometimes can be very chaotic, but often out of the chaos comes information that is on paper—the form in which your communication will finally be presented. Some of us have difficulty dealing with chaos, but in generating information, chaos can give us new ideas and insights. It may also lead to confusion, but at least the message is written. After all, whether called brainstorming, prewriting, or invention, the process is a way of placing ideas on paper so you can look at them and begin to shape them.

Try this demonstration for example. Without paper of any kind take the letters A, B, C, D and say out loud the combinations of these four letters. Do not repeat the letters, just give the 24 possible combinations.

How far did you get? Seven? Ten? More than ten shows that you concentrate very well and retain information. Most of us have trouble with more than seven; we begin to lose track of the combinations we have already mentioned.

Now take your pen or pencil and write the 24 combinations, timing yourself.

Once you see the pattern that the combinations take, the task is very easy and should have taken you five to seven minutes at the most. What is a nearly impossible feat of memory, becomes a routine exercise because writing expands our senses, and places the information in front of our eyes. As Walter Ong explains in *The Presence of the Word,* writing makes us a visual culture and we depend on the written word for reaffirmation of the truth that was spoken. The visual makes the abstract real in our minds. The visual also makes abstract information manageable by giving it a concrete form that we can manipulate physically.

Thus, the brainstorming practice is one that not only gives you thoughts in physical form, but it also helps you generate ideas. Out of that chaos, often insightful ways of perceiving old material come through.

WRITING AND BRAINSTORMING

Another very easy and useful brainstorming exercise is to write for ten minutes as fast as you can without taking your pen off the paper. That may not seem like a long time, but remember that you must write constantly, no pauses to go back over your material for corrections, no looking out the window for inspiration, no pondering the choice of words. If you get to a spot at which you cannot find the right word, leave a space. If you are stuck and cannot think, then repeat the last word that you wrote. If you have no place to go from there, then place three words at random on the paper. Place three words that go with those three, and so on. Soon you will have either a long list of words to triple, or you will have jarred something lose in your mind and started the momentum. When the well seems to have run dry, try writing from another perspective. Free writing exercises fill up the page with material.

Let me demonstrate the technique for you. In preparing a draft for this chapter I decided to write for 15 minutes in a more or less focused way about what kinds of topics should go into a book on writing well on the job. This is what came out:

- purpose of the report
- limits of information
- audience preferences
- hobby horses
- attitude of reader to poor writing
- effect of poor writing on reader
- effect of unclear sentences on read(er)
- errors and lost time and money—rods in Nuke reactor
- variety of sentences
- formats and readabîlity—sample here
- presentation of information
- tact, saying no with style and grace
- courtesy, what courtesy makes your reader feel about you
- clarity, what a well thought-out idea written clearly says about you as a professional
- directness, getting to the point makes your information clear
- consider attitudes, yours, your reader
- expectations of the writer, is this letter going to change the course of Western Civilization?
- will this report, letter, memo be readable?
  (FIVE MINUTES)
- can I expect results, should I?
- has the message been clouded by jargon, claptrap, gobbledygook, bureaucratese
- can I cut words out?
- can I sound knowledgeable with this
- can I use verbs for information and clarity
- concrete words and ideas
- organization,   step-by-step

                dramatic   chronological, emotional

                simple—complex

                complex—simple

                spacial

                     space   i.e. rt. to left

                                     left to write

                time

                area

                logic, key words
- presentation of material—clear, readable, concise
  (TEN MINUTES)
- reader—awareness of the complexity of the subject of writer; awareness of what is good writing
- good writing appears to be effortless, it appears to be something that makes perfect sense to anyone who reads it and it can express the most complicated ideas & information in a clear readable fashion

- effective writing—has impact because of the preparation that goes into it—
- effective writing has impact from its appearance of authority & factuality, its correctness, its presentation
- concentrate on content first
- the package is not important once it is opened—it helps someone to open the book, report, letter, but the reader needs the information, a 4-color cover cannot make shoddy preparation clear, readable, and direct.
- get to the point quickly in business writing

(FIFTEEN MINUTES)

That's 15 minutes and when I finished writing my arm was tired, as it should be if I had written the whole time, not taking the pen off the paper for long.

I hope that the point of this kind of brainstorming is obvious. The material must be put in a concrete form, on paper. Once it is there, it can be expanded, contracted, improved, or discarded as inappropriate. Brainstorming articulates the idea that seemed so promising as it rolled around in your head all week, but looks a bit ragged on paper.

Again the place to begin, if you plan to end with a piece of writing, is with words on paper. The place to thrash out your thoughts is on paper, or if you are lucky enough to have one, a word processor. (See Chapter 12 for a detailed discussion of their use in the Automated Office.)

Brainstorming often helps you to go beyond your perceived limits. It can give form to half-formed ideas that, unless they are on the paper, can lull you into thinking they are fully formed. Once the idea is on paper, it can be shaped. Since we are still a culture heavily influenced by the visual, and since we do not stress the powers of memory and the techniques of oratory as much as in times past, an idea is much easier to work with if it is written. If not, we will have difficulty later with its written presentation.

Putting ideas on paper helps you remember the ideas you had; otherwise you may forget them in order to go on to another item. This forget-remembering sets you free to concentrate on the next item that you wish to discuss. It is the same principle used to manage time in business, the making of a schedule or list of items or tasks. The process works well for writing and promotes efficient production every time you write, no matter how short or how routine the communication is.

FORCED ASSOCIATION AND GENERATING IDEAS

The free writing demonstrated here is not the only way to brainstorm. Arthur Kilcup, a retired owner of an advertising agency, suggests that you put down all the words you can think of that have something to do with your topic. Then when the well has run dry, force yourself to put two words that go with it next to that word. Take that one and force two more words. For example:

- brainstorming—thinking, revealing
- revealing—considering, open
- open—see, understand
- write—wrong, rite
- rite—steps, ritual

Notice that the pattern breaks after *understand*. That really is the intention of this technique, to *force* an imaginative leap. You begin to associate words with the topic faster than you can record them. The break in the pattern indicates a break in routine associations and the creation of fresh ones.

In this short space we have put ritual together with brainstorming, an association that we ordinarily do not make. But once the words are visible, we can make the connection that we need to perform some sort of brainstorming ritual to generate a topic for discussion, or to discover a new way to perceive of it. These two methods, free writing and the forced association, may seem mechanical ways of starting off, but that is exactly the point. Writing most often begins with that kind of hard work.

JOURNAL WRITING AS A
METHOD OF INVENTION

It is a gross misconception to think that writing always begins with a flash of insight. More often than not the flash of insight comes along during the process of writing. One of my students began a journal entry, which is a form of free writing, by asking herself about a report on subways that she had to write. Writing that she had not opened her eyes to the subway during all the years she had been riding them, and in fact had taken them as some vague event in her life that had no fresh qualities, she wanted to find a way to discuss the subway ride freshly. By the end of the entry she had discovered, through the act of writing, that the way she wanted to discuss the subject was not as an ordinary ride after all. Instead, she used the word *trip,* and that generated some light, a revelation of sorts. The approach to the topic was that of an adventure, a trip in the greatest city in the world on a form of transportation unique and fascinating, if we look at it that way.

QUESTIONS FOR STOKING
THE FIRE

Asking yourself a series of questions can stimulate the thought process, and force you to think about the subject in a new way. It is similar to walking the same street from the same direction every day, then walking the same street from the opposite way and opening your eyes and ears to the objects and sounds as if they were no longer ordinary. Physically changing your point of view has often the effect of changing your mental image, and thus your mental point of view.

For example, if you wanted to write about abstract notions such as inflation, honesty, communism, or freedom, the following questions might help in two ways by getting you to see the concepts from a different point of view and by forcing you to come up with some idea of the subject.

- What color is it?[1]
- What is its shape?
- What would this shape look like in motion?
- Give a distorted, biased definition of this concept?
- Imagine that this word did not exist. What kind of tribe, group, or nation would there be without the word?
- What would this word look like if it were a place?
- What animal does it remind you of?
- What person is it like?
- Who would fall in love with it if it were a person? What would they have for children?
- What does it smell like?
- What does it sound like?
- When you touch it, what does it feel like?
- You have heard of a French kiss, what kind of kiss would this make?
- Pillsbury has its doughboy, CBS has a big eye, what kind of logo would this abstraction have?
- How would you address a letter to this abstraction?
- If you were to meet this abstraction in your office, what would your reaction be?
- Imagine this abstraction pitted against another, bigger one. Name the bigger abstraction. Name a smaller one.
- What would this abstraction do if it were President?

The considerations we have discussed so far—purpose, limiting, audience, brainstorming—belong to the invention process. *Invention* is necessary in business writing even more than in creative or journalistic writing. It is essential to be able to look at a routine report freshly. It makes your reader see the information more clearly, and, therefore, use it more efficiently and with greater understanding. It also allows you as the writer to understand and control your thoughts more easily and quickly. Instead of wandering back and forth in your office, or making busy work for yourself so that you avoid the job of writing, these preliminary exercises free your creative energy so that you can get something on paper and out of your head, on a page where it can be shaped and arranged. The four techniques discussed, and several others presented in Chapters 6, 8, 9, and 10, provide an effective and productive beginning.

Arrangement is the next element of classical rhetoric. Then comes style. We will deal with arrangement in the chapters on memos, letters, and reports. We will take up matters of style in the next chapter. While reading the suggestions that follow, remember that they are part of the shaping process. Brainstorming preceded that step.

[1]These questions are similar to ones found in Peter Elbow, *Writing with Power*, N.Y.: Oxford University Press, 1981, pp. 81–93.

SUMMARY

- Clear writing reflects organized thought; poor writing often sets up a verbal fog between you and the reader.
- An effective message has an identifiable purpose.
- Consider in detail your audience's needs and expectations.
- Limit the scope of a written document through: time, space, geography, demography, economics.
- Brainstorming, or invention, is an essential step in committing thoughts to paper. Use these techniques: free writing, forced associations, a journal, questions for generating ideas, and others discussed in Chapters 6, 8, 9, 10.

EXERCISE 1–1

To determine your purpose, answer these questions:

1. What should happen as a result of this message?
2. Is this for the records?
3. What information is given or requested?
4. What recommendations should be included?
5. What impression should the reader have after reading it?
6. Is writing the best method, or would the telephone or a personal visit be better?
7. Should this entertain, instruct, direct, or influence?
8. Should this tell or sell?
9. Should this invite cooperation?
10. Should this offer consultation?
11. After reading your writing, what could the reader do that is not desirable?
12. What decisions should be expressed?
13. What actions must be specified?
14. What needs to be included? What excluded?

EXERCISE 1–2

Here is a letter that turns down a customer's request for a refund or replacement of a defective product. The customer's response to that letter follows. Comment on the way both letters address the subject, as well as the reader.

July 5, 19XX

Dear Ms. Frie,

We are in receipt of the brush you returned to us for refund due to your dissatisfaction with our product.

We are aware that many various types of hair exist and due to this, we have over the years developed a very extensive line of hair brushes to suit the individual needs.

Our products are of the best quality and we have a one-year guarantee for all our items if damaged during normal use.

Your mistake in judgment in choosing a brush for your hair is not our responsibility and does not warrant a refund.

Very truly yours,
FAIRHAIR PLASTICS, INC.

Angela Herberg

P.S. We have, under separate cover, returned your brush.

July 8, 19XX

Dear Ms. Herberg,

We are very disappointed with your company policy. There was no question as to the quality of the hair brush. Since stores do not allow a customer to try out hair brushes there is no way of knowing how well the product will work.

Ellen is fifteen years old and spent one week's allowance to buy the brush. I assured her that most business would refund money if a customer was not satisfied with the product. I was mistaken, unfortunately. However, we all must learn from our mistakes. We will never purchase another Fairhair product and we will never recommend them.

Sincerely,

Eliza Frie

EXERCISE 1–3

Use the questions in this chapter to determine the purpose of a letter, memo, or report that you must write.

EXERCISE 1–4

Analyze an audience that you do not usually write to. Identify them by answering the questions presented in the chapter.

EXERCISE 1–5

Create several audience analyses. For example, use the same information, but decide how you would present it for: your immediate supervisor, your division, the company president, people outside the company.

2

The Force of the Right Word: style and writing for the working world

The qualities of good writing remain the same whether you are a systems analyst or short story writer. Communication of an idea is difficult enough without needless complications. Good writing should be clear, direct, and simple. Unless we are aware of these qualities, in a piece that is well-written, we may overlook them. If we are struck by the ornate character of a piece of writing, and not the message it intends to communicate, we have an indication that what we are reading could use some changes.

RECOGNIZING THE QUALITIES OF GOOD BUSINESS AND PROFESSIONAL WRITING

Writing done for the world of work is judged by the same criteria used to assess the quality of novels, stories, and plays. The difference is the intent of the communication. Most work-related writing does not, and should not, try to entertain through narrative. It should present facts clearly, directly, and simply. Many writers in business and institutions make the mistake of using a style that is more complicated than is necessary.

Effective writing is simple.

However, simple does not mean simplistic, since complex relations require that all shades of difference and agreement be laid out. Advances in electronics and other technologies have taught us that simplicity is more efficient, dependable, and economical; and, therefore, better. In writing, too, simple is better.

Good professional writing gets to the point.

Words are used purposefully, not wastefully. A letter, memo, or report is written with specific goals, and each word, phrase, sentence, and paragraph contri-

15

butes toward meeting that goal. Directness and clarity make the difference between writing that communicates, and mere words on the page.

In addition to clarity, simplicity, and directness, business and professional writing:

- addresses a specific audience.
- exhibits a clearly defined purpose that serves as the foundation for the entire piece.
- uses language that is clear, concrete, quantitative, and familiar.
- employs tact and courtesy, as well as directness.

ACHIEVING CLARITY THROUGH INTELLIGENT WORD CHOICE (CONNOTATION- DENOTATION)

Any piece of writing gains clarity through carefully chosen words. Words have discrete meanings, or denotations. They have a range of meaning, or connotation, as well. Knowing the range of a word's meaning—its connotations, in addition to its denotations—allows us as writers to choose the right word that not only fits the *purpose* of our writing, but is also appropriate for the *audience* we address.

As our culture moves more and more toward the specialization of professions, and even finer specialization within a discipline, we have generated special words to discuss what we do and think. Many of these words cannot be avoided in writing. They are an irreducible core of technical terms that must be used. In a discussion of computers, *hardware, software, bytes,* and *bits* cannot be avoided. However, the rest of a written message can be made clearer for your audience if a simple word or phrase is chosen, not a simplistic expression.

To give writing power and clarity, use language that is understandable. Since the volume of technical terms is growing, this suggestion for improving your writing seems a good place to begin a discussion of the critical process of writing. The exercise below will test your versatility and also develop your ability to express the same idea less ornately. As William Strunk and E. B. White observed in *Elements of Style,*[1] "Rich, ornate prose is hard to digest, generally unwholesome, and sometimes nauseating." They directed, "Do not be tempted by a twenty-dollar word when there is a ten-center handy, ready and able. Anglo-Saxon is a livelier tongue than Latin, so use Anglo Saxon words. . . . " Of course White and others who suggest the simpler words as a way to achieve clarity and vigor in writing, know that no word in the English language is good or bad, only appropriate or inappropriate for the context. When and where to use a particular word or phrase depends on you as the writer

---

[1]From Strunk, William Jr. and E. B. White, *The Elements of Style,* 3rd edition. New York: Macmillan Publishing Company, 1979. Used by permission.

developing an ear. To sharpen your ear read books and essays. In this way you become better able to judge when a word is appropriate for your context, purpose, and audience. I would like to tell you that developing this sense—when a word is correctly used in context—is easy to learn and master. It is something we recognize first, before we are able to put it into a document. In other words, both learning and mastering the part of the craft take some time and experience.

Look carefully at the following list of words taken from reports, memos, and letters. You see these words every working day, and you may have read them in the work of writers you respect. However, they could be simplified. Try to find a simpler word or phrase that substitutes for the word or phrase in the list. For example;

finalize	—	end
visualize	—	see
absolutely essential	—	essential

EXERCISE 2–1

Supply a simpler word or phrase for each of the following examples:

absolutely essential	utilize
plan ahead	visualize
advance planning	impact (verb)
initiate	affect (verb)
activate	effect (noun)
endeavor (noun)	function
endeavor (verb)	implement (verb)
facilitate	personnel
ambulatory	prioritize
abeyance	parameters
finalize	human resources
hereby	enclosed herewith
eschew	merge together
i.e.	alternative
e.g.	vital
pursuant	attractive
herein	alleged
herewith	actualize
and/or	program
input	software
output	package
interface (verb)	viable
hands-on	upscale
technology-intensive	game plan
scenario	outplacement
maximize	bottom line
	mode

The first three examples in the list seem innocent enough, but if we look at them closely they all exhibit a common attribute of institutional writing. Each of the examples presents the same notion in both words, a tautology, or needless repetition. "Absolutely essential" only requires "essential" for its complete meaning. The word itself implies its absolute quality. English is full of words that cannot be qualified in this way. "Pregnant" provides a good example. A woman cannot be somewhat pregnant. "Plan ahead" and "advance planning" are other examples of the writer losing sight of the meaning of the words themselves. "Plan" implies the future, making the use of "ahead" and "advance" repetitious. Such repetition also marks an inflated style; but the reader could also take such writing as an insult to intelligence, as well as an indication of the writer's inability to control thoughts.

Another group from the exercise list is the legal jargon that has oozed into ordinary letter, memo and report writing. "Abeyance," "pursuant," "alleged" are fine words, but can often confuse your reader, presenting an air of legalism which sometimes creates an undesirable distance between you and your reader. Such legalisms in ordinary writing present an attitude which is less than friendly and cooperative. The words *"suspend"* or *"set aside"* for *abeyance, "according"* for *pursuant, "believed"* or *"charged"* for *alleged* work better in ordinary communications by giving your writing more power, an appearance of openness, and a friendlier, more relaxed tone.

Just as put-offish are the "-ize" words in the list: *finalize, utilize, visualize, actualize.* More than 30 years ago George Orwell in his classic essay, *Politics and the English Language,* discussed these soft verbs cloned from nouns. While some such verb makings are creative and fresh—Midasize for auto mufflers, Calvinized for Calvin Klein jeans—most are unnecessary dilutions of the solid Anglo-Saxon words that they intend to replace. No one finishes or ends a task anymore, they finalize the plans. No one uses a tool, they utilize existing technology. We no longer see or make, we visualize or actualize. Although the "-ize" words sound as if the writer knows what the message is all about, the words are part of our great national disease, Language Inflation, the persistent use of bloated verbiage. Such language presents an impressive front: but, like a stage set, it is only front—no substance is behind.

The cure for the disease requires an awareness of the power still available in ordinary language. Even the most casual survey of newspapers articles and prose fiction reveals the use of common words, but in an uncommon way. Powerful writing, good writing, has that quality. It can be read by anyone in our ideal audience, that is, anyone intelligent but uninformed.

Law is not the only profession that affects the words we use in writing. Computers have also made their contribution to the language of the workplace. *Input, output, software, and/or,* and *interface* are some of the exercise examples that have shouldered out solid, workable language to describe the exchange of ideas and information between people. *Interface* is the most mechanical of the terms. Originally electrical engineers used the term to describe the physical connection between two electrical systems of different designs, so that they could be joined. For example, two different telephone companies with different equipment needed a compatible device to make them work together. That was called an interface. The same thing is used in computers so that an IBM can exchange data with a Burroughs or a Univac. That seems like a very mechanical word to use to describe the interactions of people. Those sentiments already exist in even clearer form in the words *reconcile, exchange, communicate,* or in the phrase *"create harmony."*

The word-splice of *and/or* is the linguistic equivalent of a plus-minus sign ($\pm$). That is a very helpful symbol if you intend to show a range of tolerance in a specification of machined parts, or the accuracy of statistics. But the word *and* means, if we are talking about two items only, both; *or* means one or the other. Not both. What is your reader to think if confronted with such an ambiguous situation? Of course the reader has to stop and figure out what the choices could be. The and/or device, intended to shorten a sentence, does that very well, but it creates possible confusion by placing ambiguity before your reader. Ambiguity is fine if you are drafting a screenplay, or writing a short story, but in writing for business and industry, ambiguous terms place a barrier between your reader and the idea you want to communicate.

The words in the exercise list can provide a lively discussion at work or at home. The list is not complete, and you might add examples. Take some papers from your file and look at the inflated or ambiguous words you may have used every working day. Try to think of direct, clear alternatives for them. Thinking of words in this way develops in you as a writer the important habit of considering every word carefully. Nathaniel Hawthorne, in a review of Edgar Allen Poe's short stories, commented that the stories were powerful and unique because "every word tells." Each word carefully placed and chosen makes writing clear, readable, and vivid.

DEFINITION AS A MEANS TO CLARITY

Since we have discussed the powerful effect of strong, common, ordinary words, we should also mention briefly the definition of words. Often you must use technical terms, words that capture the meaning of much material related to the topic which would require several words or long phrases to convey the same information. It makes sense to use such terms, but they must be defined at some place in the paper so that your reader will know exactly what you mean when you use them.

The more a field deals with technical information, the more you must be sure that the words you use are defined carefully for your audience. If you are an engineer, that does not mean that every engineer will grasp your meaning immediately. Your message may be read by electrical engineers and mechanical engineers, as well as by business managers and information processors. Such diversity in your audience requires you to define terms that might be unclear to your readers.

What should you define? Words that need definition in a report can be classified in four ways. The first of the four needs the least definition.

Types of Words That Require Definitions

- *Common words in ordinary context.*   If you use the word *table,* for example, in its ordinary dictionary sense, no formal definition is needed. Such a word for a common thing requires little clarification. A more complicated word, "trust," for instance, might need some added definition if it is used in other than its ordinary sense to designate a free and honest relationship between people.

- *Common words in an unusual context.*   Everyone knows what a collar is. However, if you are referring to a piece of machinery like a lathe, the collar is not so familiar and must be defined briefly as a metal restraining ring with an adjustment screw for tightening. At an airport the apron is not what you put on when you go into the kitchen to create dinner; it is the paved surface that surrounds the taxiways and runways used by aircraft.

- *Unusual words in a common context.*   This category could be avoided. However, it falls under the use of the jargon of your profession in order to signal not a pretention, but rather your knowledge of the language of the field. No self-respecting biologist writing for other biologists calls a *Chrysanthemum leucanthemum* a daisy or a *Taraxacum officinale* a dandelion or a weed for that matter. Such words can be defined very briefly as I just did, by providing the familiar equivalent in a parenthetic expression.

- *Unusual words in an unusual context.*   Words of this type often require extended definitions; that is, the terms are often complicated and demand definitions that are illustrated with examples and cases. Before we understood the unconscious mind, Freud had to define the term extensively because it was a term no one knew, and referred to something no one had encountered. Terms like that are often the subject of articles and books used to explain the terms in full. Often textbooks of subjects such as accounting, electrical engineering, or linguistics can be considered, in part, as definitions of unfamiliar terms.

Once you have identified the type of word or words you must define, the incorporation of the definitions into a piece of writing can be made in at least four ways: in context, in the text, in the notes, or in a glossary.

By far the most effective methods of definition come in context and in the text. If it is possible, define a term with a word or phrase near the term as it appears on the paper. That is the most desirable method, because it allows your reader to continue the flow of the thought without pausing. If, however, the term requires a bit more explanation, a sentence or two can be effective without interrupting the flow.

The use of footnotes for definitions can clarify statements when properly phrased. If highly technical information needs to be defined, often that is best provided in the notes. Be careful, however, not to abuse this method of definition. As is often the case, inexperienced writers use a footnote to carry on part of a discussion, creating confusion for the reader as well as the appearance of incoherence. Notes used sparingly can provide definitions which present helpful examples and illustrations.

A glossary is a vocabulary of specialized terms with short definitions limited to the special uses. As a means of providing brief definitions for a large number of terms, a glossary works well in long articles and books. However, in short reports or a letter avoid a glossary because its use could seem pretentious and heavy-handed to your reader.

Analysis of Definitions

We have mentioned what needs to be defined, but we have not analyzed a definition. To do that we should first try to define a simple term. Take a piece of paper and write a sentence defining the word "spoon" as it is commonly used as a noun. Try to make your definition as brief as possible. Even a phrase will do. Some responses from participants in my seminars and classes have been: "An eating utensil." "A wooden or metal bowl at the end of a long stem used for eating." This definition question was used by the Army during World War I as part of its Alpha series of tests to determine who went to the frontline trenches and who was enrolled in officer candidate school. The responses above, sadly, would have landed the writer in a muddy trench along with thousands of other doughboys.

What the Army was looking for was an individual who could write a definition that had its two necessary parts, conforming to the classical formula. Any formal definition has these parts:

$$species = genus + differentia$$

The "species" is the term itself. The "genus" is a word or phrase that identifies the word as a member of a larger group, and the *differentia* shows how that particular term differs from all the other members. So a simple term like "spoon" required a *genus*, such as "eating utensil," but also required a *differentia* that included enough information to show differences. That is why "used for eating" would not do. That definition works equally for knives and forks. The type of food, liquids and semi-solids, like soups, stews and mashed potatoes, had to be included to make the definition different from that of knives and forks.

Of course the Army now uses a different method of finding candidates for leadership training, but the idea back then was to find individuals who could identify and analyze the complexity of the ordinary. When I use this exercise the members of the group who provided inadequate *differentia* look defeated. I explain that the test question was more relevant in the early part of this century when schools taught definition as a part of the core courses in rhetoric. Definition was the key to Socratic rhetoric. In the dialogues of Socrates, the method of questioning responses was an exercise in defining and identifying the *differentia* in discussions of abstract terms such as beauty. We have let the classics slide a bit, but their discoveries about the art of human communication still hold true.

EXERCISE 2–2

Now that we have discussed definition, try your hand at the word *quixel* in the following sentences. Think of a word or phrase that would be a useful cognate for the term.

1. This new *quixel* is expected to improve our record keeping by 50%.
2. Several different *quixels* are offered for sale by IBM and Burroughs.

3. No matter what the doctors say, this *quixel* will become part of every section of this institution.
4. Ten years ago it took a small room with special temperature controls to house a *quixel* with the capacity of this desk-size *quixel*.
5. The *quixel* has made memory and thinking obsolete.
6. The *quixel* keeps people from developing their true personalities.
7. He finally got all of his adolescent irresponsibility out of his *quixel*.

In trying to find a suitable word that covered all seven instances you might have discovered the way words are defined in most languages. Of course, you could grab your dictionary and look up a word to find its meaning, but *quixel* does not appear in the dictionary. Chances are the first use of the word might have suggested from the context a word like "process," the next "computers," the third "procedures." Four, five, and six could also mean "computer". At this point the first and third examples, reviewed and reevaluated could fit into a meaning for "computer." Such changes reflect the dynamic, living, nature of words which are defined by their context and use. Satisfied that computer is the correct choice, an attempt at the seventh sentence creates confusion.

How can "He got all of his adolescent irresponsibility out of his *computer?*" The use of the word computer in the context does not make sense. He can take out frustration or immaturity on the computer. He can even act irresponsibly with the computer, but his adolescent irresponsibility belongs to his personality, and that is something that cannot be programmed and fed into the machine as input, at least not yet.

What word, then, would work with the seventh example that would also apply to all of the others? What word, as we know it , would work just as well in all seven contexts without distorting its meaning, and thus causing a great deal of confusion on the part of the reader? What about *system?* As a cognate for *quixel,* system fits. It is so broad and general that it incorporates each of the concepts suggested by the sentences.

In trying to find a word that works well for each context, we have defined the word actively, generatively. We generated a word that fits its common uses. In this case the seven uses were slightly different, but nevertheless they were ones that fit the meaning of the word. Words are defined in this way in normal practice. A word and its use exist, and we set about to define its meaning. Since language is living, this accounts for changes in the meanings of terms, as well as of their connotations and denotations, over time.

If we think of words as having an existence that preceded dictionaries, and one that is constantly shaped by conversation and writing, rather than one that is carved in stone and set for all eternity within dictionary pages, then we begin to picture language and the definitions of words as tools. In the same way a traveler uses a map to drive from New York to Chicago, so a writer uses a dictionary to learn what words are available when he comes to a crossroad of meanings. James Joyce, the modern novelist, considered the dictionary to stand for tyranny since it was regarded by most people to be immutable and static. He listened to the language as it was used, and knew it to be living, changing, and fecund. We do not have to be novelists or geniuses to perceive of the language we use the way Joyce did, even for routine types of writing.

Now that many routine communications are done for you by others in a large organization through form letters, writing demands a creative ear and eye. All communications that are not routine require arrangement and consideration for your audience that call for a unique response. This is the antithesis of much of business which is characterized by acts and procedures that can be repeated over and over. A form letter does not fit every professional situation. The kind of awareness we have mentioned would help you respond in writing to the unusual situation, and allow you to generate a new form letter to fit an emerging general pattern.

Connotation and Denotation

Since we have taken up definition, another consideration in choosing the proper word is the connotation-denotation of the word. Denotation is the literal or dictionary definition, the most precise and objective meaning that is possible. Connotation, on the other hand, is the word and its associations. Take, for example, the simple word *home*. The denotative meaning of the word mentions a shelter, a valued place, a haven of happiness and love, one's close family, original habitation. However, the connotations of the word call up a far warmer picture. In the Northeast it may be a cozy snowy night by the fireplace, kids all around reading magazines on the floor, dad smoking a pipe, mom reading, the dog curled up next to the kids. Such a picture might not be yours exactly, but ordinarily the connotation of the word suggests warmth and security, safety, and happiness. The words of a popular song have a line—a house is not a home—which reenforces the connotation of home as a lot more than just a shelter.

In writing, the awareness of the connotation and denotation of words is an important factor in the choice of words. For example, words that have mostly denotative meanings for hospital personnel, both medical and nonmedical, have strong connotations for patients and their relatives and friends. The words "hospital," "operations," and "cancer" often explode in the ears of ordinary people, but working in a hospital removes the mystery of such words for most employees. The same is true for auditors for the IRS. They audit all day, every working day. Most of us, on the other hand, are never audited, but the word itself causes a great deal of anxiety. The IRS was so aware of this that they instituted a writing program for their auditors that reminded them of the connotations of their letters to taxpayers.

"We knew that, in many situations, we must be firm and specific, as well as courteous.

"Our language, however, often communicated more than firmness. It caused many honest taxpayers to feel that we thought *all* taxpayers were searching for ways to defraud the Government, and to wrongfully avoid paying their taxes. Many construed what we thought of as firm statements to be threats" (*Effective Writing, IRS Workshop*, p. 17).[2]

---

[2]Washington, D.C., no date. Doc. #T22.19/2:W93/3.

EMPHASIZING DIRECTNESS
THROUGH ACTIVE VERBS

Solid, or active, verbs such as plan, begin, merge, build, launch, end, determine the difference between mere lists of words on paper and effective documents. Verbs bring life and power to writing, a vitality often lost in the officialese of offices, institutions, or government. The verb provides a firm foundation for the rest of the sentence. Without it an idea, no matter how powerful, falls flat. Ordinary corporate and government writing—what Richard Lanham calls "The Official Style"—present bland, often mushy and muddled ideas that may be quite simple. But a collection of empty adjectives—important, vital, viable—and soft verbs—is, be, utilize—can reveal unintended meaning. Often these soft verbs are so ill conceived that they are grammatically incorrect; they may be plural when the subject of the sentence is singular, or they may have the improper ending to indicate past or future tense. This occurs not because the writer is ignorant of grammar, but because flabby verbs have such diffuse meanings that the writer can easily lose the idea in a tangle of modifying words. And if the writer loses control over the meaning, the reader will certainly have a difficult time deciphering the material.

Active verbs add vitality and power to prose. They provide a strong base for clearly constructed sentences. Carefully chosen verbs also make a document easier and more efficient to read and understand. In spite of this, most writing on the job accents the noun and uses an "is" or a "there is" or "there are" construction (rather than emphasizing a strong verb). As a result, the noun style often leaves the Earth for a while, even though, as many working people confess, the style sounds very impressive. However, as we will see in the following exercises, such a style can result in imprecision, wordiness, lack of clarity, misunderstanding, and downright falsehood.

Carefully selected verbs provide an alternative to the doughy prose. After all, English is so rich in verbs that it is wasteful not to use them. Think, for example, how many different ways you can verbally express the act of moving your body from one place to another. You can *walk, run, amble, skip, hop, saunter, jog, shuffle, lope, stroll, cruise, bump, stumble, canter, trot, sneak, slip, slide, skate, glide, dash, sprint,* and so on. Of course, each one of these verbs has a different connotation, and that is why each verb has the potential to say a great deal.

The first step in building up your awareness of verbs is to identify nouns that have been cloned from good solid verbs. We can recycle these soft nouns easily, and give strength and power to some of our old, flabby sentences.

EXERCISE 2–3

Change the following nouns into usable verbs:

1. administration
2. employment

3. authorization
4. performance
5. illustration
6. implementation
7. documentation
8. information
9. determination
10. conclusion
11. identification
12. negotiation
13. development
14. medication
15. producibility

Take, for instance, the third example, *authorization*. We usually see the word in a context such as, "Authorization was granted by the vice-president." Why not recycle that noun into a verb—*authorize*. Once we have a clear verb, we can build a sentence: "The vice-president authorized $70,000 for an extension of the research project." Or take *information* as an example, "Please send me the *information* on the project." With a solid verb the sentence becomes "Please *inform* me of the goals and expectations of your project."

Of course, you might ask, the use of a solid verb requires that you have specific information to communicate. That's right. The use of the verb as the foundation of the sentence assumes that you have something to say. The unwritten message that your reader gets from verb-based writing is: "This person really knows what the topic is all about." The reaction to noun-based writing is: "What the hell is he trying to say to me?" or worse, "What the hell are they trying to hide?"

If we can form any general rules from this they would be:

- Use verbs instead of soft nouns for clarity and strength.
- Avoid overuse of the verb *to be*.

Use of the verb *to be* often indicates a noun-based style.

"Is" Sickness

To diagnose a piece of writing for symptoms of "IS" sickness, select a random page or several pages. Count the number of sentences in the sample, and then count the number of "is" verbs. If the number of "is" verbs divided by the number of sentences results in a fraction of ½ or 50%, the "is" sickness has taken a firm hold in your writing. At 75%, major surgery could save the communication. One hundred percent results in death by boredom. Such writing can be unreadable even to you, the writer.

To cure "is" sickness, replace "is" with another verb. This may take some time initially, but once you form the habit you can construct sentences around the verb from the start. If you heed no other advice in this book but to begin building sentences on verbs, your writing should improve quickly in clarity and strength.

PREFERRING THE ACTIVE VOICE OVER THE PASSIVE FOR COMPLETENESS AND CLARITY

Since we are discussing verbs, we should mention the use of active voice and passive voice. Active-voice verbs indicate the subject acting and require something, an object, acted upon. Passive-voice verbs indicate the subject acted upon, and usually place the actor in a phrase beginning with the word *by*. For example:

Active Voice: The cook dropped the pot of hot soup.

Passive Voice: The pot of hot soup was dropped by the cook.

EXERCISE 2–4

Change the following passive-voice constructions to active:

1. Following the emergency procedure, the stranded elevator was evacuated by the chief of security.
2. The film was developed by an alternate method.
3. The policy guidelines must be revised within a week.
4. A sharp increase in infection was noted.
5. One hundred work hours a week are saved by this method.
6. Admissions should be scheduled by the clerk.
7. Implementation of these guidelines should begin as soon as possible.
8. The regulations require that the research must be conducted by technically qualified personnel.
9. Under the new regulations, a minimum standard for competency will be set by the state.
10. Feedback would be hampered by the complicated forms.
11. To avoid the possibility of confusion, the reports should be signed and dated by all persons involved.

The use of the passive voice marks modern baffle-speak, or the official language of business, institutions, and government. It is an outgrowth of the legitimate uses of the passive voice in science and occasionally in journalism. The passive voice is used correctly when:

The doer or actor is not as important as the action. For example:

John Smith mixed equal parts of HCL and H_2O as the first step in the experiment.

This sentence, according to the requirements of objectivity in reporting test procedures, provides a less desirable construction than:

Equal parts of HCL and H_2O were mixed as the first step in the experiment.

In writing about a scientific experiment, objectivity becomes a major goal. The passive voice shifts the focus of the sentence, grammatically, away from the actor toward the action. Of course, in writing about an experiment, the experiment itself is the most important event.

Another alternative to the passive voice would be:

Mix equal parts of HCL and H_2O as the first step in the experiment.

This variation is an active voice construction and may sound very familiar to you. It is the command, the imperative mood of directions. Each of the three methods has a use appropriate, of course, to the audience, the purpose, and the context of your written statements.

The passive voice is also used correctly when the person responsible for an act cannot be identified easily.

For example, a newspaper article might state:

"He said a city bus has been assigned to transport the walking injured to a hospital for examination and treatment."

We would have difficulty finding out who issued the order. In this case we could, but the act is more important than the individual who authorized it.

As another example, this newspaper account of a bombing raid uses the passive voice in the third sentence.

"People panicked and fled. Cars careened through the streets. The wail of ambulances was heard for hours."

Though the first two sentences are active-voice constructions, the actors are vague—people, cars. To reenforce the impossibility of identifying both the individuals who fled as well as the drivers of the cars, the information is reported in an indirect way. But the wail of the ambulance, as the story implies, fell on the ears of many of the people in the city. We cannot identify them either. In this instance the people cannot be identified and the use of the passive voice is an acceptable method of communicating the observations of the reporter.

Who Kicked the Dog? Danger of the Passive Voice Habit

But what of the habit of using the passive voice? What if we wanted to find out who was at fault, or who was in charge? Take Example 4 in Exercise 2–4 above: "A sharp increase in infection was noted." Who did it? From the sentence, we cannot be sure. The passive voice eliminates the doer of the act, and that is its biggest danger. The passive voice can, in the hands of a deft speech or report writer, cover up the actor.

The particular person may not be identifiable, so, as an alternative, we could identify a group to which that person might belong. For example: "Researchers noted a sharp increase in infection." This is an active voice alternative. The active voice makes us aware of the reality of an act being done by someone or some thing. Ask yourself: Who did it? Or better yet: Who kicked the dog? Don't be afraid to flush out members of your organization who hide behind the passive voice.

Advantages of the Active Voice

On the other hand, the active voice also makes writing more readable by making it more dramatic. Its structure creates, as closely as is possible in words, a picture of the action even if the action is a mental one. In Example 5 above, most of us would have written: "One hundred work hours a week are saved by this method." Compare the blandness of that sentence (which does answer the question Who kicked the dog?) with this alternative: "This method saved one hundred work hours a week." Not only is the active-voice construction clearer and crisper, it is more economical. It saves two words, or approximately 18%, by reducing the 11 words to nine. An 18% decrease in an entire paper saves time and money in writing, typing, printing, and mailing. A hundred-page report reduced by 18% becomes 82 pages. If you have no other reason for using the active voice, savings of time, money, and energy should prove convincing.

The active voice has obvious advantages over the passive. It is:

- more readable
- more accurate, closer to the event
- more economical
- livelier
- more human
- more aware of verbs
- more concerned with the doer of an act
- more conscious of the details of an action
- more knowledgeable—who kicked the dog?

A drawback in using the active voice concerns the use of personal pronouns. The sixth example in Exercise 2–4, "Admissions should be scheduled by the clerk." can easily be written as: "The clerk should schedule admissions." That may be fine for one sentence, but in writing policy guidelines the repetition of the word *clerk* becomes monotonous. We want to use the pronoun *he.* But what if clerks are male and female? An inexperienced writer uses he/she, a solution that is worse than the problem. The word *clerk* could be repeated rather than slashing the sentence with a bisexual pronoun. A more credible alternative would be to use the plural, clerks, and use the pronoun *they* which carries no gender designation. Or you could do what others have done, like Peter Elbow in his book *Writing with Power*, use *he* in some sections and use *she* in others. Personally, the pronoun referring to clerk, though it may be *he,* appears to mean in that context a reference to the genderless word *clerk*. However, the best judge is the attitude of your reader or audience toward the use of *he* as a universal pronoun to mean men and women. After all, men and women exist, and no amount of fiddling with the language will change human physiology very much. Only our perceptions of ourselves change. We could avoid the whole problem by using the passive voice, but for more powerful writing, make every attempt to shun the passive voice as a solution to the gender question. A better alternative would be to focus on the actions the person must do, as you did in reworking the example, "Feedback would be hampered by the complicated forms," to read, "The complicated forms hamper feedback."

EXERCISE 2–5

Use the active voice to express the following ideas. Rewrite these sentences, building each on an active voice verb.

1. Necessary flow times and milestones are plotted and coded to correlate with a Work Breakdown Structure (WBS), and are the result of interaction among Manufacturing, Engineering, Test and Material personnel.
2. Production Detail Schedules are then generated.
3. High concentrations of airborne solvents within the work area are required by regulations to be removed.
4. Upon completion of the design, the subsystems will be fabricated and tested to insure that they meet all design requirements.
5. Meeting schedule dates will be coordinated between the customer and XYZ Corporation.
6. The interrelationship of all tasks, consultant interfacing, and manpower assignments will be coordinated and approved by Mr. Johns.
7. MNO-North America Corporation is serviced by a network of computer systems and services provided by Data Systems Corporation and Kalk Data, Inc., both subsidiaries of the MNO Corporation.

CONCISENESS FOR
ECONOMY AND VITALITY

In addition to using the active voice, another quick way to provide vitality to your writing is to make it more concise. A concise expression is economical and clear. The official style we use in business and institutions patches ideas together with bailing-wire and chewing-gum phrases. Many such phrases are out there and we use them, forgetting the more vivid words and phrases they replaced. Conciseness is allied to the words you choose. Although the official style of business and industry favors the bloated style, it is imprecise and longwinded. Nowhere is this inflation of the language more evident than in the use of prepositions, the short words that alone have little concrete meaning, but gather their meaning from the words they join. Prepositions bind thoughts together and give the reader signals concerning the meaning. Official style phrases have inflated ordinary working prose so that it sounds more important than it is, or worse, sound pretentious and phony when something important needs expression. Exercise 2–6 below will jar your memory, dulled by the drone of officialese.

EXERCISE 2–6

For each phrase below, write one word that is a clearer, more concise, and more economical expression. For example:

> in terms of—concerning
> in the event—if
> at the present time—now

due to the fact that	in the event of
during the time that	in the field of
in the neighborhood of	as to
subsequent to	as to whether
prior to	in a number of cases
through the use of	by means of
in order that	in order to
the manner in which	in a position to
in lieu of	in the case of
adjacent to	dotted-line responsibility
to the extent that	back-of-the envelope estimate
at this point in time	it has been brought to my attention
as of this date	submitted his resignation
state of the art	in conjunction with
in terms of	in short supply
within the realm of possibility	for the reason that

These phrases may baffle you at first. "What is wrong with them?" you ask yourself. "I use them all the time. I see them in every letter, memo, and report that crosses my desk. These phrases are the mark of an educated professional," you say. All that is true. These phrases are also the spongy foundation of bureaucratic language and techno-speak. Nothing is wrong with them, but their effect, however, makes the message that is written less clear and precise. Such phrases avoid direct confrontation with communication by hedging. In research findings and in scholarship, hedging gives the reader a sense of fairness and objectivity. But the trappings of objectivity without its substance and even-handed discussion produce confusion and disbelief.

EFFECT OF LANGUAGE
INFLATION

Such writing is needlessly wordy. In the examples above, the phrases "at this point in time" and "at the present time" could, and should, be replaced by the single word *now*. The phrases are the result of legal-speak that has crept into ordinary discourse after the endless television hearings on Watergate, Koreagate, Abscam, and the daily parade of officials interviewed on television. Commericals act the same way, puffing up language with what classical scholars called hyperbole—best translated into ordinary, plain English as *hype*.

The worst impression that an imprecise and inflated piece of writing can present is that of *hype*. Your reader resists your information and your point of view immediately. In any technical field you must use the technical terms. Engineering, physics, medicine, and business—each has a vocabulary that makes informed discussion possible. Such special terms also indicate that the writer knows the field and expects that the reader is acquainted with the basic jargon of that study. However, the need for a complicated and specialized language requires even more that the rest of the communication be written clearly. One place for improvement is in the use of prepositions to make clear, concise connections between the elements of a piece of writing. For instance, an inflated phrase like "in the event that" can be replaced by the powerful two-letter word "if." The substitution of one word for four results in a 75% reduction. Think of the economy. Fewer words and letters to type makes a physical difference. But the real economy comes when we realize that "if" gets to the point, making the sentence clear and direct.

EXERCISE 2–7

Try rewriting the following examples to make them more concise.

1. After intense research and investigation, we decided that several corporations, businesses, and firms offer appropriate and/or reliable laboratory equipment which will upgrade the efficiency of our facility.
2. The employee health-care benefits program offered by most businesses in the New York area is a health-care program which now includes many health benefits not covered under previous programs.
3. The diagnosis of a problem resulting in surgery constitutes a crisis in a patient's life. The stress is increased if it is the result of a malignancy.
4. Several factors contribute to the run-off problem, and these factors will be discussed throughout this report.
5. There are several points that I am concerned about in regard to this problem at the present time.
6. It is the responsibility of each and every department head to properly arrange the affairs of this organization in such manner that each salaried employee, including himself, will receive the full vacation to which he is duly entitled.

7. We solicit any recommendations that you wish to make and you may be assured that such recommendations will be given our careful consideration.

8. The unanticipated rebound in consumer spending in the third quarter of 1979, which more than regained the ground lost in the second quarter's decline, has combined with the shift in monetary policy announced in early October to change significantly the configuration of economic activity during 1980, although the full-year results will not differ greatly from the forecast issued in September as part of the long-range plan background material.

Dr. Fox, the Professionals, and Inflated Language

More than eight years ago in the United States several psychologists wanted to see how a group of educators, social workers, and psychologists would react to a guest lecturer, Dr. Fox. The doctor lectured for one hour, and answered questions for another half-hour on "Mathematical Game Theory as Applied to Physical Education." After the lecture, the audience was asked to fill out a questionnaire on the substance of what they heard and to rate the effectiveness and expertise of the person who delivered the lecture. The results showed that the participants considered the lecturer to be very knowledgeable about the subject, and the lecture clear and stimulating.

Dr. Myron L. Fox was, however, no more a doctor than the old snake-oil salesman on the Texas frontier. He was an actor hired by researchers to fabricate a lecture, using articles from the *Scientific American,* and to present the material with double-talk, illogical references, meaningless statements, and unrelated comments, punctuated with jokes.

The Dr. Fox Hypothesis, according to J. Scott Armstrong who tested the phenomenon on written communication, might be expressed: "An unintelligible communication from a legitimate source in the recipient's area of expertise will increase the recipient's rating of the author's competence."[3] In other words, the pithy aphorism that made the college rounds and now shows up on small desk posters in offices—"If you can't dazzle 'em with knowledge, baffle 'em with B.S."—has some validity. It is a symptom of our inflated language, as well as the effect language has on us as readers and listeners.

The Language Inflator

Often we are delighted by the sound of what we hear, and the substance is either lost or misunderstood. If it is misunderstood, most folks think they are poor readers. To demonstrate the tendency of readers to accept the words as presented, let's see what happens when we intentionally and randomly generate a phrase. To do this we'll use a Language Inflator.[4] All you have to do is think of any three-digit number; 437

[3]"Unintelligible Management Research and Academic Prestige," *Interfaces,* Vol. 10, No. 2, April 1980, pp. 80–86.

[4]This is my updated version of "The Systematic Buzz Phrase Projector," attributed to Philip Broughton. It appears widely in *Newsweek,* May 6, 1968, p. 104; Richard M. Eastman's *Style: Writing and Reading as the discovery of Outlook,* 2nd Edition (New York: Oxford University Press, 1978), p. 107; and in *Effective Writing for Executives* (New York: Time-Life Video, 1981), p. 85; and elsewhere.

will do nicely. Then apply the number to the columns below—first digit to column A, second to B, third to C.

A	B	C
1. professional	1. executive	1. package
2. alternate	2. effective	2. analysis
3. maximized	3. management	3. utilization
4. essential	4. functional	4. contingencies
5. integrated	5. efficient	5. capability
6. aggressive	6. cost	6. parameters
7. critical	7. preventive	7. systems
8. crisis	8. evaluative	8. concepts
9. intensive	9. reciprocal	9. data
0. total	0. high-tech	0. predictability

If we take our random number, 437, we get the phrase "essential management systems." That sounds very impressive when you read it. It has the ring of authority, the sound of confidence and knowledge. It has all that until we realize that we generated the phrase by chance through a randomly selected three-digit number. When we think of the term having been produced in this way, it loses some of its impact. It is made up of three very general words, terms so general that every reader might think of an "essential management systems" differently. To some it might be a secretary and a typing pool, or an arts and graphics department, or an accountant, or an actuary, or a data-processing system—computers, word processors, telephone lines. The phrase is so vague as to be meaningless. It got that way through Language Inflation. That we think such bloated phrases are good writing is a symptom of the disease of poor thinking and poor communication.

The Language Inflator can be used to create a verbal fog that cannot be penetrated. It is the language that Dr. Fox uses in a lecture to be impressive, but devoid of substance. That is a great danger to you as a writer. If you are unaware that you are writing a phrase totally devoid of substance, you cannot impress your reader. The writer must always be aware of his intention, even if that intention is to misinform, confuse, or to generate vague, unclear references.

A good writer tries to be as clear as possible, but realizes that readers react to confusion in two general ways. One, readers do not like to admit misunderstanding what you have written. To do that might suggest their own lack of knowledge, in addition to their reluctance to challenge authority. Second, if your readers misunderstand you, they might assume that you are "good" with words, or that your ideas are so complex that they cannot understand your ideas, or that you communicate ideas well.

Under most circumstances complex ideas require clear, direct, and uncomplicated expression. In writing, complexity works against clarity—not that modern style cannot accommodate intricate ideas; it does. But if you have to read a paragraph several times to figure out what it says in order to understand the substance of a concept, then that is a clue to you that the writer has lost control of the language.

In some instances the vague, the general, the impressive surface without substance is used intentionally. Most advertisements do not come out and say: "Use these designer jeans, they will change your social life." The ad writers use words like "action" and "look" to demonstrate that the jeans make you desirable. In political speeches, the language is intentionally vague so it is as inoffensive as possible to most people. It has the sound of authority without communicating the substance of authority. The danger in each of these instances is that the language can be used to subvert the attitudes of people who believe in the power and authority of the written word.

Power of the Printed Word

Let me give a nonpolitical example of the power of the printed word, and how that power is magnified when it comes from a reliable source.

The billing department of a national pet products manufacturer made an error in an invoice. One of the company's warehouses in California needed 20 batteries, the kind that are about the size of an ordinary office desk and weigh about 500–1000 pounds, for its lift trucks. Since the company was national it stored all its batteries in a warehouse in the Midwest; for the sake of the narrative, call the place St. Louis. When generating the original invoice, the key punch operator accidentally entered 2,000, not 20, into the computer. The faulty invoice went through to the St. Louis warehouse. There the dispatcher noticed that 2,000 was a tall order for his operation to fill with the trucks in the lot, so he rented several tractor-trailer trucks to carry the load to California. He was overjoyed to clean out his stock of batteries, because he needed the floor space for other items. The order filler, the managers, and the truck drivers never questioned the order. Why should they? The computer does not make mistakes, and, besides, if there were a mistake, someone would have caught it. The programmer forgot to put a flag by an order of this size for that item, however.

The warehouse manager, dispatcher, biller, order filler, drivers, managers in the computer center, the warehouses, the motor pool—all acted. No one questioned the validity of the order. But somone should have phoned and asked California whether they wanted 2,000 batteries, instead of ordering a fleet of trucks.

This example is common. As a demonstration of the power of words, it is dramatic in its ordinary occurrence. It would have been just as easy to use the story of the nuclear power plant that ordered 10-foot rods for its reactors and was delivered 10 rods, each a foot long. That misinterpretation cost millions; the mistake, according to *The Wall Street Journal,* was so costly in time, money, and prestige that it is still classified information.

Ambiguity In Writing

Just imagine a sentence of only five words. Imagine, too, that each of those five words could have only two meanings. In this instance, if each of these words could cause confusion, then the possible number of meanings goes up with each word. But it does not double each time; it goes up exponentially. For the first word, 2; for the second word, 4; for the third, 8; for the fourth, 16; for the fifth, 32. So in a sentence of only five words, each with only two possible meanings, we generate not 10 possibilities for confusion, but 32. It makes sense to try to eliminate as much compounding of confusion as possible.

One way to eliminate possible confusion is to be careful of the placement of information within a sentence. The position of one word in a sentence makes a very great difference in meaning.

EXERCISE 2–8

In each of the sentences that follow one phrase is in a position that is inappropriate to the intended meaning. The result makes each sentence have ambiguous meanings. Rewrite them so that the meaning is clear, not ambiguous.

1. To start the equipment, the starter switch must be flipped.
2. The new billing policy was put into effect in two months that improved collection performance by 50%.
3. The director spoke to the manager with disturbing harshness.
4. Based on the recent evaluation of the hospital by the Commission, the safety director revised the shift change procedures.
5. While on the floor, changes in the vacation schedule were discussed by the employees.
6. Having gone through years of training, the pathologist's findings carry special significance in court cases.
7. The maintainability goal is to accomplish all scheduled internal inspections within 30 minutes and to perform 80% of all corrective actions on the machine.
8. To receive full credit for laboratory tests, the forms must arrive at the insurance company within 60 days.

In general, the phrases above are misplaced. Their meanings dangle before the reader who, having read English for some time, expects phrases to be placed physically as close as possible to the words they modify. Other languages use inflections—endings of words—to indicate where they belong in a sentence. Thus, placement in such languages is not as crucial as in English.

To avoid placing modifying phrases improperly, read a sentence aloud, trying to make the ambiguous connections your reader might make. If you can anticipate such connections, you would not allow a sentence such as number three above to stay in your work. You would edit the sentence so that, instead of implying that the manager possessed disturbing harshness, you would write: "The director spoke with disturbing harshness to the manager." This revision eliminates ambiguity and makes the sentence clearer and more direct.

Ambiguity in expository writing, such as the writing done on the job, can lead to confusion, duplication of effort, and misunderstanding. Because of this, avoid ambiguity in business writing. Ambiguity is useful in creative writing, where it adds depth and suggests other connections in the reader's mind. Ambiguity used skillfully expands a reader's perception but when used clumsily it ridicules serious meaning. Straightforward prose is more useful for business letters, reports, and newspapers.

We have discussed many techniques that can improve your style if you consciously apply them now to the next letter or report you have to write. The use of active verbs, the active voice, the choice of plain English words, the use of concise words and phrases, the correct placement of modifiers can quickly improve your writing by making it clearer, more direct, and more readable. Another element that can improve your writing, starting now, is the use of concrete, quantified, specific language.

CONCRETENESS VS. ABSTRACTION

In commenting on a contemporary's theories of religion and philosophy, Ralph Waldo Emerson said that the individual did nothing more than build "air castles." The image Emerson evoked mocked a language and an idea that had no connection with the Earth. It was about as much protection from attack as a castle in air would be.

Albert Einstein searched for the concrete solution to many abstract problems. He was reported[5] to have said,

> "Is not the whole of philosophy like writing in honey? It looks wonderful at first sight.
> But when you look again it is all gone. Only the smear is left."

As a physicist concerned with the physical world, he would select concrete, specific language in which to discuss such a system. You would think that a manager in the working world would do the same, but practice favors, wrongly and for the wrong motives, an abstract expression in instances where a concrete one would do.

In William Strunk, Jr. and E. B. White's classic *The Elements of Style*[6]—the "little book" used by numerous writers—the authors demonstrate the greater impact of concrete language over the abstract:

ABSTRACT	CONCRETE
A period of unfavorable weather set in.	It rained every day for a week
He showed satisfaction as he took possession of his well-earned reward.	He grinned as he pocketed the coin

Even the most causual comparison of the two versions of these sentences shows the vividness and power of the concrete sentences to make us picture the event, to see

[5]Quoted by Timothy Ferris in a review of *Reality and Scientific Truth: Discussions with Einstein, von Laue and Planck* by Ilse Rosenthal-Schneider (Detroit: Wayne State Press, 1981) in *The New York Times Book Review,* July 19, 1981, p. 12.

[6]Strunk, William, Jr. and E. B. White, *The Elements of Style,* 3d edition. New York: Macmillan Publishing Company, 1979. Used by permission.

with our minds. Concrete language makes us see rain and a smile; abstract language does not.

Concrete, specific, quantified language provides your readers with perceivable references, avoiding the air castles.

In the exercise below, make each statement more concrete and specific. Assume any details you think necessary to make the sentences present a clearer picture.

EXERCISE 2–9

1. In the event of a personnel interruption, the vital functions of this facility will be executed by administration and management.
2. The manager purchased several pieces of word processing equipment at an inflated price.
3. The director responds to the needs of subordinates.
4. EDP has eased many functions of various institutional departments.
5. Funds for marketing research are coming from unexpected sources.
6. To control costs and assure schedule compliance, BancService Co. will utilize corporate systems already planned and in use on other programs.
7. Project costs will be collected utilizing the company's existing cost reporting system.
8. Status review meetings will be held bi-monthly and a report of these meetings will be issued per enclosed schedule.
9. Utilizing existing disciplines and techniques the customer and MNO, Inc. are provided the means of controlling the timeliness and quality of program data.
10. Laboratory technicians must maintain performance and skill levels compatible with standard practice.
11. Supervisory personnel must meet customer needs within the parameters of institutional capabilities.

LEVELS OF ABSTRACTION

Before we discuss some sample responses to the exercises, let's examine the levels of abstraction that nouns fall into. The first level of perception interprets the object wordlessly as a constant, shifting pattern of atoms and electrical charges in a state of flux. From this constant shifting pattern, still perceiving wordlessly, the object is not the word but the thing our perception abstracts or chooses from that pattern of atoms and electrical charges.

The next step names the object—"Joe." We must understand that the name is not the object; it stands for it but omits many of its characteristics. Discussing the separation between the thing itself and the words used to represent it Linguist Alexander Korzybski made this pithy statement: "The map ain't the territory."

In the next level "Joe" becomes a *person. Person* represents the characteristics—arms, legs, hair, teeth, eyes, nose . . .—which we have abstracted from specific individuals. We can show the differences if we label them by numbers: 1, person; 2, person; 3. . . . Characteristics of specific people—a mole on the left cheek, for example—have been left out. The next level sees Joe not as a person but as an *employee*. When we consider Joe as an employee, we see only those characteristics that he shares with other people who have a job.

We can also consider another step up the ladder of abstraction to *personnel*. When we use a term such as this we refer only to what is shared with all other members of a group. Any differences, no matter how broad, are ignored.

If we take another step up, we can consider Joe as a *human resource*. In this light the human being has almost lost humanity to the perception of people as a commodity to be used in much the same way we refer to natural resources such as water, coal, and trees.

Considered as a commodity, a human resource can easily be considered as the next step up the ladder—WEALTH. At this level all references to the original object, that we considered wordlessly, have been blurred beyond recognition. If we wished to find the original pattern of energy in the word *wealth,* we could not without some extremely good guesswork and a great many clues. Deducing *man* from *wealth* would baffle even the most astute paperback detective.

Going through these levels of abstraction is not a purposeless exercise, but rather demonstrates the effect that words have on our thinking about and ordering the world we live in. In the post-modern era of electronic data processing, marketing, and other services, we have lost touch with the concreteness of the physical world. When we do that, the mental world becomes more and more like those air castles that Emerson described more than a hundred years ago.

But concrete language can be dangerous to use. If you have little or nothing to communicate, or what you have to say is trivial, specific language would reveal that naked truth. Perception through words can be manipulated, and contemporary business and industrial writing often results in a vague, cloudy, inflated construction that cannot be deciphered, often not even by the writer of the message.

Take the example—"The Hospital purchased several pieces of operating room equipment at an inflated price." On the surface we get the message that the hospital paid too much. Put country simple, they were ripped off. But other questions come to mind. What did they buy? How much did they pay? How much should they have paid? The answers to these questions require that the writer know some specific information. For example, the kind and number of pieces purchased, the price paid, and the price commonly charged on the market. To rewrite the sentence more concretely we might say: "Memorial Hospital purchased a respirator and a dialysis machine for use in its operating room for $3,000 and $5,000, or approximately 15% more than other hospitals in the area have paid. For instance, University Hospital paid $2,500 and $4,200 for the same models." Although this revision is longer and required two sentences, it provides the reader with substance instead of a vague message. The original construction would have been, we hope, augmented with other sentences to make it meaningful to the reader.

The fuzziness of language as the result of the common use of abstractions instead of concrete terms has been with us for a long time. The classic discussion of it appears in George Orwell's "Politics and the English Language,"[7] published right after World War II. Two passages from his essay might help to demonstrate further the effect of abstract and concrete language.

In the following sentences used as examples in "Politics and the English Language," decide which qualities in each are desirable and undesirable. Write your observations. Decide which example is the more concrete.

> "Objective consideration of contemporary phenomena compels the conclusion that success or failure in competitive activities exhibits no tendency to be commensurate with innate capacity, but that a considerable element of the unpredictable must invariably be taken into account."

> "I returned and saw under the sun that the race is not to the swift, nor the battle to the strong, neither yet bread to the wise, nor yet riches to men of understanding, nor yet favor to men of skill; but time and chance happeneth to them all."

While considering these examples, try to picture the idea or the event. For example, think of a "contemporary phenomena." Try to see that thing as it "compels a conclusion." Just what, exactly, did the contemporary phenomena do to bring the conclusion about? What was the phenomena? In group discussions of this example I ask people to select a contemporary phenomena. Some of their responses include: basketball, taxes, work, dating, marriage, ice cream, meetings. The abstract quality of the phrase becomes clear when we realize that the two words could represent such a variety of events. When we try to translate the abstract back to the concrete, we have difficulty.

The second example, though a bit stiff for its use of *nor yet* and *happeneth,* has terms that we can picture—race, battle, bread, riches, favor. Only time and chance are abstract, but we have the race, battle, and so forth to give us a picture of the meaning of the abstract concepts. In the first example the word chance appears as "a considerable element of the unpredictable." The six-word phrase has replaced a single word, and the idea of time has been eliminated.

If you have not been able to guess already, the two passages are discussions of the same idea. The first, as Orwell tells us, is his translation of the second into Modern English. He selected the second passage from Ecclesiastes, and since it was written in Middle English more than 500 years ago that explains the happeneth. Nevertheless its concrete language allows us after so many years to understand the meaning of the passage. The first example appears so impressive on the surface, but on close inspection, its meaning slips away. Gathering its message is almost like catching the fog or the wind.

[7]Excerpts copyright 1946 by Sonia Brownell Orwell; renewed 1974 by Sonia Orwell, and used with permission of A. M. Heath & Company Ltd. for the estate of the late Sonia Brownell Orwell and Martin Secker & Warburg Ltd. Reprinted from "Politics and the English Language" from SHOOTING AN ELEPHANT AND OTHER ESSAYS by George Orwell by permission of Harcourt Brace Jovanovich, Inc.

DANGERS OF ABSTRACT
LANGUAGE

You may comment that the first Orwell example sounds so authoritative and so much like what you are used to reading and hearing. That may be true, but the use of abstract language has a very real danger for you as a writer, in addition to its effect on the reader. If you write without concrete images, you will soon lose the thread that connects the message you are creating in words on paper with the real world. You will have written in honey, or built your castle in air. Successful writers discuss abstract notions, but they know that the reader must be able to see the ideas. Concrete, specific, detailed language is the best method.

The next exercise demonstrates the danger of abstractions in writing memos, letters, and reports in the workplace.

EXERCISE 2–10

These examples are the productions of working executives. Rewrite them. Of course, you need to add concrete detail, so for the purpose of this exercise, assume any quantitative information that makes the sentences plausible.

1. A detailed knowledge relative to problem areas and approaches to their solutions has been accumulated, with the result that advanced activity programming by individuals in middle- and top-mangement positions, and the latter in particular, prior to a promotion situation or their initial assumption of their new responsibilities, is indicated and in fact recommended.
2. User charges are based on the amount of square feet for which each service department commits to pay and as such are fixed for the department but may vary for the individual in the event of a change in the agreement as to how time and/or space is to be shared within a department.
3. The creativity of an expressive analyst lies in his/her capacity to regenerate this potential transitional life space rather than the possession of any particular artistic skill.
4. It is our considered opinion that your collateral posture should be one of quality and professionalism and should be carried off in just such an environment, as opposed to one with all kinds of bells, whistles, and balloons.
5. Dick has been 30 odd years with the Company and has, with great distinction, progressed through a series of important promotions during his career. He is a special type of individual with an unusual blend of qualities. He will be missed. We wish him well in his retirement.

Example 2 above caused a three-hour discussion at a major New York City institution. Several people attended the meeting, including those in charge of accounting, building administration, instruction, as well as the computer analyst, researchers, and staff representatives. At a low figure of $12.50 an hour for an administrator, the misunderstanding and confusion caused by the vague memo cost the institution $35 for each individual there. But that is only the surface cost. Consider the amount of work left undone while these people struggled to clarify the policy and the cost per hour doubles. So the generation of a vague sentence written in haste could end up creating a great deal of confusion and costing much time, energy, and money.

The extra time that it takes to write a clear, concrete, and readable message pays for itself in the elimination of possible confusion and misunderstanding that might follow.

The sentence might be rewritten:

The amount charged to each service department is based on the number of square feet it has agreed to use. Though the rate for each department is fixed, the agreement can be changed when the use of the space or the amount of time it is used changes.

Or, more briefly:

Each department is charged by the square foot at a fixed rate. When the needs of the department change, the agreement can be changed.

The first example in Exercise 2–10 is the product of a vice-president of a large corporation—insurance, we'll say. He has a very responsible position and feels that the language he uses should reflect the importance of his position within the company. The hidden message in his memo draws attention to his position as a big wheel, so the language is puffed up. His goal of sounding official and important wiped out what should have been the real goal of the message, and that was a directive concerning the way promotions were handled. Instead of inflating the message to make him appear in control, he should have written to the point. The hidden goal and the actual goal of the message are both thwarted. In an effort to sound important, he sounds pompous and inept, and makes a simple message complicated to the point of incoherence.

The example, on closer inspection, reveals a sentence that is not only unclear, but incomplete. The core of the sentence reads "The result is indicated and in fact recommended." The result is "advanced activity programming"—which could mean the promotion, or the instruction. It is so vague that no reader could act on it. And that is the real test of a working communication: can someone act on it without telephoning you or coming by your office to find out what you wrote. A few minutes of thoughtful effort can make any message clear, direct, and useful instead of inflated, confusing, and unreadable. We can take the example and make it at least two sentences. "We have gathered information on the problem areas of our organization, as well as a list of ways to solve them. Because it is indicated, I recommend that managers who are about to be promoted, or who have just received promotion, receive instruction in both the problems and the methods of solution."

Language that is concrete is language that is vigorous and readable. Use words that your reader can picture. Don't say a weather event if you mean rain. Don't say a promotion situation, if you mean promotion.

Direct, concrete language generally makes writing more forceful. However, a direct message may require some tactful tempering to avoid bluntness and offensiveness.

SUMMARY

- Effective writing is simple, not simplistic.
- Good professional writing gets to the point, addresses a specific audience, exhibits a clearly defined purpose, considers the context, and uses tact and courtesy.
- Intelligent word choice provides clarity.
- Define common words in an ordinary context, or in an unusual context as well as unusual words in a common or unusual context.
- Place definitions in context, the text, the notes, or a glossary.
- Formal definitions follows this formula:

$$species = genus + differentia$$

- Strong verbs bring life and power to professional writing.
- Overuse of "is" represents a symptom of weak writing. Cure "is" sickness with strong verbs.
- The active voice makes writing easier to read and more energetic.
- Use concise expressions for economy and to combat inflated language.
- Ambiguity creates confusion in professional writing, and the proper placement of words helps eliminate it.
- Concrete, quantified, specific language allows an audience to picture an event, which makes the document vivid.
- Abstract language in memos, letters, and reports often creates confusion and needless expense. Language that is concrete is language that is vigorous and readable.

3

Unwritten Messages in Writing

When we analyze the writing process we reveal three main parts:

WRITER   →   MESSAGE   →   AUDIENCE
   | | |
PREWRITING   WRITING   READING

If you are familiar with the communications model used graphically to represent any information transaction, this appears simpler. The communications model is more abstract because it accommodates anything from gestures to multiplexed computer transmissions:

SOURCE   →   SIGN   →   MEDIUM   →   RECEIVER ⎫
   | | | ⎬ FEEDBACK
ENCODE   TRANSMIT   DECODE   ←──────────⎭

This model serves to demonstrate that the idea you have goes through several processes before the reader receives it.

If we consider the beginning of the process of thought as similar to the invention of an idea, then our discussion of brainstorming fits into the initial stage of writing. In business and professional writing the information you must communicate exists, making the invention process center on the arrangement and style of ideas, not the creation of them from nothing.

Once you have the notion you wish to communicate, you must find a suitable form that will be understood by your reader. In writing, that process is your selection of words. The word is made up of symbols, or letters, which are themselves symbols for sounds. The written word then, represents the sounds of the spoken word which we as a society agree stands for a thing, an action, a quality, or an emotion. When we speak or write, however, most of us do not consciously consider words this way. But we must give some thought to this element of writing if we are to write with consideration. After all, a successful piece of writing starts with an idea. We select words that fit it closely and put them on paper. The reader sees them and, if we have done our job carefully, the thought in the reader's mind will be very similar to ours.

Our explanation is, of course, simplified, but it serves to illustrate that consideration in writing requires you to write for an audience. No piece of writing that has consideration is done in a vacuum, and writing that gets results has the quality of consideration.

What constitutes consideration in writing is a tone that incorporates tact, courtesy, and naturalness with the demands for clear and direct communication. In exploring the unwritten messages of writing, consideration represents an essential factor.

BARRIERS BETWEEN WRITER AND READER

Barriers to clear communication exist. Your efforts to write often must surmount not only the subtleties of language, but physical and psychological hurdles as well.

Obviously the writer and reader must share knowledge of the same language for any understanding to result. And the reader's unfamiliarity with professional terminology might prompt the writer to explain terms in greater depth. The writer should also seek any difference in cultural attitudes and beliefs that the reader might have. Whatever those differences might be, the writer should not patronize. That usually insults an audience. Remember that most people can understand complicated ideas that are presented clearly and directly. When you cannot accurately identify a specific audience it is safe to assume that you write for a knowledgeable, general audience.

Usually the barriers of language can be diminished through careful word choice and clear organization. Logical arrangement of a message helps the reader comprehend information and ideas quickly. If material is not presented in an order that can be easily followed and understood, the reader will have difficulty acting on it. Such disorganization lacks consideration and fails as practical, work-oriented writing.

Even though it is often his job to read your letter, memo, or report, you must still exhibit your understanding of the reader's needs by maintaining his interest. But your effort to interest the reader need not imitate a newspaper or magazine article that "hooks" readers with dramatic headlines and leads which stimulate curiosity.

Material that is understandably arranged, grammatically correct, and factual usually interests people by virtue of its straightforwardness. Your control over the words you choose exhibits, through the language, your consideration for the reader.

Another obvious unwritten message is the physical appearance of documents. Send a document that is cleanly typed and correct, as well as spaced on the page for easy reading. Smudged paper, sloppy typing, and arrows to show that a line belongs somewhere else suggest to the reader that you thought so little of him and the message you wrote that you did not bother to have the page redone. A small effort speaks loudly in your favor, and if you take the trouble to type, at least make sure it's done correctly and readably. Other physical conditions that might distract your reader remain beyond your control. Often your reader is in a cramped, poorly lit office. Your message must also compete with scores of other papers on a person's desk. That is why you should demonstrate your understanding of the reader's expectations by getting to the point early in your document.

You can also anticipate a reader's reaction by placing yourself in his shoes. Understand the information you must write for him from his perspective. With that understanding, present information that is direct, clear and useful. You don't need a degree in psychology to make yourself aware of another person's reaction to controversial or new subjects. Present them accordingly. That is, try to recall what convinced you, then use that information to convince your audience. As writers we often make the mistake of assuming that a reader is on our side from the start. Some readers may not like an idea no matter who expresses it. That psychological barrier requires more than writing. Often it demands an entire campaign to alter attitudes.

DEVELOPING AN APPROPRIATE TONE

Tone, an important element of consideration, reflects the writer's attitude toward not only the material, but also the audience. In daily behavior people can be angry or hostile, excited or depressed, unsure or confident, shy or cocky. Through a carefully constructed tone writing can reflect those qualities and more as well. An inexperienced writer often ignores the impact of tone on the reader.

A carefully thought out and selected tone can reduce the likelihood that you will insult, offend, or put off your reader. An effective attitude seriously considers the effect of the material from the reader's point of view. For example, if you sell word-processing equipment, sell the machine on its merits, but don't list the merits of the machine without first placing those strong points in a context the listener can understand. In other words, if your machine has a memory capable of retaining the equivalent of 300 pages, consider the reader's perspective. Suggest to the customer that the memory capacity will reduce the filing space needed in the office, because the disks are much smaller than 300 pages. Also point out that the memory enables the faster retrieval of documents, and more accurate reproduction of revised text than is possible with retyping.

A reader-oriented *attitude* considers the tendency of people to get interested when the conversation focuses on themselves. People love to talk about themselves. However, this attitude differs from a tone that patronizes. Any poorly considered tone risks that.

Tone in writing is important because it sends unwritten messages. It is to writing what body language is to conversation. Tone as the attitude toward the material and toward the audience demonstrates your enthusiasm and invites your reader to share that interest.

Determining Tone Consistent With
OBJECTIVES

Often the workplace is more like a psychic battlefield than a place of coopera-tion and harmony. Given the military-like organization of most large businesses and the authoritarian style of management in this country, it is no wonder that when we write for work, we want to write so we do not draw any fire. Many do not want to draw attention to themselves for any reason, good or ill. While that may be a proven business survival technique, it certainly promotes a stilted perspective for writing or for personal fulfillment, for that matter. The tone of writing reveals any reticence and, as we discussed in Chapter 1, this stems from the fear of being wrong—or worse, being thought a fool.

All too often this defense against error or embarrassment, this retreat to the foxhole of your desk can get in the way of your purpose when you write. Since writing is an active process, such defensiveness hinders the creation of documents that communicate directly. In that sense the essential *purpose* of writing evaporates into the secondary purpose of defending yourself and your dug-in position. When this takes place, the *tone* flashes your defensiveness and sets up doubt in your reader's mind—or worse, insult and challenge.

To demonstrate the importance and impact of tone, let's examine an example. The following paragraphs are taken from the conclusion of a report which evaluates cost, purchases, and sales of an institutional food service. Notice the way the writer of this report addresses both the subject matter, as well as the reader of the report. Comment on the tone and the attitude toward the reader and the audience.

> While a number of cost reduction actions are responsible for our evading the full force of the hurricane character of inflationary winds, the most prominent being menu and product changes, along with aggressive purchasing practices.
>
> At about this time, I suspect the reading of this report is becoming as painful as the writing, and so with a measure of compassion born of empathy, I'll move along to a few concluding nonstatistical comments.
>
> We trust that those in our department who are constantly called upon to plan, execute, and control anew, glean the satisfactory results from this report, that their efforts made possible. To those whom we are accountable to, we trust this report lends some insight as to what has happened in the department during the fiscal year, and why it has happened. We are also hopeful that those to whom we are account-able, have the benefit of other reports that the industry may generate, that would extend to them the means to fairly measure our performance, against those institu-tions of similar endeavor and ranking.
>
> If any of the readership has questions stimulated by the reading of the attached statistical analysis charts, we would be most pleased to render our opin-ions.

Before discussing the attitude of this report, were you able to find the major grammatical error in the example? The first paragraph, right. It is a group of phrases and clauses masquerading as a sentence. It is in fact not a sentence at all, but only parts of a sentence. The group of words has no verb. As we mentioned in Chapter 2, the foundation of good, clear writing is made of verbs. The writer of this report unintentionally demonstrates so little familiarity with the material that he provides an incomplete thought. In this instance, the incorrect grammar communicates not ignorance of the language, but a lack of confidence in the material. In most of the business papers that you write or read, grammatical errors indicate something other than illiteracy. The untutored reader, however, will focus on the error, rather than on the reason for it.

An incomplete thought indicates a lack of control over the material. It also represents the writer's fundamental misunderstanding that someone must read the material, an unwillingness to communicate for fear of sounding foolish, or ordinary carelessness. Any of these contributes to an undesirable unwritten message.

In addition to the grammatical error, the conclusion of this report strikes an unacceptably combative tone. If we look closely at the first block of words, the use of wind as a metaphor might be desirable in the hands of a better craftsman, but here it only draws further attention to the incompleteness of the sentence. Had the writer adjusted his attitude the real message could have come across. Instead of a challenge he could have written: "We were able to reduce the effect of inflation through menu and product changes, as well as through aggressive purchasing." Stated directly, the message is not only clear, but communicates the writer's confidence in the actions that were taken. The *tone* changes from doubt, challenge, and self-consciousness to professional assurance.

The next block of words attempts to make a transition or bridge between the body of the report and the concluding remarks. But it is inflated, and sarcastic. The references to the readers' pain and the writer's pain might be funny in casual conversation about the general character of business reports, but it falls flat on the page. Avoid jokes, unless you have the natural talent for delivering them. The effect in this case creates an undesirable *tone* by revealing the writer's flippant attitude toward the reader and his uninterested and pained attitude toward the subject of the report.

The next block of words is punctuated incorrectly which makes reading and understanding difficult. But more importantly, the use of "we trust," injects an air of superiority on the part of the writer. The message seems to be handed down to a subordinate rather than up or across to peers or superiors. The use of the phrase "those to whom we are accountable" further compounds the undesirable effect.

Worse, the sentence, which in an inflated way suggests that the reader get reports from other institutions and "fairly measure our performance" with theirs, directs the reader of the report, rather than informs him. What should be here in the place of a challenge is the summary of the other reports and a comparison. After all, the writer wants to say that, given the effect of rising prices and costs due to inflation, he has done very well, in fact better than similar operations in institutions of comparable size and prominence. Instead, the challenge communicated doubt.

And the last group of words, contrary to its invitation to discuss the substance of the report, further requires the reader to do the writer's job by attaching charts with no explanation or evaluation.

Most business reports are more mindful of the tone than this one. In fact the example taken as a whole would say to an uninvolved observer that the report was an exchange between individuals who were having more than a passing argument. This report almost amounts to a declaration of war, its tone is so brusk and uncooperative.

To determine an appropriate tone look back at the purpose which you identified at the beginning of your writing. Notice that the purpose might be to change the attitude of the reader, or to present some unfavorable information. The tone selected makes a difference in your reader's reaction to your information. Use these questions to identify an effective tone:

- What is my attitude toward this message?
- What is my attitude toward this reader?
- Is this information "good" news or bad?
- Can parts of this be left out? Should they? Why?
- Is this reader one who will respond to levity—or one who will be offended by it?
- Is irony appropriate?
- Can overstatement or understatement be used appropriately?
- Does the tone fit the intended meaning?
- Have I underestimated the reader's intelligence?
- Can the reader misinterpret my attitude?
- Could information be arranged differently?
- Could facts be selected and presented more appropriately?
- Has the material been presented simplisticly or simply?
- If jokes are included, are they left unexplained?
- What is the reader's attitude toward the material?
- What is the reader's attitude toward me, the writer?

A Reflection of Professional Ability

After thinking about the questions concerning tone, it may have occurred to you that the ability to provide useful answers requires your command of the material in the report. Exactly. The report is the surface presented to the outside world. Its unwritten message reinforces an individual's knowledge and expertise. Clear writing demonstrates the writer's ability to handle information and present it clearly because he has the professional background, experience, and know-how to do so.

Straightforward writing inspires confidence in your reader that the material you present is accurate and thorough, as well as useful and clear. Of course, professionalism in part is made up of those qualities. Writing has the power to enhance your professional standing.

Writing is organized thought and as such requires planning. Most readers expect that in the finished product. Well-written material does not call attention to the writing itself, but to the information. Poor writing shoulders its way into the awareness of the reader, obliterating the purpose of the report. It tells us that the writer is not really in control of the material. And if that is the case, he may not be in control of other elements of his position.

The company vice-president's long-winded memo on the promotion procedure which we quoted in Chapter 1 forces us to question his ability to understand and communicate information.

The written word has the unprejudiced capacity to reveal muddled thought and illogical presentations for all to see. That may account for the common fear of writing that we discussed in the section on brainstorming. But writing also shows the depth of ability and talent as well. This book strives to plumb this depth by giving you an understanding of the process of writing and the impact of the written word on an audience. The more effective your writing, the greater will be your audience's perception of your ability to act professionally.

Because writing requires understanding of a topic beyond the capacity merely to perform, but to absorb, arrange, and interpret facts, many business people consider it a good barometer of your ability. In addition to job performance, an individual capable of writing clearly presents a more desirable manager than one who can perform, but writes unintelligibly.

The reality of weekly, monthly, quarterly, and other types of reports, memos, letters, and policy statements represents an integral part of every job. New electronic technology makes your ability to write even more important because ordinary memos and reports will soon be distributed over computer networks that will replace interoffice mail.

POSITIVE EXPRESSION

Hyped-up, unsubstantiated optimism has no place in serious writing. Using extravagant expression undermines credibility. However, any piece of information can be written in a positive manner and still cling to its validity.

For optimistic writing to be effective, you must demonstrate your control and understanding of the facts. Presenting ideas and information in a positive way requires a realistic picture of the subject, in addition to mature familiarity with the material.

Almost any message can be written positively.

In writing this way make sure that the bad news you might bring is still presented truthfully. For instance, if we were to write: "No solution to inflation has been found," we have written a negative version of a truthful statment. In a sense it is negative because it cuts off elements of the truth, and presents only a partial picture. If, instead, we write: "Businessmen, bankers, and government officials continue to search for the solution to inflation," we present a fuller picture of the situation, and add the valid information that the search for a solution continues.

Positive expressions, on the other hand, can lead to blandness, the written equivalent of a vacant smile at a party. Worse, it can foster deception. Instead of writing "tax increase," a call for more taxes was dubbed with this example of deceptive newsspeak: "revenue enhancement." The phrase disappeared after several protests against its obvious deception.

Effective use of positive expression requires you to have a familiarity with the connotations of words. (See Chapter 2.) *Cancer,* for instance, has negative connotations, as does *inflation.* Careful selection of the context in which you use these and other terms determines the impact of your message. Of course, the positive expression of an idea depends on the message, your purpose, and the demands and needs of the audience.

Positive expression demonstrates your knowledge, confidence, and optimism to the reader. Compared with negative statements, information expressed in positive terms has a more favorable impact on the audience when the reader perceives the whole expression to be valid and truthful.

In addition to its desirable impact, positive expressions demonstrate that you have placed yourself in the reader's shoes. Analysis and consideration of the needs and expectations of the audience allows you to develop a reader-oriented tone that we mentioned. Then you see your ideas from the reader's point of view and ask, "What's in it for me, or what do I get out of this?"

Such analysis reveals the self-interest of people. In writing just as in conversation, most people like to have attention drawn to them and to have the conversation focus on them or their interests.

To achieve a reader-oriented letter, mention the reader by name. Of course, this is a mechanical device and its effect can diminish if you use it too frequently. Do not overdo this because the opposite effect will occur and your reader will sense your patronizing attitude.

You can also put off a reader if you toot your horn too loudly. The reader might perceive boosterism as arrogance. Instead, present words so the reader can see that your accomplishments are good for him. For example, your company had for years led the way in aggressively establishing flexible hours for employees, an approach now used by hundreds of companies across the country. You suggest to a possible investor in your firm that you are a concerned, civic-minded organization, and your program of flexible hours demonstrates not only responsible management and sound business practice, but community involvement. You communicate a confident attitude and spirit rather than a solicitation of recognition and praise for your own accomplishments.

Positive, reader-oriented expression allows you to demonstrate your maturity, confidence, and knowledge. Being positive does not imply that you must sugar-coat bad news—in fact, bad news goes down easier if you express it right out. For example, here is the first paragraph of a second-quarter report from the president of a manufacturing company:

> The company reported losses for both the second quarter and the first half ended June 30, 19XX. These unfavorable operating results are principally attributable to the continuing, unprecedented market weaknesses in our key business.

The rest of the report goes on to explain how the declines in sales of the major business operations—freight cars, railroad equipment, and truck parts—affected the company's profits and its ability to pay the quarterly dividend to stockholders.

Such bad news is not easy to express positively, but the explanation conveys a sense of confidence that the company is strong and will ride out what is also expected to be a long slump. The report presents this positive statement before its closing paragraph:

> ... we are continuing tight controls over working capital and capital expendi-
> tures, and feel confident that we have established sufficient financial flexibility until
> the recovery in our railroad and truck market occurs.

Ending on an optimistic note, optimism based on the realities of the company's operations and its markets instills confidence. This company has assessed and prepared for the storm ahead and will weather it, not because it has wished away conditions, but because it has approached its problems realistically, and shared its situation and prospects with its stockholders.

In the recommendations that a national service organization uses to guide customer-service employees, the same holds true. Give the bad news first and then the explanation, if necessary. Bad news and the expression of "No" can be done in a tactful, courteous manner, one that recognizes the humanity of the reader.

IMPACT OF THE MESSAGE

Suggestions for the positive expression of ideas assumes that you as a writer have thought about how your reader will react to the written message. Much of the impact of the message, of course, depends on what you say and to whom.

That brings us back to the analysis of *audience*. Go over the questions you asked yourself in the beginning to determine the nature of your audience. This time, however, your answers might be more sharply focused because you have a specific piece of information to communicate. For example, when a nuclear plant experienced a serious cooling system leak, the utility that operated it sent an explanation of the shutdown to its customers. In addition, it increased the rate to offset the cost of the oil used to generate electricity while the plant was shut down. The customer newsletter, sent along with monthly bills, begins:

> You may have asked yourself why your electric bills should go up while the Reactor 1
> nuclear plant is down. Why, in other words, should you pay for what some people
> claim were our mistakes?
>   The fact is that Reactor 1 is out of service not because of our mistakes. It is
> out of service because it needs new pipes in its heat exchangers, a refueling,
> maintenance work, and certain plant modifications which had been planned some
> time ago.

If the reader takes this beginning at face value, the message is that the accident in which 100,000 gallons of river water accumulated on the floor of the plant's containment building, and in the reactor cavity as well, had nothing to do with the shutdown. The shutdown was planned. If that is so, even the most ill-informed reader would ask why the rate increase was not also planned. Had the writers of this letter for the utility considered the impact of the message they could have known that customers would not react favorably. In fact, lawsuits followed to bar the utility from passing on the costs of the leak damage to its customers.

Though the impact of a message should be considered, the facts should nevertheless be presented. The utility could have explained the problem directly, and suggested that the rate hike was necessary to protect the generating facility, and to safeguard the people in the area from contamination. Further, the accident provided an opportunity to make necessary repairs that were dictated by the technological advances that were designed to solve the leak problems experienced by all generating facilities, conventional and nuclear. Had they explained that the costs would be shared equally by shareholders, the company, and the consumers, then the message that a rate hike was warranted might have been accepted more easily. The message sent to all utility customers with their bills appeared to burden consumers with the cost and responsibility. Of course, consumers responded negatively.

INJECTING TACT, COURTESY, AND NATURALNESS

This section may seem presumptuous in its attempt to "teach" the social qualities of tact, courtesy, and naturalness in writing. You might say that someone either has those qualities or they do not. That is not entirely true, since each one is an attribute of a civilized person, what used to be called "good manners." What, you ask, do manners have to do with the business world?

Manners in writing still say a great deal about you as an individual and about your company as an organization. For example, if you work for a home appliance manufacturer, you know that competition and buyer resistence can be fierce. Customers have heard many pitches. If you were to build up your product by pointing out how bad the Grubart Blender is, then you have not only employed poor marketing judgment, but you have also presented your case tactlessly. By denigrating your competitor you tell the customer that you do not have enough positive information about your own product to let your case stand on its merits. The section above concerning the positive expression of information also holds true here. A positive message is often tactful.

Tact in writing often requires nothing more than common sense. For instance, one of your co-workers fouls up the order for the week and you must file a report to explain the additional cost and inventory to your regional supervisor. An admission of the mistake and a straight-out explanation of how the inventory and costs will be evened out over the next few weeks would be a more sensible way to explain the situation—that is, if the co-worker is ordinarily accurate and conscientious. A tactful message says you are in control and that you can solve problems that arise in the course of the day's business.

Like tact, *courtesy* is allied with polite behavior and manners. Why bother with such archaic remnants of a distant past? Doesn't business deal in bottom lines and results? Yes, but to get results, business deals with people. Even if the business is making steel or sewing needles, somewhere in the company's dealings people are intimately involved. But more than that, the report, letter, or memo that you write becomes the record of what happened. It remains long after emotions that might be less than civil have passed. So to avoid a letter that returns to haunt you, it is good practice to write with courtesy and tact.

To do this, you often have to write indirectly, which contradicts our earlier assertion that good writing is direct and to the point. The approach you take depends on the context of the message and the circumstances that surround it.

Many people speak to you directly and make themselves quite clear in face-to-face conversation. Sooner or later you get a written message from one of these individuals and you have to check the name to be sure the writer is the same person that you know from business conversations. Too many people feel that in writing they must be formal, that is, stiff. Professional situations call for writing that is readable and natural, not wooden or stilted. For example, "Please be advised that" in writing sounds too much like a butler presenting the master of the house with some bit of news about the burst pipes in the basement. In other words, people that you meet in the course of the day don't talk that way and it sounds phoney when we read it. Would you feel comfortable saying, "enclosed herewith"? Then feel the same discomfort in using wornout phrases that come between you and your reader.

Again no one can teach naturalness, and if you are a stiff and formal person, then stiffness is naturalness for you. But the point here remains, naturalness says that you are a human being. Good communication needs that basic of all qualities.

SUMMARY

- An awareness of the process of writing and an understanding of the barriers that come between the writer and the audience provide a basis for business papers.
- Writing that is tactful, courteous, and natural helps overcome psychological and language barriers that impede a reader's understanding.
- An appropriate tone—one that is consistent with the purpose and also meets the audience's needs—is essential to effective writing. Tone is the writer's attitude not only to the material, but to the reader as well.
- Clear writing often reflects professional ability, and most business people equate the two.
- Experienced writers anticipate the connotations of the words they choose, and judge a word's impact on the reader.
- Positive expression of facts and ideas offers another method to get results through writing.
- Tact, courtesy, and naturalness lead to considerate and effective writing.

EXERCISE 3–1

Rework the following sentences. Present a positive tone and maintain the substance of the message.

1. No cure for cancer has been found.

2. Periodic review of progress and status will be held to ensure that no problems go unrecognized and without proper corrective action.

3. Your failure to respond to my suggestions on the policy revision statement was a surprise to the committee.

4. To avoid errors and mistakes in the processing of laboratory tests, please provide accurate information on the request form in the proper space.

5. We have made every attempt to insure that your stay in the hospital will be as free of pain and anxiety as possible.

6. Prompt payment of your bill saves us accounting expenses and brings our operating costs down.

7. To offset these detractors of product quality, we have introduced our Pride of Workmanship Program (rework costs and attendant schedules delays).

8. While the elderly, the poor, the handicapped, the unemployed do not represent the affluent society, they should not be treated as second-rate citizens.

9. Under no circumstances would the project to preserve the historical blacksmith shop be allowed to flounder and die.

10. As far as the collateral material costs are concerned, it is not possible at this time to estimate costs until volumes for each of the components referenced are more closely defined.

EXERCISE 3–2

1. Select a piece of writing you have received recently. Determine the writer's attitude toward you as a reader, and toward the material. Does the piece you choose exhibit consideration? Rewrite the selection, making its tone demonstrate consideration.

2. Look for the unwritten message in something you wrote in the past year. What did your writing imply that you did not intend at the time? Rewrite the document, making it more direct, courteous, and tactful.

3. Choose three ads in the newspaper. Explain the unwritten messages of each.

4. Select flyers, memos, or letters you have received that are stiff and overly formal. Rewrite at least two of the examples in your own voice, injecting naturalness into them.

5. Find an example that takes a tone inappropriate to the purpose and the audience. For instance, it may be too informal with someone that is unknown to the writer. Revise the selection to make its tone appropriate.

EXERCISE 3–3

As company vice-president you have just been informed of a rather sticky situation. A vendor who has supplied you for 10 years with raw materials for the manufacture of nylon parts has recently sent material that does not meet the specifications that you require for your process. Your assistants have called several times to find out what has happened and have gotten the runaround. You have called yourself, but could not get through to the company's manager, or its owner. You have been told by their subordinates that they could not answer your questions at that time.

Write a letter to this company's owner explaining your situation and your reluctance to discontinue a business relationship that has been, up to that point, very harmonious and friendly.

EXERCISE 3–4

Your company has just received a request from the local chapter of the union that represents about 75% of your employees. They want to use the company parking lot for a fund-raising benefit bazaar and festival. The proceeds are to go to a community charity for crippled children. Though you want to let them use the facility for such a worthy cause, your lawyer has informed you that if you open up the plant grounds for such an affair, your insurance premiums will double. However, the company owns a vacant lot three miles away from the plant on the edge of town that could be used for the festival with the lawyer's blessing.

Send a letter explaining the situation to the union leaders. Be as tactful as possible.

4

Clarity and Directness
with Hard-Working Sentences
and Paragraphs

When someone mentions grammar, we generally remember our grade school English teacher—a stereotypical matron who spoke a strange dialect of our language, peppered with odd references to coordinate clauses and subordinating conjunctions. Grammar, we were taught, was a set of immutable rules to be followed to the letter.

However, the true nature of grammar has less to do with authoritarian rules and more with the possibilities of a language and its intuitive expression. Grammar, usually linked with syntax or the order of words, presents a plan for our language to give words a meaningful context. Look at these words:

> our of plan language a presents grammar

Each of the words has some meaning for us, and we even begin to rearrange them to gather more, but we need a coherent order to provide a context. That, as far as grammar is concerned, is one of the main ways we limit meaning in writing.

Even if we had no books, we would still have a grammar since it is the agreement that native speakers have about what is expected when we communicate. As William Irmscher observes, in *The Holt Guide to English* (N.Y.: Holt, Reinhart, Winston), grammar is "the structure of speaking and writing that we accept—an internalized set of principles that enable us to use the language—whether or not we are students of grammar."

SENTENCE PATTERNS

Since the early 1950s transformational grammar has provided us a different, more streamlined way of looking at the structure of English. Instead of the way you might have learned, this concept describes basic sentence patterns, or sentence kernels, as the building blocks of our language.

Pattern I
Subject—Intransitive Verb—(Adverb)   She eats (loudly).
Pattern II
Subject—Transitive Verb—Direct Object   The grocer bought the fish.
Pattern III
Subject—Transitive Verb—Indirect   The Navy awarded us the contract.
          Object—Direct Object
Pattern IV
Subject—Linking Verb—Predicate   Opportunity is bright.
                    Adjective
Pattern V
Subject—Linking Verb—Complement   Smith is the leader.
Pattern VI
There—Verb, usually linking—Subject   There is a tavern in the town.

English sentences use one or more of these patterns to express ideas. The first three of them carry more force because they build on strong, active verbs. The last three present weaker messages, because the verbs do little work.

Types of Sentences

We use the basic patterns to form each of the four types of sentences:

- declarative—statements
- interrogative—questions
- exclamatory—expressions of strong feeling
- imperative—commands or requests

Each of the four types can be simple, compound, or complex. For example:

Simple   The manager runs the program.
Compound   The manager runs the program, and the vice-president directs the manager.
Complex   After each manager submits a report, the administrator prepares an evaluation of the month's operation.

Often a change in the function of a sentence, from declarative to imperative, for instance, requires a different word order. When you ask a question, the order of words also changes, with all or part of the verb coming before the subject. Making a command or request affects word order. Usually the imperative sentence omits the subject, allowing the reader to infer "you."

Coordination and Subordination

The linking of sentences containing similar or associated information demonstrates for the audience the relationship between the ideas. Coordination commonly ties ideas of equal value together, while subordination presents more of the contrast between two statements, one of which is grammatically stronger.

Coordinating conjunctions:	and, but, or, nor
Subordinating conjunctions:	because, since, if, although, unless as, as though, so, after, before, while, when
Correlative conjunctions:	either . . . or, neither . . . nor, both . . . and, not only . . . but also, so . . . as, whether . . . or
Adverbs as conjunctions:	also, besides, indeed, however, nevertheless, anyway, on the other hand, for instance, thus, so, accordingly, then, later, finally

(See Table 4–1 for a fuller list of transitions.)

The careful use of coordination and subordination improves the quality of the presentation, making the information clearer to the audience. For instance, look at this group of sentences:

> Both the Navy and ZAR Corp. have been at the forefront of developing this new technology. ZAR Corp. has recently been awarded a study contract for digital automated test program generation work.

Table 4–1   Transitions and Their Uses*

TRANSITIONS		USES
1. and or, nor also moreover	furthermore indeed in fact first, second . . .	For adding something
2. for instance for example for one thing	similarly likewise	To add to, illustrate, or expand a point
3. therefore thus so and so hence consequently	finally on the whole all in all in other words in short	Tally consequences, using minor points to emphasize a major point

TRANSITIONS			USES
4.	frequently occasionally in particular in general	specifically especially usually	To qualify a point or illustration
5.	of course no doubt doubtless	to be sure granted (that) certainly	To concede a point to the opposition
6.	but however yet on the contrary	not at all surely no	To reverse direction
7.	still nevertheless notwithstanding		To return the thought to major argument
8.	although though whereas		To attach a concession to one point
9.	because since for		To connect a reason to an assertion; to show effect-cause relations
10.	if provided in case	unless lest when	To qualify or restrict more general ideas
11.	as if as though even if		To suggest hypothetical conditions that strengthen and clarify a point
12.	this that these those who whom he she	it they all of them few many most several	Relative and demonstrative words (adjectives and pronouns) tie things together, pointing back as they carry the reference ahead

*Table of Transitions, Copyright 1983 by Kenneth Friedenreich, Ph.D. Used by permission. From *Impact and Style in Business Communications.* Los Angeles: Educational Services Division, Brentwood Publishing.

Some relationship between the two statements is implied. When we combine the sentences by using the proper conjunction, the relationship becomes clear.

> Because ZAR Corp. has been at the forefront of developing this new technology with the Navy, it has recently been awarded a study contract for digital automated test program generation work.

When you coordinate and subordinate sentences effectively, you give your writing clarity and power.

Modifying Structures That Distinguish Mature Writing

The patterns and types of sentences we have discussed rarely appear in our writing in their barest form. The sentences we write and speak present more subtlety, often combining two or more patterns and usually expanding the meaning with modifiers. We can easily identify adjectives and adverbs, the most common modifiers. In addition to words, phrases and clauses also act as modifiers, as in:

> The support beam that leaned near the doorway fell loudly when a sudden vibration shook the building.

In this sentence *support* and *sudden* are adjectives, *loudly* an adverb, *near the doorway* an adverb phrase, *that leaned near the doorway* an adjective clause, and *when a sudden vibration shook the building* an adverb clause. The *beam fell* forms the basic pattern or kernel of the sentence, and two other kernels, *(beam) leaned* and *vibration shook,* appear as modifiers in subordinate clauses.

Most of us became aware of these structures in public school and college. Native speakers of English, even if they cannot name them, use these structures easily and unconsciously in communicating. After all, a child of six forms a sentence of greater grammatic complexity than the example when he shouts to an older child, "If you don't stop that, I'll tell mommy."

Why mention all this here?

Simply because, when we write, most of us continue to use only the modifying structures that we were able to speak when we were five or six. A mature writing style employs other structures of modification—the noun cluster, adjective cluster, verbal phrase, and absolute. When used appropriately, these structures promote credibility by demonstrating knowledge to the audience. Regular use of these four help careful writers to inject variety and vitality into their work.

Noun Cluster   As its foundation the noun cluster uses an appositive, a noun which explains another noun or pronoun in the sentence. The appositive becomes a noun cluster when modifiers cluster around it:

> Avco, *a major developer of computer technology,* employs more than 20,000 professionals.

An inexperienced writer would have spread the information inefficiently over two weak sentences.

Adjective Cluster   An adjective and its modifiers form an adjective cluster:

The management program, *first on the list of alternatives,* presents efficient quality control at reasonable expense.

Here the mature writer makes a strong single statement where another would have written two weak sentences.

Verbal Phrase   Because the verbal phrase begins with the -ing or -ed form of the verb, it is the easiest of the structures to identify and to use.

Two pilots, *flying at different altitudes,* tested the emergency systems for 20 minutes.

The focus on the act of testing demonstrates the power of a verbal phrase to clearly show the reader what information is most important. As with the other structures, an inexperienced writer would have written two unfocused sentences.

Absolute   The most sophisticated structure in English, the absolute, is grammatically independent of the sentence it joins. Rather than modifying any single word or phrase in the sentence like the other modifying structures, the absolute modifies the whole sentence. Experienced writers who publish the books and short pieces that we read use the absolute regularly. For instance:

The plane circles, *its tanks empty.*
The crippled ship, *its deck awash,* steamed into port.

Of course, the absolute could be presented as a separate sentence. However, that would weaken the association between words, expressing the idea less efficiently and economically.

The four structures of modification give your writing force, while allowing you to show differences in emphasis. The mere use of these structures does not immediately make you a better writer, though expressing ideas with them provides your writing with emphasis, clarity, focus, and economy.

The conscious use of noun and adjective clusters, verbal phrases, and absolutes places you firmly on the road to efficient and effective writing.

STRUCTURING PARAGRAPHS FOR DIRECTNESS

Because it is linear and your reader captures your meaning one word at a time, writing needs development. We use paragraphs to indicate blocks of information that go together, since most thoughts are too complex for a single sentence. Most thoughts come to you in a lump, not one word at a time, and you need to shape them so the reader can follow along with you.

Briefly, as Aristotle observed of plays, every piece of writing has a beginning, a middle, and an end. Applied to a paragraph, that requires you to provide your reader with:

1. An *entrance* to the thought, idea, or information;
2. A *tour* through the details;
3. And an *exit* to the next idea, or out of the whole idea if it is the last paragraph.

This basic structure gives your reader a clearer and more easily understood presentation. Carefully constructed paragraphs invite the audience to read the pages of your text.

Some of the same methods that you use to develop a whole piece of writing work well with paragraphs. Use one or a combination of the following approaches to organize ideas and to make your point clearly.

Detail   In expressing a general notion, add to it with specific details. Include only the important ones, arranged according to time, space, logic, or another pattern that clearly brings your idea to the reader.

Example   To demonstrate an abstract or general concept, use an appropriate example. A carefully selected example helps the reader to picture your point.

Definition   Develop blocks of ideas with a formal definition in which you provide your reader a larger category, or *genus,* for the term, as well as appropriate differences, or *differentia*. Informally, a brief explanation of how something works or a listing of its major parts helps define concepts and tangible objects for the reader.

Classification   Like definition, placing a notion in a class of similar notions provides your reader with points of reference to aid in understanding.

Comparison and Contrast   Once you have placed an idea in a class, explaining it requires comparison and contrast. Ask yourself what it shares in common with the others, and what is different about it. Providing clear similarities and differences emphasizes ideas for your reader.

To discover some of these points, try the brainstorming techniques we have used. (See Chapter 1, as well as 6, 8, 9, and 10.) One or all of the techniques should uncover fresh perceptions of the similarities and differences.

Analogy   Like comparison and contrast, analogy helps dramatically to flesh out ideas for your reader. Its value in technical writing is less than that of comparison and contrast because analogy loosely associates ideas rather than connecting them logically. In argumentation, analogy furnishes a weak explanation.

Cause and Effect   An appropriate cause of effect supplies your reader with answers to how, why, or what for. The development of ideas this way exploits your reader's natural curiosity, as well as the desire to know the answer to a question.

Good writers use a combination of these methods to express thoughts clearly and coherently. Also some mechanical ways of developing ideas enhance the quality of your style, making your message more understandable and direct. For example:

- The repetition of a key word or phrase can provide a familiar landmark for the reader, emphasizing and recalling points made earlier. However, repetition can also provide a lullaby, putting your reader to sleep. Cultivate an ear for the kind of repetition that leads to boredom; then avoid it. Reading aloud helps to identify sentences that need to be reworded.
- Parallel sentence patterns carry your reader through your thoughts and provide a dramatic form of development as well. Here, too, overuse of the same pattern appears dull and could dampen your reader's interest.
- Transitional expressions provide mental connections for your reader. Use of these words also furnishes a clear context, reducing the possibility of misunderstanding.

SUMMARY

- The three basic sentence patterns which achieve effective messages build on strong, active verbs. The other three present ideas weakly.
- A mature, professional writing style uses appropriate coordination, subordination, and modifying structures—noun cluster, adjective cluster, verbal phrase, absolute. The four structures provide writing with emphasis, clarity, focus, and economy.
- A effective paragraph provides the reader with an *entrance* to the thought, a *tour* through the details, and an *exit*.
- Develop paragraphs through: detail, example, definition, classification, comparison and contrast, analogy, or cause and effect.
- The repetition of a key word or phrase, as well as the use of parallel sentence patterns and transitional expressions, enhances the quality of writing style.

EXERCISE 4–1

The following sentences exhibit many different kinds of grammatical and stylistic problems. Some of them are sentence fragments, others try to say too much. Some present ambiguous modifying phrases. Revise each of the examples to strengthen the sentence.

1. Do not mail cash, stamps, third party checks, or checks made out by anyone other than the registrant or a family member having the same last name are not acceptable.
2. Ask about our free tire warranty.
3. The city is overcrowded, unemployment running high and no unemployment office, with filthy and unhealthy conditions of living.
4. Appearance may often be the major factor in selecting the type of structure to be used as where a bridge which will fit in and actually contribute to the appearance of an area.
5. World Premier! A spectacular drama of love and adventure that races to it's climax at the World Olympics.

EXERCISE 4–2

Read the following sentences. Rearrange the order to improve the development of the idea and to show relationship between statements. Some sentences can be combined to further improve the sense of the paragraph.

1. Electronic data storage uses less space.
2. The system is now being tried in several divisions.
3. Some resistance to the change can be expected only because it is new.
4. For several years the hospital has studied the use of computers in record management and storage.
5. Malpractice suits require records to be kept for long periods, often for 20 years in the case of births performed at the hospital.
6. The administration is committed to electronic record storage since it saves time, money, and labor.
7. Computers are wonderful machines which can process thousands of items of information in seconds.

EXERCISE 4–3

In the following examples, revise the sentences so that they exhibit development. Some of them need nothing more than transitional elements and others need a complete overhaul.

1. You should neuter your cat. The reason for this is to protect your cat from future illnesses of this sort. This will also protect her offspring from the same illnesses.
2. No representations are made by us that our service will or can render equipment absolutely free from occurrence or re-occurrence at any time of such items as failure of the elevator to level off at landings, eccentricities in the operation of hatchway doors or car doors or any defects not ordinarily revealed by the customary inspection and testing methods used by us.
3. Prepare a form 10–37B for each document change. The form should be prepared with a No. 2 pencil. The left-hand column should be left blank. All dollar figures should be in thousands of dollars. All completed forms need to be verified by the financial officer.
4. The service area will include the entire country. It is a considerably large area. The area is rural. Some low-income residents live some considerable distance from the facility. In these cases, the people may be provided with bus transportation. Some may have car pools.
5. This procedure was designed to provide greater emphasis on individual accounts by our regional offices. By giving regional managers increased responsibility and flexibility for selecting, scheduling, and reporting work, greater use is made of our regional office staff and reports are issued soon after preliminary work is completed.

EXERCISE 4–4

1. From business letters, memos, and reports find five sentences that can be improved. Write the sentences and your revisions. Explain what you did and why for each example.

2. Analyze the development of a paragraph taken from a document that you thought was particularly clear. Explain how the material was developed and give an explanation for the effectiveness of the paragraph.

3. Find a group of three or more sentences that do not hang together. Explain the lack of coherence or unity in the selection, and rewrite the sentences to form a well-organized paragraph.

4. Compare two paragraphs you select as good and poor examples of development. Explain the similarities and differences between them, and revise the poor example.

Part II

Types of Writing: Memos, Letters, and Reports

PRELIMINARY
CHECKLIST

Deciding Whether to Write, Use the Telephone, or Meet in Person

Write   if you need to communicate the same information to many people.
if a record of the exchange is required.
if the information is complex, requiring the reader's concentration.
if the message contains a great many details, numbers, facts, drawings, and similar data.
if rules, regulations, company policy, or custom require a written statement.
if you want to emphasize an idea, proposal, or procedure.
if you want to capture someone's attention.

Telephone   if you need to speak with an individual or small group.
if you have a brief message to communicate.
if you need a simple question answered.
if you would like information sent to you.
if you need to check dates, times, places of meetings or appointments.
if you need to negotiate from a weak position.
if you want to make personal contact or set up a personal dialogue.

Meet   if you need to talk with several people in a discussion.
if you want to make the message personal.
if you want to show that you are involved.
if you want to discuss information.
if you want information, figures, drawings or other visual data explained to you or by you.
if you want to make a presentation of ideas or completed work.

Follow Up   if you want the information exchanged orally confirmed or
in Writing   verified.
if you need explanation or clarification of information gathered on the phone or in the meeting.

Don't Write   if you do not want a record of the conversation.
if the information is routine and insignificant.
if you have nothing further to say.

5

Useful Memos
of Substance and Selectivity

The memo is the most common of all business communications. It can take many forms, from casual scrawls on scraps of paper to lengthy documents similar to reports. Ordinarily, though, a memorandum is a brief reminder or a short record of an agreement or exchange.

Effective memos are short and emphasize major points at the expense of details. Since you usually direct a memo to a person you know, it should exhibit a natural, informal tone. But remember, too, that a memo is also a written record.

Most organizations have forms for memos and, more likely than not, they have circulated some guidelines for their use within the company. Nevertheless, the following general list should help you write a clearer, more direct memo no matter what the format.

Effective memos:

- place important information first
- address a carefully defined audience
- focus on only a few topics, ideally just one
- identify options
- communicate the importance of the information
- employ an informal tone
- provide clear directions

Use a memo to:

- provide a written record of an oral transaction or communication
- serve as a short report on a minor project, or on the progress of a major one
- create a source for future reference
- document actions

DECIDING WHEN TO USE A MEMO—NOT CRYING "WOLF" IN THE OFFICE

Since it is a short, written communication, the memo is often abused more than properly used. For example, in your eagerness to move up the corporate ladder you distribute memos to everyone from the chief executive on down. Often inexperienced individuals engage in this blatant use of them as a personal newsletter.

However, the result is just the opposite of the intention. Most recipients of memos that do not concern them become irritated at receiving what amounts to organizational junk mail. Don't waste their time in your effort to appear busy or hard working. Spend your time on necessary and important paperwork. Though some of your colleagues generate memos, snowing paper with their signature throughout the organization, they soon earn a deserved reputation for generating wind. Few, if any, people give much attention to their material because so much of it wasted their time in the past. They have been crying "wolf" in the office. The more you bring attention to routine and ordinary matters, the less impact you will have when you need it. Sure, you will attract attention initially, but attention that casts you as unprofessional.

Memos have the reputation in many organizations as the dumping ground for just about everything that goes on. Much of that abuse rests on the common misconception that the quantity of paper somehow equals quality. Instead of sending a direct message, many people write a small pamphlet where a paragraph would do. For instance, take this memo from a lawyer to a hospital administrator. Figure out what the lawyer wanted to say, and determine the impact of this memo.

```
Memorandum to: Mr. T. R. Ewing
               Assistant Administrator
          re: Hospital Pharmacy

    I am writing this memorandum in response to your request
that I put in writing the opinion I previously gave to you with
respect to Section 6816 of the Education Law relating to
Pharmacy.
    You inquired as to whether a pharmacist was authorized, in
filling a prescription, to provide a lesser quantity than
that provided in the doctor's prescription.
    Section 6816 of the Education Law, so far as material to
this inquiry, provides, as follows:
    "Any person, who, in putting up any drug, medicine, or food
or preparation used in medical practice, or making up any
prescription, or filling any order for drugs, medicines,
food or preparation puts any untrue label, stamp or other
designation of contents upon any box, bottle or other package
containing a drug, medicine, food or preparation used in
medical practice, or substitutes or dispenses a different
article for or in lieu of any article prescribed, ordered, or
```

demanded, *or puts up a greater or lesser quantity of any ingredient specified in any such prescription,* order or demand than that prescribed, ordered, or demanded, or otherwise deviates from the terms of the prescription, order or demand by substituting one drug for another, is guilty of a misdemeanor***"

I have italicized that part of the section which is pertinent to the inquiry and requires interpretation.

It may well be argued that the italicized words relate solely to the component ingredients of a particular prescription and not to the total volume or quantity thereof. However, it is my opinion that the purpose of this Section is to restrict the pharmacist from varying the prescription written by the physician and that the words "greater or lesser quantity of any ingredient" would encompass not only the component elements of a prescription but also the quantity of the entire prescription. It appears perfectly obvious that with respect to many prescriptions the furnishing to the patient of quantities in excess of those prescribed may have injurious effects and, of course, limitations are prescribed on prescriptions which may be re-filled and the numbers of times they may be re-filled.

There would be nothing in the statute to prohibit such practice unless it falls within the range of a greater quantity of any ingredient specified. If this interpretation is placed upon greater quantities it must also apply to lesser quantities. While it may be argued that the filling of a prescription with a lesser quantity than prescribed is less likely to present any harm to the patient there may well be circumstances where injurious consequences may result.

In any event, it is my opinion that the pharmicist is under obligation to conform to the directions provided by the physician in his prescription and not to substitute his judgment for that of the physician. There may well be situations where excessive quantities may be prescribed and under such circumstances it would certainly be good practice for the pharmacist to bring this to the attention of the physician for appropriate correction. There may also be other circumstances which would make it inadvisable for a physician to prescribe large quantities of a medication subject to deterioration before they can be utilized. These, however, are matters within the province of the physician and not the pharmacist. If the practice of certain physicians in repeatedly prescribing excessive quantities of medication is apparent this may be brought to the attention of the Center's medical staff for appropriate consideration by the physicians on that staff.

F. A. Logan
General Attorney

cc: Mr. B. G. Pillar
    Mr. A. R. Dorson

This memo wastes the administrator's time. T. R. Ewing only wanted F. A. Loophole to answer the question with a yes or no. Of course, the lawyer eventually did that, but it took the administrator a while to find it. In fairness, the memo is also for the record, and in that sense the task could have been handled a bit differently. The lawyer could have attached a short answer to a formal explanation that included an interpretation of the law and a justification for the conclusion.

As it reads, the administrator cannot act on it quickly. Here the establishment of a record overshadows the thrust of the main purpose—action. Of course, a lawyer must cover all bases, but that does not excuse wordiness and inefficiency. After all, the report in memo form merely states a rather conservative interpretation of the law in order to guide hospital policy concerning the change to uniform doses in the dispensing of medication.

POSTERIOR COVERAGE AS A MOTIVE FOR WRITING

A rule of thumb, let's call it the Law of Posterior Coverage, divides the number of memos you write to cover your actions by the total number of memos to get a fraction or percentage. If the number ends up to less than 10%, don't worry, from 10% to 50% indicates something is wrong either with your own performance or more likely your supervisor's ability to perform. Above 50%, revise your résumé and start looking for a way out of the organization.

Memos written to protect yourself present a good barometer of the health of your company, and your place in it. You might find that you are writing many memos to protect yourself from an individual in the organization who uses memos to build himself up at the expense of fellow employees. Office politics of that kind require other skills on your part.

Another way to protect yourself is not to write a memo in anger. Words spoken are forgotten over time. But writing remains for as long as the paper exists. When you sit down to express dissatisfaction in writing, do not send it right away. Wait a day or so, then read over the message to see if that is really what you intended to say.

DETERMINING WHO SHOULD RECEIVE A MEMO

On memo forms, spaces for the date and the memo writer present no problem. The subject line, sometimes designated by the Latin preposition *re,* acts as the title and should be direct and businesslike. Newspaper-like headlines or cute titles are inappropriate because they make your message appear frivolous and unprofessional. For example, something like "Rub a Dub Dub" makes a silly title for a reminder to assembly-line supervisors that employees who handle white furniture parts must keep their hands clean.

After the "Subject" line, the "To" line creates problems. Effective memos address specific persons or groups. For example, if you are confirming the discussion of a project and are asked to present the results of the meeting for a committee of 30 people, then you need not list all the people, merely address it to members of the committee. The memo might also be distributed to individuals by the demands of company policy or organizational protocol. Avoid over distribution, that is, sending the document to every official in the organization regardless of policy or their need to know. If you have doubts about whether or not the memo should go to a particular individual in the organization, ask. Your immediate superior or an experienced secretary should be able to tell you if the memo has been distributed that way before. The larger the organization, the more specific are the guidelines for distribution. Large banks, for example, prepare secretarial handbooks that also list the usual types of memos each division receives.

As we mentioned, if you send copies of your memos to the wrong people, you will certainly attract attention. However, the unwritten message you send in this case might be that you are overly ambitious or, worse, that you are uninformed or do not care about guidelines.

As we move closer to the automated office, and as more and more computer manufacturers offer machines that provide for electronic mail—that is, sending messages through computer terminals—the ability to send memos rapidly to everyone in a company increases. Instead of writing a memo, having it typed and either photocopied or mimeographed, we can have the computer network distribute the material electronically. Though a great time and energy saver, it has the potential for wide abuse. Initial uses of the system in some Wall Street firms produced a flood of memos that ordinarily received minimum distribution.

The technology of electronic communication has a double edge. The power of the computer also magnifies the errors of the writer. The new technology demands greater knowledge, sophistication, and ability when writing on the job. (See Chapter 12.)

Like a letter, the tone of a memo should be professional and businesslike. That does not mean formal, since their nature calls for informality. You do not have to include, and in fact should not include, small talk or messages of goodwill.

Memos need not be signed like a letter. Most people simply place their initials next to the name at the top. Some sign their names in full. Others leave memos unsigned. All three are acceptable. Barring company policy, follow your own preference.

Hand-Written Memo

Although most memos should be typed, you can write in longhand if you desire a personal touch. Make sure that you write legibly if you do this.

Tone in Memos

The tone of any piece of writing reflects not only the attitude of the writer toward the subject, but also toward the reader.

MEMO SITUATION CHECKLIST		
Situation	Advantage	Drawback
to everyone in the organization	rapid and simultaneous distribution of general information	Make sure the information is really for everyone
to specific departments	directs information to the group concerned	can slight a group or include one that is not involved
to positions	emphasizes the authority of a tier of employees—managers, supervisors, etc.; sounds official	can be impersonal, stiff
to several specific people	More personal, like a small meeting; efficiently informs a small group	can make some recipients defensive
to one person	personal; only one person gets the information	can make information sound too personal
from group or committee name	clarifies positions and goals set by the group; sounds official	a broad distribution might create conflicting responses
from department or office name	sounds official and authoritative	sounds impersonal so be sure the content is appropriate
from position or title	impersonal; stresses the company organizational structure; formal	places distance between you and the reader
from first name or titleless name	personal	can diminish some of your authority

Memos written for circulation outside the company should be thought of as letters. (See Chapter 6.)

A memo that remains inside the organization should have a natural, straightforward tone. It should reflect the informal working relationship that you have with the people in your office. Formality in a memo to a colleague appears awkward, as if you had something to hide. Formality creates distance. Examine this memo to determine its tone:

                                              September 15, 19XX
TO: All Members of the Company
RE: Restatement of Package Inspection Policy

   There are many aspects to a good security program; package
inspection is one of them. Accordingly, the Company reserves
the right to inspect all packages coming into or leaving the
premises. Packages, for this purpose include such items as
briefcases and shopping bags. Responsibility for inspection rests with Security personnel.
   Everyone's cooperation is essential to making this and all
security programs effective. I know that you will continue to
help in our efforts to maintain a safe environment.

This memo concerning the inspection policy is direct. The tone appears natural and, considering the nature of the message, almost friendly. A different attitude could have created a far different effect. A tone that reflected the actual concern over the increase in theft and the alarming frequency of assaults and other crimes on company property could have created undesirable misinterpretations.

This memo could have read:

   The rise in theft and the increase in the number of reported
assaults on the company property compel us to enforce strictly the policy that all employees wear their ID badge at all
times while in the facility. Also, all packages, handbags,
briefcases and shopping bags will be inspected on leaving the
building. Anyone who does not cooperate with the security
guards on this will be subject to severe penalty, perhaps
leading to termination.
   Your cooperation is required to keep the company safe from
violent crimes and to protect our capital investment from
theft.

Written this way, the memo certainly offends almost everyone. Here the tone reveals a siege mentality, and implies that all persons in the company are violent criminals or thieves. Obviously, the exaggerated tone of the rewritten memo presents a point—the information you send out in a memo is always taken personally, even though it may be sent to all members of the company. The chance of misinterpretation is immense.

In a related instance, a consulting engineer sent around a memo that caused the same effect. He had noticed that his engineers spent a lot of time around the coffee machine and the water cooler, engaged in what appeared to be idle conversation about baseball or inflation. He wrote a memo informing the entire staff of his observation, and suggesting that they get back to work. The memo hit like a torpedo. To restore office morale he had to meet personally with each staff member to reassure them that they were doing a good job. The implication that he was goofing off angered each employee. All said that the work was getting done and that it was quality perform-ance. It took the consulting engineer almost a week to repair the damage done by a memo that had an ill-conceived tone.

Remember that a human being reads the memo you write. Tact and courtesy temper the impact of bad news or disagreeable information.

Excluding Irrelevant Information

Memos often present unnecessary, uncalled for, or just plain irrelevant in-formation. If you usually include everything, and leave it up to the reader to sift through for relevant facts, you have put in useless information. More important than wasting the reader's time, you run the risk of providing facts and information that might suggest a conclusion or action you do not want.

To check for this, look at each sentence in the memo and ask yourself what the sentence contributes. If your answer reveals that the sentence adds little or nothing, cut it out, or combine the needed information with another sentence.

Including Important Information

If you have carefully determined the *audience* you are writing toward, and you have identified your *purpose,* that should define the important information. Take the analysis of audience and purpose and check through the memo to see if the important points appear in the memo itself. If an item is missing, write it in.

Ask yourself if every word and every sentence contributes clearly and directly, without offensiveness, to the point. If you can say that every word belongs, then you have written an effective memo.

EVALUATING MEMOS—WHAT TO LOOK FOR

One effective way to improve the quality of your memos is to look critically at some you have written, and some that you have received. The first step in the evaluation of a memo requires criteria that you will use to identify the strengths and weaknesses in any example. The Letter, Memo, and Report Evaluation here provides those criteria and aids in evaluation.

This scorecard will help reveal parts of the memo that could be improved, rethought, or rewritten. It also furnishes a device to analyze the message and identify its use of style, language, and organization.

The ability to evaluate and analyze a completed memo establishes criteria for critical judgment that help you write memos more directly, understandably, and professionally.

The evaluation criteria form three groups—Language, The Whole Memo, Letter, or Report, and Style. In scoring each of the 15 items, give 0 for an average rating, +1 to +3 for above average, and –1 to –3 for below. Any example, of course, has both strong and weak portions. Because of the negative values for scoring, these cancel out each other. In other words, a total score above 0 indicates a generally useful memo. Any score below 0 suggests that the example needs an overhaul for some or all of its elements. The high score of +45, and the low of –45 represent limits. Scores that approach those extremes, a +39 or a –42, for example, indicate that the evaluation has not been carried out seriously enough.

The point here is not the total score. The scorecard, in fact, has no line for a total. Instead, the evaluation emphasizes the identification of parts of the example that need further work.

Letter, Memo & Report Evaluation

Take any sample writing, and rate it according to the items listed below. Score each from –3 to +3. Use 0 for an average rating.

Language
  1. Clear (unambiguous words)   1._____
  2. Concise (brevity without sacrificing clarity)   2._____
  3. Correct (accurate word chosen)   3._____
  4. Concrete (words the reader can picture)   4._____
  5. Courteous (positive attitude, reader oriented)   5._____

The Whole Memo, Letter, or Report
  6. First things first (answers who, what, when, where, why, and how
      as soon as possible)   6._____
  7. States purpose clearly   7._____
  8. Addresses reader appropriately   8._____
  9. Includes important details   9._____
10. Excludes irrelevant information   10._____
11. Illustrations are relevant to the purpose   11._____

Style
12. Variety of sentence lengths   12._____
13. Concise phrases   13._____
14. Clear reference of modifying phrases (no unintentional ambiguity)   14._____
15. Active voice used   15._____

If, for instance, the scorecard indicates positive numbers in the Language section, and negative numbers in four of the six items in the second part, that reveals the need for additional work on the organization and arrangement of the material. For the device to be useful, however, the scoring must be candid and straight-forward.

What to Look for in Each Category

The Language section concerns the selection of words, and how well they are suited to the whole message. The next deals with the organization and arrangement of the whole document. The last section of the evaluation form on style focuses attention on the effective presentation of facts and ideas.

Each of the 15 items concerns a different element of writing. In each of them you look for a particular demonstration of word choice, organization, or style.

1. *Clear*. Look for the selection of words and phrases that are not intentionally ambiguous. Also look for words that are direct, sentences that do not beat around the bush, and phrases that avoid language inflation. If a few exist in the example, rework what is there to supply clarity.

2. *Concise*. Look at the way an idea is expressed. If the writer uses few words and the meaning is clear, then the memo is concise.

3. *Correct*. Unlike the Cheshire Cat in *Alice in Wonderland* who proudly asserted that a word meant whatever he wanted it to, a business memo or report must use the standard definition of the word. That should not imply that you chain yourself to the dictionary, but rather that you should consult it as a source of common agreement on definitions and spelling.

4. *Concrete*. The use of concrete rather than abstract language makes information and ideas clear to your reader. It also enables you to keep the information under your control. For instance, you might request that your office staff attend a "social function" on Friday afternoon, when all you wanted them to come to was a brief birthday party for your mail clerk who has just turned 50. In addition, concrete language allows you to refer to the memo later and quickly understand its contents. Confusion results from a letter or memo that you cannot understand. When you look at the "from" line and see your own name there, despair replaces confusion. If you wrote it and you cannot understand it, then the abstractness of the language needs to be revised with concrete, specific expressions. If not, no one will understand your meaning.

5. *Courteous*. Check the memo for positive expressions. Remember that positive statements result in favorable reactions from the reader more often than negative ones. In another section of the book we noted that the negative expression, "no cure for cancer has been found," cuts off the reader. Stated positively, "Research-ers continue to look for cures for cancer," not only makes a valid statement, but also gives the reader the impression that even though the answer has not yet emerged, someone works on it.

   A courteous message also considers the reader's needs, and directs information toward him. For instance, you might solicit contributions to the annual charity drive with this: "I am interested in having our division of XYZ Corp. top all others in the company in contributions to the BCD Fund." Such an expression seems self-serving. Something that meets the reader's needs might appear this way:

"Your contribution to the BCD Fund demonstrates your concern for the sick and needy."

6. *First Things First.* As a general rule, come straight out with information in order of importance. Don't beat around the bush. Brainstorming helps determine which idea is the most important.

7. *Purpose Clearly Stated.* Make your reader aware of your goals. If you wish someone to consider your plan to reorganize the division, come out with it. However, if you know the idea will meet stiff opposition, build up to it with inductive development—organization that moves from the particular to the general.

   Give high marks to a memo that exhibits a clearly identified purpose. To check for this, see if the "Subject" line tells what the memo is about, and if the memo itself bears out the expectations raised there.

8. *Addresses the Reader Appropriately.* Here look at the "To" line. Is the document written to one person or to many? Is it written with the idea that it will become part of a file? Also determine whether or not a memo has the appropriate attitude toward the reader. (Refer to the "Memo Situation Checklist.")

9. *Includes Important Information.*

10. *Excludes Irrelevant Information.* These two elements of a memo go together. You can get high marks on the inclusion of information and not on the exclusion. Since you feel uncomfortable with flat statements, you hedge, adding heaps of detail to strengthen the position. Remember that relevant information clearly supports conclusions and recommendations.

11. *Illustrations Relevant to the Purpose.* Be sure graphs, tables, and drawings are useful in illustrating the purpose for the audience. Select the examples that best illustrate your point. In scientific reports, follow the scientific method, and be sure that you choose representative examples. Slanted impressions can result from examples that are exceptions.

12. *Variety of Sentence Lengths.* A well-written memo employs a variety of sentence patterns and lengths. Variety allows the reader to understand the information quickly and without confusion.

13. *Concise Phrases.* Writing gains power and clarity through words and phrases that get to the point. For example, "because" easily replaces "due to the fact that," adding strength to your message.

14. *Clear Reference of Phrases.* A business communication should not leave much room for interpretation. It should not be ambiguous. Unlike short stories, poems, plays, and novels which depend on ambiguity for depth of meaning, a business communication strives for one clear meaning.

15. *Active Voice.* The active voice adds power and directness to your writing. Use it. Remember our question—Who kicked the dog? An active voice construction will always help identify the person or group responsible for an action.

You can use the scorecard to analyze your own memos, or ones that you receive. The 15 categories apply to any memo, and should help to identify the strengths and weaknesses in your writing. They should also enable you to pinpoint the trouble spots so that you can revise the memo, or avoid the weaknesses in the next memo you write.

Another memo follows. Evaluate its effectiveness with the scorecard. Read the memo over completely, then give it a score for each of the 15 items.

Interoffice Communication
October 17, 19XX
Memorandum to: Steven Green
From: Will Smith
Subject: CPR—mobile Problems—CCU and OR

Your concern regarding the subject Medikco equipment is certainly justified. To have a $33,000.00 new piece of equipment not functional makes no sense. Upon receipt of your 9—23—XX memo I contacted:

A. The Manufacturer
B. Our Accounting Department
C. Our Messrs., J. Bogst
                H. Nelson
                M. Steel

and have determined the following.

1. We indeed own the CPR—mobile located in the 15th Floor CCU. This piece of equipment was delivered in mid February, 19XX and paid for in May, 19XX.

2. The unit in the Operating Room was repaired on 9/26 and is now in working condition.

3. The current leakage problem as discussed in paragraph two of your 9/23 memo has caused our detection system to alarm. Other factors do exist to create this problem.

4. Medikco has stated that the 460 microliter "leakage" as described is normal and that no specifications in our order required more sophistication. A review of the order verifies this.

5. Medikco offers a 3KVA 115 to 115 isolation transformer to correct this situation. They claim that this addition will reduce leakage to less than 100 micro amps, which I have been advised is acceptable. The cost of this transformer is $500.00 delivered and installed.

6. The reason the unit cannot be used in other locations than the CCU is only because of the wall plug. Mr. Cando and Mr. Passbuk both have told me that an adapter would solve this problem. This could be done "in house."

In view of the above, I suggest that:

A. We wait for delivery of the table on order.
B. Purchase the isolation transformer.
C. Have maintenance supply the adapter.

Your comments?

_____
Will Smith
cc: Dr. Albert
    J. Bobst
    H. Nelson
    M. Steel

Before you evaluate this sample memo from Will Fixit to Steven Grene, you may have had a feeling that something in it was not quite right. The scorecard helps to reveal weak areas so they can be strengthened.

Your score for the first section on Language might give the sample average marks on clarity. The memo is generally direct, and words do not have any unexpected ambiguity. It is, however, not startling in its clarity, but more like the windows on your car before a wash, not as clear as they could be. For this we might give the memo a 0, indicating it is neither poor nor spectacular in its clarity.

The next three categories—Conciseness, Correctness, Concreteness—also deserve average marks. The writer could have said "working" for the phrase "now in working condition." The infinitive "to alarm" is used incorrectly since its transitive implication requires, in this context, an object. We ask, "Alarm who, or what?" The correct use would express the idea as, "The leakage problem tripped the alarm."

You might have given a mark as low as a –3 for First Things First because important items do not appear near the beginning of the memo. Obviously Steven Grene must act on the information in the memo, and it should be presented with that in mind. The order of the memo determines what the reader will know first, and also what should be done first.

The inverted order used in many newspapers is effective in memos. (See Fig. 5–1, The Inverted Pyramid.) Such an order assumes that your reader is busy and has hundreds of pieces of paper to sift through weekly. With that in mind, ordering information for the reader makes you stand out as a manager. A disorganized memo implies that you have a slight grasp of the material. At best it indicates you are overworked and have no time to write a clear memo.

Take the extra time and, if necessary, the extra expense of retyping the message so that it can be acted upon efficiently. In the case of this sample memo, Grene had to call Fixit for an explanation of the memo, using time that could have been directed toward correcting the problem.

In addition to the order presented in this memo, the score you might have given to the statement of *purpose* could also be as low as –3. A clue that the writer was confused about the purpose of the memo appears in the "Subject" line. Is this memo really about "CPR-mobile Problems?" More accurately, it is "Correcting CPR-mobile Problems" or even "Identifying and Solving CPR-mobile Problems." The title gives away Fixit's failure to determine the purpose before he wrote the message. Answering questions about the purpose helps you avoid a similar situation.

The memo is, of course, addressed to Grene, but it will also go out to the people who will have to do the work. So the message, though not literally addressed to the maintenance and engineering departments, should have them in mind. Ultimately, Grene must get the word to them to do something. That brings in the next three items.

This sample memo would have properly gotten low marks for its use of details, its inclusion of relevant information, its exclusion of irrelevant information, as well

The Inverted Pyramid

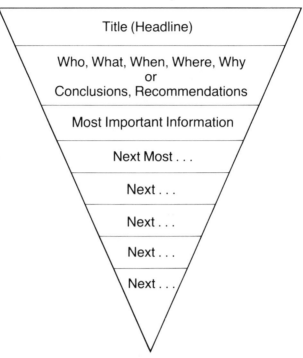

Title (Headline)

Who, What, When, Where, Why
or
Conclusions, Recommendations

Most Important Information

Next Most . . .

Next . . .

Next . . .

Next . . .

Next . . .

FIGURE 5–1   The Inverted Pyramid Structure of Writing

as its selection of the details to include. This memo appears to include *all* the details, as if Fixit pulled the CPR-mobile file and put the information down in the memo as he leafed through the bills, order forms, repair and warranty information, and the bills of lading. That might be fine for information at the end of the report, or information in an appendix. But the reader must know what to do, who to contact, and when the problem will be corrected.

After going this far in the memo, you should have already identified the weak points—order of presentation, identification of purpose, determination of audience needs, selection of relevant details. Even though these items are part of a memo that is average in its use of language, these elements that pertain to the whole message make it an ineffective total message.

The style portion of our example is also average. However, when we examine those elements we realize that the problem does not lie with the use of the active voice over the passive. Instead, the whole memo is weak because the information itself is presented with little regard to its organization, and to the use the information serves.

Before you write your next memo, review the 10 steps necessary for writing. That will help you avoid the problems demonstrated in our sample.

SUMMARY

A memorandum is a brief reminder or short record. Effective memos emphasize major points, exhibit an informal tone, and are short. Use memos to document your own actions; avoid using them as a personal newsletter.

Direct a memo to the proper person or group. When you address a colleague make the tone reflect the informal working relationship that exists among people in the office.

Include relevant information; exclude irrelevant details. Evaluate your writing for careful word choice, organization of ideas, and style. Memos that produce action use clear, direct, courteous, and unambiguous language. They get to the point, and use strong, active verbs.

EXERCISE 5–1

Evaluate the following memo, using the scorecard. Identify strengths and weaknesses. Pay special attention to the organization as well as the style.

```
November 11, 19XX
MEMO TO: Mr. M. Watson
         Mr. G. Johns
FROM:    Mr. W. Harvey
RE:      ELECTRONICS FOR LEASING
    As mentioned in my note of 11/8/XX, I met with Mr. Cod of
E.E.C. today. Please refer to your copy of my letter to E.E.C.
of 10/29/XX and the problems noted thereon. These "problems"
resulting from our meeting of October 26. The result of my
meeting with Mr. Cod:
    1. As you know, we received a lump sum of $140,000.00 at
       start of agreement. The monthly payments of $5,140.00
       represent the guaranteed yearly minimum total, broken
       into 12 equal payments (as originally agreed and re-
       quested).
    2. Daily Rental Information—this information is kept by
       the attendant and forwarded to E.E.C. accountants on a
       weekly basis. The report notes:
       -total sets available in Big Data Corp.
       —  ''   ''   rented.
       —  ''   ''   on "complimentary" bases.
       —  ''   ''   out for repair (these are replaced with
       working sets).
    3. Collection & Loss Information—on a 2 week cycle, the
       E.E.C. accounting people have a "run" of accounts re-
       ceivable which includes monies uncollected and/or un-
       collectable. I am advised that we are running between 7%
       & 8% uncollected which E.E.C. feels is not too bad. It
       seems most uncollected monies are due to a) death of
       patient, b) complaint set did not work or was never
       turned off, c) had address or name, d) just plain no pay.
```

These reports will be forwarded to us as soon as possible. I'm having them come to me, for now, to see if they will be of use. We used to get these (sent to Mr. Parsens) but was stopped as being too cumbersome. I'll keep you both advised; probably ask for a meeting to review and discuss.

    4. Attendant—is, as set up by E.E.C., an "independent contractor" paid a salary plus incentive bonuses. He is insured, through E.E.C. and we will receive a copy of same.

Re: console unit for turn-on—E.E.C. will check out cost and advise.

That is it, to date. We should meet as soon as we get the first update. Please advise.

EXERCISE 5–2

1.   Select three memos you have recently received, and three you wrote. Decide which of your examples get to the point effectively, and which do not. What is the weakness of the ineffective items? Rewrite them.

2.   Compile a list of jargon and inflated phrases used in memos around your office. Translate the terms into direct, plain English.

    Show your list to others in your organization and get them to comment on and discuss the terms with you.

3.   In memos you have received or written, identify sentences that use passive voice, as well as inflated words and phrases.

    Rewrite the sentences, using the active voice. Write the inflated terms in plain English. Incorporate these changes into a revision of the memo.

4.   Examine a memo of your own or one you have received from outside your organization. Find the main ideas. Rearrange the information according to the order you think best fulfills the purposes of the writer as well as expectations of the reader. Try to form a generalization based on this particular case.

EXERCISE 5–3 (ADVANCED)

Read the following eight memos carefully. Evaluate two of them, using the scorecard. Write a memo in which you present to each of the two writers the results of your evaluation. Make suggestions that would improve the memos they write in the future. Along with your comments and recommendations, provide a revision of each memo that demonstrates the improvements you suggest.

Example 1

Date: February 4, 19XX
For: My Insco Associates

Subject: Group Life Insurance—Imputed Income

    In certain cases, the amount of inputed income included in the W-2 reports of Insco employees increased significantly for 19XX. Imputed income calculations, which are based on total group life insurance coverages (including spouse's benefit), reflected the following changes in our group life plan in 19XX:

- For the spouse's benefit, an increase in the basic monthly benefit formula, an improved cost of living feature and removal of the remarriage forfeiture provision.
- The availability of additional coverages, including new monthly income options and a fourth level of insurance.
- A decrease in contribution rates, at all ages.

Since insurance amounts are tied to compensation levels, individual coverages were enhanced by salary increases occurring during the year. Additionally, employees celebrating quinquennial birthdays are charged with increased imputed income rates under applicable federal regulations.

We have closely examined the actuarial and legal aspects of the 19XX imputed income calculations, and we are satisfied that the procedures which were used are appropriate.

I trust this information is a helpful aid to your review of your 19XX tax status.

Example 2

February 12, 19XX

IMPORTANT NOTICE

TO: All Faculty
FROM: C. M. Bush, Dean
SUBJECT: Flu

You are probably aware that there is a lot of flu around but may not be aware of its extent. Students have begun calling our office to ask assistance in obtaining extensions of paper deadlines and making arrangements not to be penalized for missed examinations.

We checked with Dr. Saul Benson, Medical Director of the University Health Service (Infirmary). He said the Health Service is seeing more than 40 students each day with flu symptoms. This particular flu is characterized by high fevers. Recovery takes a week to ten days. Because of the high fevers it would probably be unreasonable to expect much intellectual work to be completed during that period. Unfortunately, flu is highly contagious, especially in a dormitory living situation.

None of us, especially the students, need this additional complication in our lives. Please do what you can for students with the flu by being flexible about deadlines for written work and by arranging either make-ups or forgiveness for missed exams or laboratories. (If your course grading system allows one missed exam, this can be it.)

Thank you for your cooperation.

Example 3

TO: Ms. Jean Alexander
FROM: Pure Researcher

In view of Mr. Joel's extensive research experience including work in instrumention, computer design, programming and management, he will be very invaluable to the
present phase of the research project since it is imperative
that we change from manual to fully automated control of our
instruments to enable unattended and sustained test programs continually throughout the night, weekends and holi-
days. [This should result in a more rapid determination of
the sequence of UK due to the increased massing of data during
the unattended periods of which would otherwise not be util-
ised, minimise loss and deterioration of precious samples,
maximize on shutdown opportunities for routine maintenance
to optimize equipment efficiency, exercise more precise
control of repetitive tests, etc.] For example, he will be
modifying the electronic recorder of the amino acid analyzer
for fluorescence measurements, and interphasing the high
pressure liquid chromatographic instrument to a micro-
processor. In order to facilitate the handling of massive
data obtained from amino acid sequencing he will assist us in
writing computer programs to handle, reduce, store and
correlate the data. Furthermore, he will assist us on various
aspects of the project with minimal supervision. Also his
duties will entail routine maintenance and trouble-
shooting in our laboratory set-up.

Example 4

January 28, 19XX

TO: ALL OUR TENANTS

Due to the necessity for Metro Power Company to replace the
steam to the hot water supply and heating supply in our
building.

We trust that this inconvenience will begin at about 8:00
AM and end by 12 Noon on *WEDNESDAY MORNING JANUARY 30, 19XX*.

Please understand that we will offer all the assistance and
cooperation to make the shut down period the shortest
possible.

Example 5

TO: Residents of 40 Park—19th Floor
FROM: Neil James
SUBJECT: Move to 10 Park—17th Floor

An ugly rumor has been circulating recently to the effect
that, within the next several years, staff located at 40 Park
would be kicked out of the office we have grown so fond of and
exiled to some remote northern location. A careful check has
revealed a few shreds of truth are to be found within this
malicious tale. Specifically, renovation has begun on the
17th floor of 10 Park and according to current plans, we are
scheduled to be moved to 10 over the weekend of October 4/5 or
11/12 (19XX).

While further instructions will no doubt be available
shortly before the move, we can make the relocation process
simpler in a number of ways by removing all the papers, books,
bicycles, etc. which have accumulated over, under, around
and through our desks and other furniture, and discarding the
things we no longer need. It may also make sense to forward
certain items to central storage facilities out of the city
if they will not be needed for an extended time.

Your supervisors may have some more specific suggestions
about what kind of clean-up makes sense for your unit but in
the mean time, please start thinking about how we can get rid
of our surplus materials before early October.

Many thanks.

Example 6

TO: Residents of 40 Park—19th Floor
FROM: Neil James
SUBJECT: Move to 10 Park—17th Floor, #3

The latest word is that our move to 10 Park will be over the
weekend of October 18—19.

Please let Nancy Wind know by September 26 how many card-
board boxes you need to pack the materials which cannot be
locked inside your desks, credenzas, etc. and we will order
them for you. These boxes should be packed by the end of day,
October 17.

For those who do not have experience in folding and packing
boxes, a special ten hour crash course will be offered the
weekend of October 11—12. Tuition refunds will be paid by the
company for those receiving grades of "C" or better.

Successful completion of Folding and Packing will result
in two credits toward the Advanced Administrative Services
Certificate. Please see me if you are interested in reg-
istering.

Example 7

December 28, 19XX

MEMORANDUM TO: Chairmen, Deans, Vice Presidents, Faculty and Department Heads

SUBJECT: Signing of contracts or making commitments on behalf of the Hospital and Medical School

It has come to the attention of the administration that members of the faculty and other personnel of the Hospital without authority have been signing letter agreements and other forms of contracts or making oral commitments purporting to bind the Hospital and the Medical School. Several of these agreements, which have not been reviewed by appropriate administrative officers or counsel, have presented legal problems to the institution. Continuation of this practice may not only result in serious consequences to the Hospital but may well subject unauthorized persons making such agreements to personal liability.

Except in those cases where specific written authority has been granted to designated individuals, *NO CONTRACTS, OTHER DOCUMENTS, OR ORAL AGREEMENTS PURPORTING TO BIND HOSPITAL AND MEDICAL SCHOOL OR ANY SCHOOL OR DIVISION THEREOF ARE TO BE SIGNED OR MADE BY ANY HOSPITAL OR SCHOOL PERSONNEL.*

In any case where a signature is required on behalf of the Hospital, its Schools or divisions, to a contract, agreement or other legal document, the same should be transmitted to the administrative offices of the Hospital or, where appropriate, to the Office of the Dean, Office of Sponsored Programs, or the administrative office of the Hospital or the Medical School. Where necessary, the documents will be reviewed by legal counsel and arrangements made for signing by authorized signatories.

T. W. Moover
Executive Director

I. A. Greenly
Dean

Example 8

June 5, 19XX

TO: All Company Employees

RE: Identification Badge Policy

It is the policy of the Company that all employees and other persons having regular business at the facilities wear photo identification badges.

Badges must be worn conspicuously at all times. They are to be worn picture up above the waist, on the outer clothing. Identification badges will be required for conducting business at the cashiers and General Stores and for entrance to selected areas.

Replacements for worn or damaged badges or for badges reflecting status changes will be issued free of charge. However, lost or stolen badges will be reissued only after payment of a five dollar fee to the Company cashier.

Effective July 7, 19XX the Security Department will assume responsibility for issuing photo identification badges.

Responsibility for compliance with this policy falls primarily on, but is not limited to, department heads, supervisory, and security personnel. Personnel throughout the Company are authorized to request unidentified individuals to produce valid identification or otherwise identify themselves.

J. Wilson Thomas
Executive Vice President

EXERCISE 5-4 (ADVANCED)

1. Write a memo to the appropriate committee in which you present a policy change. Be sure to include your assessment of the problem and justification for the change.

2. You have noticed a security report indicating a 10% increase in crime on the factory premises. In walking through the building you have personally noticed that a majority of the employees do not wear identification badges. Draft a memo to the entire staff concerning the need to wear proper identification at all times.

3. Jayne Johnson, a co-administrator, has asked you to help her evaluate a morale problem that exists in her unit. You spend the better part of Tuesday and Wednesday afternoons observing the section and talking informally with the staff.
   You find:

   - Poor lighting
   - Understaffing because of a hiring freeze
   - Cramped work areas
   - No in-service training budget
   - No promotions granted in the last 10 months
   - Pay of senior staff members barely keeps ahead of the new employee's starting salary
   - Forced overtime
   - Drab-colored walls
   - Arbitrary application of the rules by the supervisor
   - No supervisory recognition for work well done
   - No feedback from supervisor on procedures

   Write a memo to Mrs. Johnson outlining the problems you found and recommending changes for improved morale.

4.   You are in charge of a division of a large corporation. Among the operations in your plant is the chrome plating of electric coffee makers which are assembled in another part of the country. The director of quality control has informed you that the plating is only half that required by the specifications. This means that the line will be slowed down and the production will be cut in half.

   Write a memo in which you inform the company of the problem and outline your plans to solve it.

5.   You have just attended a three-day conference on a topic important to your company. Write a memo explaining what you learned there. Offer suggestions and recommendations based on the conference.

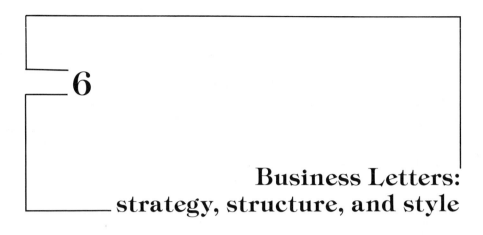

Business Letters:
strategy, structure, and style

The letter represents the most familiar way to communicate with someone outside the company. Though not as rapid as the telephone, a letter can be more effective in achieving results, in placing ideas in the right hands, or in building good will. A well-planned letter provides a record as well as a foundation for discussions and transactions. It also reflects positively on you, and speaks highly of the organization that has you on its payroll.

ADVICE FROM THE PROS

Malcolm Forbes, of *Forbes* magazine, offered his approach to writing business letters in an ad that was part of the International Paper Company's series, "The Power of the Printed Word." He placed letters into three categories: mundane, stultifying if not stupid, and first rate. In order to produce an effective letter, he recommends that you:

- call the person by name
- tell what your letter is about in the first paragraph
- be specific
- be positive, nice, natural
- keep it short
- be honest
- don't exaggerate
- be clear
- use accurate English
- edit ruthlessly

Forbes follows much of the useful formula that appears in *Plain Letters*, a pamphlet of the Records Management Handbooks series developed by the National Archives and Records Service to reduce and simplify paperwork. The government recommendations list four desirable qualities in letters: shortness, simplicity, strength, sincerity.

The 4-S Formula

Shortness

- Don't make a habit of repeating the letter you received in your answer.
- Avoid needless words and needless information.
- Beware of wordy prepositional phrases such as "with regard to" and "in reference to."
- Don't qualify your statements unnecessarily.

Simplicity

- Know your subject so well that you can discuss it naturally and confidently.
- Use short words, sentences, and paragraphs.
- Tie thoughts together so your readers can follow you from one to another without getting lost.

Strength

- Use specific, concrete words.
- Use active verbs.
- Give answers first, then explain if necessary.
- Don't hedge.
- Avoid negative words.

Sincerity

- Be human. Use the names of people and personal pronouns.
- Admit mistakes. Don't hide behind meaningless words.
- Don't overwhelm your readers with hyperbole, hype.
- Express yourself in a friendly, dignified way.

Both the government pamphlet and the Forbes suggestions describe well the qualities of a good letter. Both give sound advice and are useful for quick reference. A business letter, however, presents a writing problem that calls for a strategy, not a formula. The 10 steps we mentioned in Part I offer an approach to letter writing. Look at them again before you sit down to write.

STRATEGY FOR LETTERS

Brainstorming

Brainstorming techniques work well when applied to letters. Often you can write a draft of the letter in a 10-minute focused writing. That may help to reduce the time required to produce a letter. Starting cold, a one-page letter may take 60 to 80 minutes to complete. If you plan the letter, that time is certainly well spent.

Planning pays off in the reader's response to the message. Well-written letters create results; ones you dash off and send out in a few minutes may create misunderstanding and usually require future letters and memos of clarification. The Bureau of Land Management estimated that a poorly conceived and executed letter can cost as much as $579 to write and read. They further suggested that increased writing time often results in very large savings in reading time.

In other words, the time you spend planning—determining the purpose of the letter and ascertaining the needs of the reader—pays off. A letter written quickly may appear to save money by reducing your cost, but it reaps little benefit when compared with the costs incurred by others in reading and in generating clarifying documents.

Use the brainstorming techniques to determine the purpose of the letter. Write down clearly what you think the letter should do. Come up with one sentence that focuses the point of the letter for you. For example, you might write a letter to prospective clients: "This letter convinces executives that they can trust me to place them in top positions at powerful companies." Such a sentence, although it may not appear in the letter itself, helps you to understand what you do in the letter. You might say, "This letter should convince the manufacturer that I should get the contract for making the special fittings." Or, "After reading this letter, the bank and the regulating agency will agree that we can operate the expanded facility, and that we are worth the risk of a $250,000 loan."

Determining the Purpose

When *determining the purpose* of the letter, keep in mind the unwritten message that you send. That purpose may turn out to overshadow the others.

At the very least, the letter should say that you are professional and that you know the etiquette of business. Some companies spend a great deal of money on the design of stationery and letterhead to communicate the professionalism of the company. Don't let the content of the letter contradict that. Every letter that you send should promote your organization.

A clearly determined purpose makes the reader's job easier and, in turn, reflects well on you. Rarely will you get compliments, but you will not get irate calls.

Here is a letter from a town supervisor. The adminstrative assistant wrote the letter, and the official signed it. Occasionally you may have the job of writing a letter for someone else. Whether you sign the letter or not, you must still decide what you want the letter to accomplish. Examine this letter and write in one sentence what you think it was written to achieve:

```
                                        May 5, 19XX
Senator David Q. Jones
44 Senate Office Bldg.
Washington, D.C. 20510

Dear Senator Jones:

   As the Congressional budget continues to be scrutinized,
and cuts are made to benefits that so many of our constituents
rely on daily, I implore you to search your conscience before
voting for these proposed reductions.
   The elderly and the poor who receive benefits, the retarded
and handicapped, those who are unfortunate in finding them-
selves on unemployment lines, some for the first time in
their lives, veterans who served this country and now look to
us for help . . . can you say "no" to their needs? While these
people do not represent the affluent society, they should not
be treated as second rate citizens.
   Please give very serious consideration to the needs of
these people before you cast your vote.

                                        Sincerely,

                                        Martin Rivers
                                        Supervisor
```

Could you identify the purpose of this letter after reading it? Of course, but you may have taken some time to come up with a simple statement like: "Senator, please oppose budget cuts that would hurt the elderly, the retarded, the unemployed, and veterans." Like many letters sent to legislators, this one employs what might be unusual in a business letter, an appeal to the emotions. Most business letters appeal to financial concerns or the saving of time, energy, or money. Or they suggest a specific course of action. In this case, however, that appeal might just be the right one, considering the topic.

Even though the Senator happened to be a champion of social issues, in writing the sample letter, a more effective approach would have addressed the problem differently. The letter could have pointed out that maintaining people in their own homes through monthly income supplements costs state and federal agencies less than maintaining them in public institutions.

To achieve a stronger impact, the letter should have stated its purpose clearly.

The reader's attitudes and expectations.   Once you identify the purpose, you must identify and understand the needs and expectations of the reader. Here is a letter that demonstrates, almost painfully, what can happen when you write a letter with almost total disregard for the audience:

Dear Executive:

I am pleased to present my thoughts concerning why you should select Jones Career Company over other search firms, such as Joe Smith & Sons Career Consultants, Tom Jones Recruiters or, perhaps, a specialist in the Real Estate area.

Let's first discuss the specialist—a specialist is blocked from many real estate companies because many of these organizations are clients. This has nothing to do with the people for whom we would be searching, but it does mean that this consultant has been in and out of many of the same areas that you want to go into and, perhaps, has a jaded opinion, and does not know it cannot be done.

At Joe Smith's or Tom Jones' you are not going to get a special deal, nor special attention; they are not going to walk across the street for you—you are small potatoes to them. Even if they sold the job and you met the marketing person, that is not the person who would be performing the search. The search would be done by someone whom you have never met, who does not understand the inner workings nor the dynamics of your company.

Now, let's talk about why you should choose me, Hilda Gildersleeve, to do a search. You will select someone with whom you have a good rapport and someone who has a complete understanding of your needs and requirements. In your business and in mine, *trust* is the main asset that we have. If we do not have a client's trust, then we have nothing. Admittedly, there is much I have to learn about the Real Estate market, but I have been very successful in learning the cosmetics business, the consumer packaging industry, financial (private placement, corporate finance, analysis, fixed income, etc.).

I am looking forward to the challenge of learning why I am going to be instrumental in putting top people in a top firm.

Cordially,

Hilda Gildersleeve
Vice President

As an introduction to Hilda's professional ability, this letter starts off on the wrong foot. It promotes her executive placement service by pointing out the inadequacies of the competition. An audience usually reacts unfavorably to this, because it implies that you have little to recommend for your own service.

In an effort to push "trust" as the main selling point, Hilda's letter undermines that by asserting that the client can trust her. But she gives no concrete reason to do so. The last sentence reveals her lack of control over the letter when it suggests that she would like to know "why she will be instrumental in putting top people in a top firm."

Had Hilda considered that this letter would be read in silence by one person, she would have tried to alter its content and presentation. First, she should have asked herself how to communicate what she offers—a professional that the executive can trust. A positive presentation of her performance record for other industries and her efforts to learn the real estate business would have helped the reader to understand her ability to act as an effective representative.

Hilda's letter instead demonstrates a lack of business sense, rather than professionalism. The letter, intended to interest readers in Hilda's service, turns them off. She would have greatly benefitted from an acquaintance with a strategy for writing business letters and analyzing the audience's needs.

Other situations also call for a careful analysis of the reader's needs. For instance, assume that you are writing to acknowledge a mistake made in a customer's account with the company. The person has called, so you know that she is concerned and anxious to clear up the problem as soon as possible. You plan to check the records and alert the accounting department; you informed her of those actions in another phone conversation. After checking, you find that indeed she has been incorrectly charged, and that the error caused embarrassment and inconvenience because the computer was flagged to indicate a hold on additional charges. The person found this out while shopping the day before. You make another call to let her know you will correct the error.

When explaining what occurred, you might initially ignore the error, or shift the blame. It is easy to blame a computer. It is not human, and most people know little about them and fear them enough so that the machine's imperfection makes a satisfying culprit for problems. But resist the temptation to shift the blame. More likely than not your reader will not be convinced, since that excuse has been overused. The loss of credibility in the long run also compounds the mistake.

Admit Mistakes

Once you admit a mistake, state clearly what has been done, or what will be done to correct it. Here is an example that might follow our hypothetical situation:

Dear Mrs. Yamata,

   You are right. We did incorrectly charge you, and this in turn triggered the computer to "flag" your account. We have already instructed the financial manager to correct this immediately, and a revised account statement will be in the mail to you tomorrow.
   We apologize for the inconvenience this has caused you, and would like to invite you to be our guest for a complimentary lunch in our cafeteria on the 8th floor of our store. A certificate is enclosed, and all you have to do is present it there the next time you come here to shop.

                         Best regards,

This letter does more than present the facts; it builds good will. It also acts as a sales letter by tempting Mrs. Yamata to come back to the store. Without this attitude toward the reader's needs and expectations, the account could be lost. But more than just one account, the loss could be compounded because people talk loud and often when they have been treated badly by an institution.

Personal, Informal Tone

In a letter your attitude toward the reader has more impact than it does in other types of writing since you address the person by name. You share the silence together, communicating through the paper. Being aware of that should encourage you to write naturally. Often when people write they sound stiff and remote, but an informal, personal tone brings out positive and favorable responses from your reader. A people-oriented attitude builds good will and stimulates cooperation.

Another way to be personal and informal in a letter is to use the pronouns "I" and "we." You sign the letter with your name, so why not use "I"? And if you write on behalf of your company, use "we." Of course, if you practice using the active voice in your writing, personal pronouns become second nature. Instead of "It is believed" or "It is understood," say, "I believe" or "We understand." Also prefer a construction like "we desire" to "It is desired by this office."

Words with positive connotation make a letter personal, and show your willingness to work together. For example, *pleased, appreciate, welcome, thanks, cooperation,* and similar words capture your reader's sympathy.

In the salutation of the letter, use the person's name, and *spell it correctly,* to set an informal tone. If possible, avoid "Dear Sir." If necessary, make a telephone call to find out, for example, who handles the purchase orders that initiated your letter in the first place. A "bullet" letter, one that is directed to a specific person, creates a more positive response than one addressed to "Sir" or "Madame."

If you write to a woman and do not know if she is married, single, or prefers to be called Ms., you can use her full name, for example "Dear Jane Pelts," to avoid offending her. Then you do not appear too chummy by calling her Jane, and you sidestep using a title—Miss, Mrs., Ms.—that might be inappropriate. The previous generation used "Miss" to address professional women, but that seems stiff for the contemporary office. Even the elimination of the title is now used and accepted. However, if you feel uncomfortable writing a letter without using a title in the salutation, call to find out the reader's preference.

A letter written with a personal tone reflects well on you and your organization. The favorable unwritten message it communicates can be, and often is, more important than the content of the letter itself. The person who received the letter, for instance, may not remember months later the particular details, but that person will certainly recall the impression the letter gave. If you were personal and professional, that stands out, and the reader will be more inclined to select you and your organization over others. Every letter you send contains the possibility of unexpected, future benefits. Not only does it pay to approach every letter that you write with this attitude, but letters with a personal tone create a more pleasant working atmosphere.

CATEGORIES OF LETTERS

In the beginning of this chapter we mentioned Forbes' three classes of letters—stultifying, mundane, and first class. His descriptive ordering focuses on quality. Without giving up the notion of quality within each category, business letters fall into two broad divisions—*unique* and *routine*.

The Unique Letter

You send a unique letter in response to a specific circumstance or occasion. Such letters consume much of your time and energy, but unique letters are read and acted upon. For this reason, the planning and creativity you invest in determining the purpose and analyzing the audience pay off in results.

Unique letters can serve as reports or proposals, as sales presentations or job applications. They make special requests and provide information. Often such a letter asserts authority and responsibility.

Here is an example of a letter asking for a speaker at a convention. Since the writer knows the reader, it is informal, yet professional.

Dear Lynn:

It would give us great pleasure to have you as speaker for the Participating Conference on Directors of Volunteer Services at the Annual State Assembly to be held May 1–3, 19XX, at the Conrad Hilton in Chicago.

Enclosed is the program plan and, as indicated, our session is on Tuesday, May 2, at 3 p.m. We would like you to speak no longer than twenty minutes; then allowing ten minutes for questions about your talk before we go into the "Open Forum," in which we want you to participate as well.

We are in a position to offer you only $100 for your participation and hope this is agreeable.

We are excited about this program for it is the first for our discipline alone. You may be interested in what the Executive Vice-President stated in his correspondence to me: "By recommendation of our Education Committee and approval of the Board of Directors, we are establishing this pilot program for 19XX for the DVS as distinct from our traditional Auxiliary/Volunteer Program. This is in recognition of the growing professionalism of the DVS and despite much commonality, that your particular educational needs do differ."

Further details will be forthcoming in due time.

Sincerely,

June Smith
Director, Volunteer Services

Enclosure

The best feature of this letter is that it presents the point right in the first sentence. It provides the reader with the kind of information that she needs to know to accept or decline the invitation. It also indicates that the fee is not for service. However, the apologetic "only" might be offensive and the writer could have called payment an honorarium instead.

The paragraph that quotes the Executive Vice-President could stand some explanation. Whether in a letter or other document, quoted material should provide a condensed version of ideas, or at least an interesting expression of facts. In this case, the quotation generates more confusion than clarity. The writer would have better served the reader's needs by explaining the information rather than quoting it.

As a unique letter, it carries a personal tone, and makes the point directly.

In addition to special circumstances, a unique letter is often appropriate in an ordinary situation. Here is a letter, for example, that responds to an error in a membership registration.

```
                                      June 15, 19XX
Michael B. Goodman
4700 2nd Avenue
New York, NY 10116

Dear Mr. Goodman:

   Thank you for pointing out the error on your membership
card. There is apparently some problem in our computer's
printer program which caused your name to be misspelled. You
are correctly listed on our membership roster.
   Enclosed is a corrected card. Please accept my apology for
any irritation you may have felt at receiving the misspelled
one. We are checking into the problem, and are hoping that it
didn't affect too many members. Your pointing out the error
has been most helpful, as this is the type of problem that
would not necessarily be discovered in the course of our
routine computer audits.
   Again, our many thanks for your support and concern. Please
don't hesitate to write if you have any other questions re-
garding your membership.


Sincerely,


Membership Service Manager
```

All of that for a petty mistake in a printed card? If that occurred to you as you read this letter, you are right. This unique letter in response to a minor problem might be interpreted as overkill and, as a consequence, the tone of the letter might seem patronizing. On the other hand, the handling of the problem, which was not really a complaint, demonstrates an effort to build good will and also to provide some explanation of the problem.

As a unique letter, it makes the point, admits the mistake, provides an explanation, and outlines the steps taken to correct it.

Unique letters also call for tact. Here is an example. The personnel officer must inform an employee who has cancer about disability and pension options open to her without giving the impression that the organization is trying to let her go.

March 2, 19XX

Mrs. Lola Glaster
181 Farget Lane
Levit, New York 11706

Dear Mrs. Glaster:

Attached is an application for an ordinary disability retirement. After our recent conversation concerning your employment with the company, I explored what was available to you as an employee with only a few months employment. It appears that your many years of membership in the State Employees' Retirement System will give you an opportunity for an early pension.

It is required that you be on the payroll or on an official medical leave when you submit the application. Since it takes many weeks to be processed I would suggest that you send the application immediately. (Be sure you have a copy for your files. Also, send us a copy.) If you return to work, it can always be cancelled.

A booklet is enclosed from ERS explaining the program.

Please call me if you have any questions.

Yours truly,

Elizabeth Booking
Personnel Associate

EB:fha

The letter tactfully informs the reader of her options, even though she has been with the company only a few months. Such situations demand the personal touch, and the last line of the letter invites the reader to communicate questions and problems directly. Considering the circumstances, the letter presents a friendly, professional, and tactful response.

In other circumstances, however, you may need to write a letter that asserts authority and responsibility. For example, the new director of a county health facility received a copy of a curt memo from the medical review office, listing the auxiliary police candidates scheduled for x-ray. The note further dictated the time for the procedures. The administrator learned of the request the day the candidates arrived for x-ray. Though, as a county hospital, it routinely performed such services for other county agencies, the administrator had to assert his authority over the administration of the service. Here is a letter that he sent to correct the problem.

```
                                    January 28, 19XX
Principal Personnel Analyst
Office of Employee Medical Review
Farmingville Health Center
Farmingville, Ohio

Dear Mr. Fiddler:

    In order to alleviate any possible conflict in scheduling
in the future, the following is the procedure to be adhered to
when requesting any services from Goham Health Center:
    1.  All correspondence be addressed to the Administrator—
        cc: Department head or staff person to deliver service.
    2.  A request for service must be received before any sche-
        dule can be made.
    3.  Written confirmation of requested or revised schedule
        will be forwarded to your office.
Your cooperation is appreciated, thank you.

                                    Sincerely yours,


                                    Robert Brown
```

Notice that the tone of this letter is brusque, contrary to the advice offered in this chapter. Under the circumstances and considering the purpose of establishing authority and responsibility, the brusqueness fits well. The appropriate, businesslike coldness informs the reader that the writer is in control and that he is firmly in command—not the other way around.

Hand-Written Letters

Almost all business correspondence is typed, a complete change from the not too distant past when all letters were hand-written. But such an old-fashioned approach can provide a personal touch. A hand-written letter can capture your readers' attention and often their patronage as well.

Such a unique letter is appropriate when:

- the message is short
- the relationship is already personal

On the other hand, a hand-written letter to a casual business acquaintance, asking for routine information, can appear trite and inappropriate.

Write a letter by hand if you have determined that such an approach is appropriate to the situation.

Routine Letters

More often than not, you will write a letter in response to a routine situation such as a request for information about your company's services or products; or information concerning repair. Often routine letters:

- answer requests from charity and community groups for volunteers
- offer condolence
- acknowledge occasions like births, weddings, graduations
- recognize promotions or superior service
- accompany reports as a letter of transmittal
- serve as progress reports
- grant or refuse credit
- sell or advertise products or services
- answer complaints
- provide information through mass mailings

Here is an example of a routine response to a request for information concerning an investment fund.

Dear Friend:

Thank you for requesting more information on The Mega-Money Market Account.

Enclosed you will find a brochure which outlines the benefits of this exciting investment vehicle—as well as our Prospectus with an application which you can use to open your account.

For well over a century, The Safe Investment Group, our investment adviser, has provided millions upon millions of people with financial protection they can depend on—in insurance policies and pension funds.

And now, to help you offset the monetary erosion caused by today's economic uncertainty, the Group has developed an investment opportunity to make your money work harder for you—The Mega-Money Market Account. It enables you to enjoy the current higher earnings of a multi-million-dollar investment in the money market simply by investing $2,500 or more.

I invite you to read the enclosed brochure and our Prospectus. Then, when you're ready to open your account, just fill in and return the application provided, along with your first deposit, to begin earning our high interest. A postage-paid envelope is enclosed for you convenience.

> With the economy continuing to look uncertain, I do hope
> you will take advantage of this investment alternative very
> soon.
>
>   Sincerely,

The letter not only offers information, but tries hard to make a sale. If you spend the money to send a letter, it should at least attempt to make a sale. Here the pitch appeals to the reader's need for a secure investment that will protect the money he has and also make it grow. Since it is a routine situation, the entire letter goes out to anyone who asks for general information. That makes it a form letter.

Form Letters

These save a great deal of time and expense. Carefully constructed and expressed letters make little effort to conceal that they are mass-produced. Readers have come to expect a general form letter in response to general questions. If the case has a special twist, you can add a note in your own handwriting, or clip a short note to the letter with the additional information. This saves time and money without offending the reader.

Some form letters provide information. Others like the next example, create confusion and a sense of doubt about the company's candor.

> September 17, 19XX
>
> To Holders of Common Stock:
>
> Last year, F/P Corporation ("F/P"), a wholly owned sub-
> sidiary of FPC Corporation ("FPC"), acquired approximately
> 72% of the outstanding shares of Common Stock of the Company
> (representing 67% of the outstanding voting securities of
> the Company) from certain shareholders for $.60 per share
> ($.12 in cash and $.48 in 9% installment notes). In connec-
> tion with the acquisition of Common Stock, FPC and F/P agreed
> to make a tender offer or effect a merger on terms no less
> favorable than the terms of such transaction.
> I am enclosing a copy of an offer by F/P to pruchase any and
> all shares of Common Stock of the Company for cash at $.60 per
> share. Information concerning the position of the directors
> of the Company concerning the offer and the terms of the offer
> is set forth in the enclosed offer, and each shareholder
> should carefully consider the information set forth therein
> before making a decision whether to tender shares.
> You should note that the offer is scheduled to expire at
> 10:00 A.M., Eastern Daylight time, on October 27, 19XX unless
> extended. The offer and accompanying letter of transmittal
> include instructions for tendering your shares which should
> be read carefully. If you have any questions or require any

assistance, you may contact FPC at the telephone number and address listed in the enclosed offer. You may also wish to contact your local broker, dealer, commercial bank, trust company or nominee for assistance concerning the offer.

Very truly yours,

The purpose of the letter, to inform stockholders of a merger, gets lost in the tangle of this letter. It may be, as a lawyer who looked at this letter suggested, that the purpose was just that—to confuse the stockholders. If that is the case, the reader can certainly doubt the candor of the letter.

Boiler-Plate Letters

Routine situations may have several unique wrinkles. For that reason a mass-produced form letter is not only inappropriate, but could insult the reader. A solution, however, may lie in a boiler-plate letter. Large organizations and government agencies that have a heavy volume of daily correspondence use this approach to respond to slightly different routine situations. Simply, the boiler-plate letter is a number of predrafted paragraphs written to fit several situations. The numbered paragraphs are copied and distributed throughout the organization. Then, when a letter comes in, the respondent provides an opening paragraph if necessary, the numbers of the predrafted paragraphs in the order that they are to appear, and any additional material to a typist. The result is a custom-tailored letter from "stock" parts. This approach works well only if the predrafted material fits. Otherwise, the reader is confused or, worse, insulted.

The boiler-plate letter can be produced quickly, if you have access to word-processing equipment. The predrafted portions in this case are stored in the computer memory. The typist's job is made even more productive since this technology eliminates the retyping of the stock paragraphs. (See Chapter 12 for a discussion of writing with electronic equipment.)

Letters That Say No

Under the best of circumstances, most of us have trouble saying *no*. In conversation one person may hem and haw and beat around the bush, but sooner or later the other understands the message—*no*. No, that job is taken; no, we cannot provide the service at that price; no, we cannot allow you to use our parking lot for your charity carnival.

Writing presents special problems for saying *no* since readers do not have the benefit of your facial expression, gestures, and other body movements to inform them that the *no* does not carry personal or emotional threat. How can you write *no,* yet attain the positive reassurance that facial expressions, glances, and gestures provide in conversation? Often the body language says that you and your organization are decent people, even though you have turned down the request.

Letters that say *no* must rely on tact and courtesy. Put yourself in the readers' shoes and ask how you would react to the way you have said *no*. A sincere tone helps communicate bad news without alienating your reader.

In your efforts to anticipate reactions, do not forget your purpose. Make sure that you state the bad news clearly. In meeting your purpose and the needs of the reader:

- present the bad news
- explain your decision
- encourage the positive
- invite future activity (if appropriate)

Look at the following letters that turn down job candidates:

Dear Mr. Coprof,

    We have completed our search and have extended an offer for the position as office manager. It is with regret that we inform you of our decision.
    We had hundreds of applicants for the position, and there are many reasons for selecting some candidates over others—strengths, needs, compromises—so our decision reflects more than a simple judgment of your abilities.
    Thank you for applying, and we wish you the best in your career.

Or this letter:

Dear Mr. Reading:

    Thank you for your letter, which was referred to me by Robert Lauter. While your qualifications are quite impressive, I regret that your services do not meet the needs of our company.
    We certainly appreciate your interest in Pulp Paper Company and would like to take this occasion to wish you success in your future career.

Which of these letters states the bad news clearly, yet recognizes the needs of the reader and builds up the positive image of the company as well? The second does a better job than the first, although the letter is less specific. It directs itself to saying *no* to the applicant and also to reassuring the person that he has fine qualifications even though they do not fit the company's needs. That thin line presents an honest explanation. The other says *no*, but rubs it in by saying that the applicant was among hundreds of other people also not selected, as if that makes the rejection more palatable. The tone of the first focuses on the writer at the exclusion of the reader. If the hundreds of applicants were seen as potential customers and not as unsuccessful applicants, the writer of the first letter would have been more tactful.

Keep in mind that letters saying *no* are not the most welcome item people find in the mail box, so present bad news as humanly as possible. Instead of some disembodied phrase like "We regret to inform you . . . ," try an approach that acknowledges the reader's reaction. For example, "I know how disappointed you will be, but we cannot send the gifts you ordered before the date you requested."

EVALUATING LETTERS

The scorecard used to evaluate memos (See Chapter 5) also works to identify the strengths and weaknesses of letters. The three main categories—Language, The Whole Letter, Style—allow you to isolate difficulties in a letter so you can focus attention on them productively. The evaluation criteria also provide a checklist for effective writing.

PARTS OF A BUSINESS LETTER

Letterhead	QXR Company, Inc. 11778 Industrial Way Dallas, TX 75225
Date	August 26, 19XX
Address	Gerald S. Thoms, Marketing Director MikroChip Corp. 237 Tape Drive Boston, MA. 06534
Subject	Subject: Combined Sales Effort
Salutation	Dear Gerry:
Message	I have completed the plans for our combined advertising campaign. QXR and MikroChip will begin the full-page ads in seven trade magazines, starting October 1st.

I have contacted the advertising departments of the publications and found the current rates and available space. I have attached a list.

I think that we should try to place the ads in other magazines as soon as possible, since the Fall proves to be very active for the type of product we offer. Let me know your thoughts on this.
Could we meet in Boston on September 9 to coordinate the current project, and also to discuss the next move? Please call me and let me know if that accomodates your schedule.

Closing	Sincerely,
Signature	Ron Massey
Typist's Notation	RM/mb
Other Notation	Copy to Quentin X. Reardon Encl.: 1
Postscript	P.S. We think this joint effort has been positive and would like to continue such projects.

Letterhead, Date, Inside Address

Letters that you write for your company should be on letterhead, or your personal company stationery. This emphasizes your official position with the organization, and displays their distinctive business papers. Quality letterhead suggests to the reader, "We are pros. We are in business to stay. We invest in our letters." However, all that expensive business stationery is for nothing if the content of the letter is poorly conceived, organized, and expressed.

Every letter you send should also have a date. It helps your reader identify the message, and eases filing for future reference. One reason for writing a letter is to create a permanent record.

In addition to the date and letterhead, every letter has an inside address, or the address of the person or organization you are writing to. This can occasionally create a problem. For instance, in following up a telephone conversation with an assistant buyer, and after telling her that you will send a request for further information, you realize that her name might have several spellings. It could be Colin, Colon, Kolen, and that her first name might be Jane, Jayne, or Jain. What do you do? This might seem a minor point to dwell on, but the worst possible mistake to make in any business communication is to misspell the name of the person you address. People expect accuracy, though they will hardly notice it. However, they *always* notice if you make a mistake.

The best way to find out is to call, not the person, but the secretary or clerk who handles her calls and ask directly how the name is spelled and also the correct job title. That short call can save you not only embarrassment, but lost business.

Often, though, you may deal with a department, and the person you have worked with in the past may have been transferred, promoted, or moved. In that case, the inside address should be the name and address of the company. Beneath it, use an attention line: Attention: Mr. Bill Jackowitz, Benefits Director. This saves the letter from following the person to his new post and then back again to the department that can help you.

Subject Line

Many organizations use the subject or reference line in business letters. This serves as a title for the letter, and helps the reader recognize your subject or reference quickly. For example, if you have had several conversations with the same people and know that they are involved with many projects with your comapany, as well as with other organizations, give them a useful title or subject line like:

Subject: Mailing List Development Costs for Investment Service Marketing.

Keep the subject as short as possible, without sounding like a breathy headline from an afternoon tabloid. Generally, a good subject or title coincides with the sentence that you generated to determine the purpose of the letter.

Salutation

Custom calls for the standard salutation "Dear" followed by the person's name—"Dear Mr. Smith." In a formal letter use a colon (:) after the salutation. A comma (,) implies less formality. If the letter uses the writer's first name only, which suggests informality, use the comma.

When writing the salutation, a problem often arises. When you have meetings and conversations with clients, contemporary business etiquette allows you to use first names, though you are not on a personal basis. What do you do when writing to a person to follow up the conversations you have had? Simply drop the Mr. or Miss or Ms. and use the person's whole name: "Dear James Smith." That lets your reader know that your are not stiff and rigid in the letter, but that you both are still not old school chums. The quasi-formality of American business is best acknowledged this way. This should not offend the person, who might be a bit startled by being addressed as "Bill" in a letter.

Of course, if you are in close personal contact with the person and have exchanged letters for years, then the first name is appropriate. Since use of the first name only is a judgment call, you decide to use it or not by evaluating your relationships with the people you are writing to.

Other salutations to persons of special rank are listed in secretarial manuals and often in dictionaries. Use conventional forms to address a senator or minister, a priest or a professor, a president, a lawyer, or a mayor. If you are uncertain, check to be sure.

Avoid, if possible, using "Dear Sir" or "Dear Madame." Here again you can use the telephone to find out who is in charge of the area you are interested in.

Above all, address the person by name. This communicates interest and, according to a manual circulated by Chase Manhattan Bank, "adds warmth to a letter."

The Message

This element of business letters has three parts: introduction, body, and conclusion.

Introduction.   The first paragraph of the business letter determines whether your letter gets through, or whether it gets filed or passed along to the clerical staff. Every letter that you write should begin positively. "Your demonstration of the amount of money we could save by using your equipment impressed even our financial analysts." Then jump right in and make the point: "We would like to have a more complete discussion with you." The letter should then lead to the necessary details.

The details might cover the general topics that you want the person to address in their presentation, or they might be the items that you need specifically.

If possible, add a personal touch to the letter. Do this if you have had personal conversations with the people either during lunch, at a conference, or at professional meeting. But use the personal touch wisely and tactfully. Don't get too chummy and don't appear too easy going. After all, business letters project professionalism and are also used for the record.

The Body.   The body of the letter clearly presents information or an argument to persuade an unconvinced reader. Select important details for the body. If you have five examples, present the best *one* in the letter. If you find that your letter requires several pages, you are actually writing a short report. In doing that, you may elect to write a cover letter that introduces the information presented in the form of a report or a proposal. (See the following chapters on reports and proposals.)

Under most circumstances, a letter should be a single page. With that as a guideline, the challenge for the writer of an effective letter is to select and arrange facts and ideas as succinctly and clearly as possible.

Conclusion.   At the end of the letter suggest the next step, but be specific. If you wish to set up a meeting, for instance, say, "Would Thursday, January 21, at 10:00 a.m. be an agreeable time for us to meet in your office?" That gives the reader a specific time and place to work with. It is more effective in arranging a meeting than if you wrote: "Let's get together next week to iron out the details." If they cannot meet with you when and where you specified, they will give you an alternative. A specific time and place focuses the discussion and gets over the awkwardness of who will arrange the details.

Also in the conclusion of a letter, add any appropriate personal reference. For example, "Congratulations on your promotion."

But avoid phony expressions of sincerity and the cliches of business letters. You are better off with an abrupt end than with phony sentiment.

Try to make the end of a letter positive, even if you are writing to get someone to do something. For example, I recently got a letter that ended, "Thanking you in advance for your cooperation in this matter." I have, and no doubt many others have, an emetic reaction to such pre-thanks. It implies that the person knows that I will cooperate and thanks me even before I do the job. I am often insulted by the implication. It is better to end by saying in a tactful, positive way, "We welcome your help and cooperation in this matter," or, "We look forward to working with you toward the resolution of this project."

Closings.   Like the salutation, the closing of a letter follows convention. But you have many choices. Some closings are reserved for public officials and persons of rank:

Respectfully yours,
Yours respectfully,

But in most cases choose one of the following:

Sincerely yours,
Sincerely,
Very truly yours,
Truly yours,
Yours truly,
Cordially,
Cordially yours,

The closing can also include a reference to a special or seasonal event:

Happy holiday season,
Merry Christmas,
Best wishes,
Regards,
Best regards,

Don't be cute in closing, unless your profession—for example, public relations or clothing design—emphasizes creativity.

Signature.   Write your name clearly. A stylized signature expresses nothing more than an inflated ego. By all means be distinctive, but be legible.

Notations and Postscripts.   Make sure that the notations of enclosed material and attached material reflect the reality of the letter. Also be clear about the distribution of carbon copies, and make sure that the "bcc" notation—blind carbon copy—appears only on copies of the letter and not on the original.

Use a Post Script as a gentle reminder or to emphasize information in the body of the letter. On a business letter the P.S. can provide a personal touch to an otherwise factual letter; for instance, "P.S., I enjoyed your presentation."

SUMMARY

- According to the pros, effective business letters are short, simple, and direct. They use clear, specific language, and admit mistakes. Good letters are personal without being patronizing.
- Letters require a strategy. Plan letters by using brainstorming techniques. Ask yourself questions to determine the true purpose of the letter, and to analyze the needs of the reader.
- A positive attitude toward the reader in letters achieves results.
- Business letters are unique or routine. Unique letters respond to special situations or circumstances. They are personal statements that make a specific point. Routine letters are appropriate responses to general requests and situations. Such letters can be either form letters or boiler-plate letters.
- Saying "no" in a business letter requires tact, clarity, and directness. Present the bad news first, then offer an explanation. If appropriate, invite other actions to follow up.
- Evaluate letters for concise, courteous, and concrete language. Understandable organization puts first things first, and addresses the reader and the subject clearly. Use a clear, specific style based on strong, active verbs.
- Business letters contain these elements:
    letterhead
    date
    address
    subject line
    salutation
    message: introduction, body, conclusion
    closing
    signature
    notations
    postscript

EXERCISE 6–1

1. Find three examples of boiler-plate letters that you have received. Comment on the unwritten messages that each example sends. Rewrite each example, incorporating improvements on the way each addresses the reader.

2. Look over your files and select three effective letters that you have written in the last year. Identify points that you feel were strong. Evaluate them with the idea of incorporating those strengths into the letters that you now write. Revise another letter you have written recently, but felt it could have been more forceful.

3. Compare form letters that you have received from two companies that offer similar products or services. Discuss how closely the letters approximate the way you feel toward that company and the subject of the letter. Did the letters change the impressions that you had of the companies?

EXERCISE 6–2

1. You have just received a request from the Children's Aid Society to use the company parking area and grounds for a Fund Raising Breakfast. You call the Director of Security who informs you that the liability insurance policy forbids "games, recreation, or social affairs" on the grounds owned by the company. Though you would like to help this respected community organization, you fear the insurance company will raise its premium if you allow the grounds to be used in that way. Also you do not want to set a precedent that other charities might take advantage of in the future.

   Write a letter to the Children's Aid Society and tactfully refuse their request. A public park three blocks away might be more appropriate.

2. You are a hospital administrator. James Osborn, a professional film-maker, has written you a letter requesting permission to film the birth of his child in the delivery room of the hospital. His doctor, who has been affiliated with the institution for 10 years, has given verbal approval. Your hospital has a long-standing policy forbidding the filming or video-taping of procedures in operating room areas. However, some exceptions have been made in the past for medical films and documentaries.

   Write a letter in which you explain your refusal of Mr. Osborn's request. Include an explanation of what he must do to be granted an exception to the rules.

3. Write a letter to Dr. A. C. Greene, a nationally respected advocate of preventive medicine, in which you ask him to address your administration and staff on beginning a health maintenance program. Explain that your personnel department is in the process of adding a program much like the one Dr. Greene has advocated for almost 25 years.

4. You have just interviewed three people for a beginning office position. All of them are extremely qualified, eager, handsome, and intelligent. However, you have chosen Helen Albright for the job because she has completed two years of college in addition to her work experience. All other qualifications are equal. You would honestly like, however, to keep the others in mind since you anticipate growth in your office. Say "no" to the other two candidates, and explain your decision.

5. The Alfred Cazin family has applied to you for credit. In the credit application, and in the subsequent check of references, your credit department has found some unresolved criminal judgments against Alfred. On that basis alone you must refuse credit. However, the rest of the information checks out positively. Write a letter refusing credit to Cazin.

6. As the director of customer relations you received a call and a letter complaining that a floor polisher sold in your suburban store broke and caused $2,000 damage to a customer's hardwood floor. The person seeks damages as well as money back on the machine. You don't really believe the story, but you have not yet sent anyone to inspect the floor and the machine. You suspect that the floor may have been

damaged before, and the polisher was not the cause. Let the customer know what you plan to do to settle the complaint.

EXERCISE 6–3 (ADVANCED)

Compare the way each of these form letters addresses the reader. Identify strengths and weaknesses in them. Rewrite both of them.

Dear Investor:

Thank you for inquiring about the Contact Money Market Fund. We have enclosed our information package for you, including a prospectus which completely describes all the features of our fund.

Contact Money Market Fund has been extremely well received by thousands of investors and has grown to over $850 million. The fund invests in large denomination money market securities, which are backed by the U.S. Government, by major banks with assets over $500 million and by major corporations rated "prime" by Moody's or "A" by Standard and Poor's.

When you consider the facts, it's easy to understand why so many individuals, like yourself, have been investing their money in the fund.

- Minimum investment only $2,500
- Check-writing privilege (minimum $500); initial charge $6—subsequent checks free
- Absolutely NO CHARGES for deposit or withdrawal

Think about it. If you have excess funds tucked away, why not transfer a portion of them to Contact Money Market Fund? Simply complete the enclosed account application and mail it together with your check in the enclosed postage-paid envelope. You'll be glad you did.   Sincerely,

                                        President

P.S. For our current yield, call toll free 800–555–5348; in New York call collect 212–555–5555.

Dear Stockholder:

We are pleased to announce certain recent amendments to the Clean Energy Company Dividend Reinvestment and Stock Purchase Plan. These changes have been made primarily in response to comments received from stockholders as to how the Plan could be improved. The changes made that will be of interest to you are as follows:

1. Optional cash payments may now be made on a monthly basis. Previously, such payments could only be made quarterly. (See "Optional Cash Payments," page 4 of the enclosed Prospectus);

2. Optional cash payments may now be made of not less than $25.00 per remittance nor more than $12,000.00, in the aggregate, per calendar year. Prior to the amendment, such payments were limited to a maximum of $3,000.00 per quarter; and

3. The Plan was made available to holders of preference stock of the Company.

For those stockholders who are not presently participating in the Plan, this opportunity is taken to point out that the Plan is designed to provide an economical and simple method of purchasing additional shares of common stock of the Company at a 5 percent discount and without the payment of any brokerage commission or service charge.

Participation in the Plan is entirely voluntary and may be cancelled at any time. You may choose to participate immediately or at some time in the future by completing and returning, at any time, the enclosed authorization form to The Moebus Bank, N.A., in the envelope provided. If you do not choose to participate, do not return the card and you will continue to receive your dividend checks.

Complete details with respect to the Plan, including the recent amendments, are contained in the enclosed Prospectus which you are encouraged to carefully read and retain for future reference.

Very truly yours,

Enclosures

EXERCISE 6–4

Examine this letter. Comment on the way it addresses the audience. In a different version of this example exaggerate a strength or weakness.

August 26, 19XX

Ms. Barbara Boatwright
Executive Director
The James Carsons Foundation
1100 Glendon Avenue
Bayside, California 90099

Dear Ms. Boatwright:

Enclosed is a copy of a resolution of appreciation which was adopted by the Soundview Town Board on August 15th. Please share this with the members of the Board of Directors.

We are truly grateful to The Carsons Foundation for their benevolence in awarding this very generous grant for the preservation of the Carsons Blacksmith Shop.

I am certain this project will flourish under the guidance of George Savage and the Soundview Historical Society, as well as residents who have indicated an interest in this restoration project. The Foundation will be kept informed of the various stages of this community undertaking.

Sincerely,

Michael O. Finny
Supervisor

MOF/jp
Enc.
cc: G. Savage

EXERCISE 6–5 (ADVANCED)

Discuss the way the following letter says *no*. Point out the organization and the manner the writer uses to deliver the bad news. Write an alternative letter, or rewrite this one.

May 8, 19XX

Ms. Precious Stokes
P.O. Box 273
East Base, New York 11995

Dear Precious:

After reviewing the papers and messages you left at my office, I called the office of Elaine Lich regarding your particular problem. My questions were referred to Mr. Roghtib, who called a few days later to discuss the matter.

You must keep in mind that strict guidelines are established by the Federal government for distribution of Community Development funds, and these guidelines do not allow for variations. First of all, it was pointed out that your delay in submitting the required documentation to substantiate your application resulted in failure to meet the deadline for that allocation period. Therefore, you were advised by the County Community Development Corporation that your application would be considered in the next assignment period when funds became available.

Secondly, when an inspection is made for contractual work, the regulations clearly specify that priority must be given to repairs or alterations which will improve a residence and make it more habitable. Items such as insulation and storm windows, which your house lacks and was designated to receive, are priority items. The worksheet that was filed on your behalf indicated these improvements, however, it is my understanding that at this point you took the option of filing your own worksheet.

Mr. Roghtib further explained that appropriate steps which would lead from the sliding door unit were also a priority item and would be part of the project. Cosmetic improvements such as a deck replacement are not included in the guidelines which the inspectors must adhere to.

I could only request on your behalf that your application be given every consideration when funds are made available again.

Sincerely,

Supervisor

EXERCISE 6–6

Here is the draft of a boiler-plate letter that will go out to an attorney who had a summons served at the hospital on someone who is not associated with the institution.

Write a final version that gets to the point, but also includes the essential information.

Date _____

Name and Address of the Attorney

Re: /Title of Action/

Dear Sir:

We are returning to you herewith a copy of a summons (or summons and complaint) which was delivered to this office on (specify date of service) and purporting to constitute service under Section 308 CPLR upon _____, one of the defendants named in the above action.

After receipt of the enclosure, we had our records examined and advise you that on the basis thereof it appears that the said party is not employed by the New City Medical Center nor do we know of any circumstances under which any of our facilities could be considered the "actual place of business" of any such person. We are, accordingly, returning the summons (or summons and complaint) to you since we are not in a position to make any other disposition of it.

Yours very truly,

EXERCISE 6–7 (ADVANCED)

Here is a letter that serves as a report. Compare it to another type of letter found in this chapter. What major similarities and differences can you see between the two? Comment on the way numbers are presented. Revise this letter, making it a clear, readable reference document.

Dear Mr. Sand:

In response to a request from Mr. A. Wayne of Fone Lines for prompt response and aid in clearing a portion of the third floor of the White Plains 2 building, agreement has been reached concerning new equipment required on the fourth floor to accommodate this undertaking.

Two D4 mounting bays will be required. Because of the short intervals involved, extraordinary measures are being taken. One bay, previously ordered for special service requirements at Newburgh West is being diverted to White Plains 2 for this job, and the second bay will be obtained from the Western Company production line on an emergency basis. Funding will be arranged within the Metro Telephone Company April View by transferring $200,000 from Fastran Keys #BMAN 75554 ($100,000), GMID66638 ($50,000), and ISUF67395 ($50,000).

Please consider this letter as authorization to proceed with installation of the equipment, and to render billing to *Coneckt* Company Project No. W7827. Completion of installation of these bays is required by September 16, 19XX.

Carrier mountings identified for use with interstate carrier systems will be identified in an application to be filed with the F.C.C.

If there are any further questions, or if we can assist you in any way please contact Mr. T. Selleck of my organization on 212–555–2981.

                              Yours truly,

TJS/ec
Attachments
Copies to: Mr. J. S. King
           Mr. J. Poweg
           Mr. N. Camen
           Mr. A. DiBlanc
           Mr. J. McIntine
           Mr. A. Wayne
           Mr. J. Jenks

EXERCISE 6–8 (ADVANCED)

Read the following letter carefully. What is its purpose? What is the tone of the letter? What does the writer expect in such a letter? What should be the next step after the letter?

Identify strong points. Locate weak portions. Write a revised letter that you think meets the writer's purpose and the audience's needs more effectively.

Dear Mr. Alexander:

Thank you for your time and courtesy when I called, following-up your correspondence with Al Geffry of our office.

During our conversation, we briefly discussed ventilation requirements and air cleaning systems for your paint manufacturing facilities. In response to your request for additional information, I have enclosed several equipment catalog cut-sheets. These sheets provide information on various air cleaning devices for use within your plant. A brief description of these equipment items is presented herein.

We anticipate that the airborne contaminants in your work areas are primarily paint pigments and solvents. The equipment suited to remove these contaminants will vary, depending on the design parameters of each application.

Pigments are removed from an airstream by various dry and wet processes. The characteristics of the pigments in the airstreams and the economics of disposal or reuse, will determine the system to be chosen.

Using a dry process removal system, high pigment loadings are best removed by a high-efficiency cyclone, followed by a fabric-type filter. The cyclone will remove the heavy particles while the fabric filter will provide removal of the fines. This equipment set-up can be designed for pigment recovery and reuse, if required. Attachment A presents a cut-sheet for a typical cyclone equipment layout. Attachment B describes a cartridge fabric filter system.

Nuisance, or small pigment air loadings, may be handled by a cartridge-type filter alone. The airstream characteristics will determine the actual equipment set-up required.

Pigments in the air streams can also be removed by wet process systems. A typical wet process system could incorporate a venturi scrubber, followed by a separator tank. The venturi scrubs the pigment from the airstream with a water spray. The air and water mixture pass to a separator tank where the water and pigment are collected.

Attachment C presents a description of a venturi-type wet scrubber.

High concentrations of airborne solvents within the work area, are required by regulations to be removed. Location of the ventilation discharge from the work area will determine the solvent removal requirements from the airstream. If the airstream is discharged outside the building, sufficient dilution should occur, such that separate solvent removal is not required. If the airstream is returned back to the facility, solvent removal from the return flow will be required.

The type of solvent used will determine the removal mechanism required. If the solvent is water soluble, a venturi scrubber or similar wet scrubbing device will remove the solvent from the airstream.

If the solvent is non-water soluble, more extensive treatment is required. A wet scrubber may be used which will remove solvent in a slurry. The solvent's solubility will require a scrubbing liquid, other than water, be used.

Other alternatives for non-water soluble solvent removal are: thermal oxidation, catalyic oxidation and activated carbon absorption. Thermal and catalyic oxidation remove the solvents by incineration. Attachment D presents a descriptive cut-sheet for thermal and catalyic oxidation equipment.

Activated carbon filters absorb solvents on its surface. Regeneration of the filter recovers the solvents for disposal or reuse. Attachment E presents a description of an activated carbon filter unit.

I hope these descriptions and equipment catalog sheets will provide you with basic information on the various air cleaning systems applicable to your operations. Before any systems could be developed for use, all the system parameters, such as: airborne contaminant levels; ventilation requirements; and recovery, reuse or disposal requirements would have to be detailed.

I would like to meet with you to further discuss this information, and to review your needs and ways in which we may be of assistance. I will call to determine when we can set a mutually, convenient time for this meeting to occur.

If you have any questions or need further information, please do not hesitate to call.

Very truly yours,

ALPHABETA CONSULTING ENGINEERS, P.C.

J. Robert Ecoles
Vice President

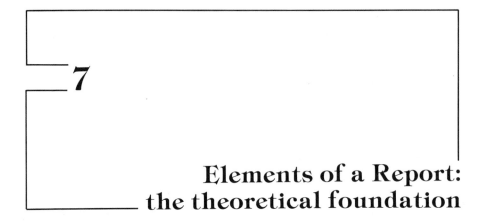

7

Elements of a Report: the theoretical foundation

A report is a written account of a project that is prepared in an organized, often formal way. It presents details more extensively than letters and memos. Because of this, reports routinely use additional elements of writing. They frequently call for the definition of terms, the description of objects and processes, the classification and arrangement of information, the presentation of interpretation and analysis, and the communication of conclusions and recommendations. Each of these items plays an important role in the writing of a clear, well-organized report. This chapter examines those elements of writing as separate items. However, no report uses any one of them exclusively, and most reports exercise a combination of elements to deliver information concisely and understandably.

DEFINITIONS

A formal definition follows this formula discussed in Chapter 2:

Species = Genus + Differentia

In defining a word—the *species,* place it in a larger group—the *genus,* and then mention the relevant differences—the *differentia.* The information that you supply must be necessary for the reader's understanding, and you should include a sufficient quantity of related details. Through the presentation of details, definitions furnish a method for developing ideas.

Strategies for Writing Definitions

Current business, scientific, and technological terms are not often found in even the most up-to-date dictionaries. Thus you are forced to create a definition of your own. Brainstorming techniques work well in providing you with the raw material to build a useful definition for a report. You can use free writing to get started. Think about the term, and generate a list of things you associate with it by writing continuously for 10 minutes. Put down anything that comes to mind, no matter how far-fetched. The random list that you create, remember, is raw material and cannot seriously be considered as the final product.

Select one of the following terms, and generate a list of characteristics:

lawyer	marketing	executive
manager	ethics	socialism
accountant	workmanship	capitalism
engineer	integrity	work
industry	advertising	money

Look carefully at the list of associations and indicate which ones are closely related, unrelated, or tangential to the term. To be part of your definition, an item must contribute to the reader's understanding.

Once you have gone through this procedure, you may find that you have assembled enough information to write a clear definition. Or you may realize that you must seek more information. After creating the list through brainstorming, you may discover other ideas equally important and informative.

After you have culled the list and decided which items are relevant, make sure that in writing the definition you:

- set up a common ground
- provide comparisons to show the reader what it is like
- avoid circular constructions, using the word to define itself

Stated simply, the *common ground* insures that both you and your reader communicate in the same ball park. You might have trouble, for example, attempting to define a computer for a Huasa tribesman—a nomadic Nigerian. Finding a common ground would be a struggle since they base time on the cycle of wet and dry seasons, and have little need for the way westerners calculate it. It might tax the imagination under those circumstances to explain how calculations made on a device that uses on and off electrical impulses indicated on silicon chips can keep track of money, information, and resources. The common ground may begin with an analogy to a book, since the Huasa use orally transmitted stories to record the past.

The common ground may be the most difficult of all the requirements for definition. But without it, you will regress, defining each genus all the way back to a general term that both writer and reader understand.

Common ground implies that definition is not *of* the term, but *for* an audience.

Defining a computer for a member of a nomadic African tribe is an exotic case, yet it is similar to the electronics engineer who must define *capacitor* or *circuit* in the computer in an effort to explain to an office manager how long it will take before the machine will be humming along, crunching numbers again.

Effective definitions expand on shared experience and knowledge to draw *comparisons* and *contrasts* for the reader. In that way you tell what the item is like, and draw a mental picture of it by evoking characteristics that it shares with something the reader knows about already. To define "heart," for instance, we might compare it to a pump. In doing so, the reader sees how it works. Add to that, "The heart looks like, and is about the size of a clenched fist," and physical characteristics emerge.

Circular definitions muddle the clarity of a report. Avoid the repetition of the term, or a form of it, in the definition itself. For example, "a person who keeps accounts" or "a person who accounts for money spent" each represent unacceptable definitions of *accountant*. Circular definitions such as this create needless confusion, and add little to the reader's understanding.

Extended Definitions

Definitions can become a substantial portion of a report. In such instances, they are really extended definitions. The following letter to company stockholders focuses on the term LIFO and its use by the company:

TO OUR SHAREHOLDERS:

Within a few weeks now we will be reporting our 19XX results over the news wire services. Since we anticipate converting a substantial part of our operations to the LIFO (Last-In-First-Out) method of inventory valuation, I think it appropriate to write and let you know of the change prior to the actual release.

The proposed conversion also requires a restatement of our 19XX quarterly results on a LIFO inventory valuation basis. This causes me to be a bit concerned that, without some advance notice and explanation, condensed versions of our results picked up from the news wires could make it appear that we had a downturn in the fourth quarter. This was certainly not the case. As a matter of fact, our net earnings for 19XX under our current reporting method show an increase in excess of 20% over 19XX.

We will not be able to release our earnings until auditing in the field is complete. When the results are in, we still expect to find a favorable comparison with our prior periods. We do not expect any material delay in reporting our results.

Essentially, LIFO is a method that allows a company to report earnings after charging cost of goods sold with current inventory cost. As the words suggest, under LIFO, last-in (higher-priced in most cases) items are used first to determine the cost of goods being sold and ultimately the profits resulting from that sale. In short, LIFO responds to inflation.

Since we are convinced that inflation is entrenched in our

> economy and will remain with us for some time, we think LIFO
> accounting is now the more appropriate method for us. We also
> believe the move will be well accepted by the investment com-
> munity as a positive step on our part. As an additional be-
> nefit, the use of the LIFO method will result in a reduction of
> federal income tax payable by the company, and in turn permit
> a reduction in comparative borrowing and interest expense.
>   In closing I would like to say that, no matter how we keep
> score, 19XX was a very good year for the company. I also expect
> that the next few quarters will be difficult, but the outlook
> for BuilTight Corporation in the 19XX's is so positive that
> we look forward to the many challenges with optimism.
>
>   Sincerely,

The letter is actually an extended definition of the *Last-In-First-Out* method of placing value on inventory. It goes beyond the limits of the formal definition to provide illustrations and reasons for using the different method—"LIFO responds to inflation" and results in a reduction of the taxes the company must pay. These two reasons may not be *necessary* and *sufficient* items for the definition, but they do add to the reader's understanding by recognizing their needs. The LIFO explanation also demonstrates that extended definitions allow you to develop ideas through the use of details and illustrations that are appropriate in reports.

Placement of Definitions

Once you have gathered the relevant information that you need to provide a useful definition, determine where to put it. A definition, we mentioned previously, can appear in any of the following positions:

- in a parenthetical expression in the text
- in a footnote
- in a glossary of terms
- in context

Each has distinct advantages.

Placement in the text allows the reader to have information when needed, and where it least distracts from the flow of the writing.

Footnotes that present definitions appear to be more scholarly and authorita-tive, but can often seem pedantic. In most business and industrial reports you do not wish to create a stuffy, pedantic impression. Even in scholarly writing, the definitions are part of the text, allowing brief comment and documentation to appear in foot-notes. (See Chapter 10 for an examination of notes and documentation.)

A *glossary* can provide an efficient way to present definitions of many special or technical terms in your report. As a list of equivalent meanings, it establishes a common ground with an introductory audience. It also allows you to form a common language among professionals, and create agreement concerning the meaning of terms.

Definitions appear in *context*. Providing necessary information as part of the flow of the writing is the preferred method for definition. Defining through context closely approximates the way most of us capture the meanings of words.

As with most solutions to writing problems, the best is also the most difficult to produce. Writing a report that defines technical terms by providing a clear context requires more concentration. It also takes more time. Taking more time runs counter to the get-it-done-yesterday mentality of American business. However, time spent to improve the quality of the report pays for itself.

In taking extra time to provide a clear definition, you recognize the needs and expectations of the audience. Though it may not guarantee clarity, it helps promote it in the report.

Jargon

Words that require definitions can be grouped as common, technical, and esoteric. Jargon falls into the last category, and it is strictly language limited to scientific or technical use. In conversation, it often serves as useful verbal shorthand. However, it is usually considered gibberish. When George Orwell described contemporary writing in "Politics and the English Language," he also offered a useful definition of jargon.[1] He wrote:

> Prose consists less and less of words chosen for the sake of their meaning, and more and more of phrases tacked together like the sections of a prefabricated henhouse. . . . There is a huge dump of wornout metaphors which have lost all evocative power and are merely used because they save people the trouble of inventing phrases for themselves. . . . Modern writing at its worst . . . consists in gumming together long strips of words which have already been set in order by someone else.

This description could easily apply to the quality of the writing in business and industrial reports. Many of the reports that cross your desk every day have that prefabricated quality. It demands little thought or imagination on the part of the writer. It also sounds mundane.

Words and phrases that sound familiar find their way into writing merely because they are available. Such phrases discourage fresh thinking about an idea. Jargon represents that tired-out use of the language, and demonstrates in most cases a lack of ingenuity in expression. At worst, jargon underscores that the words, the meaning, and the writer have never made a casual meeting with one another. The conclusion of the food service report discussed in Chapter 3 shows what can happen when words and their meanings part company.

It is easy to condemn jargon as poor writing. But it has an official ring to it, and

[1]Excerpts copyright 1946 by Sonia Brownell Orwell; renewed 1974 by Sonia Orwell, and used with permission of A. M. Heath & Company Ltd. for the estate of the late Sonia Brownell Orwell and Martin Secker & Warburg Ltd. Reprinted from "Politics and the English Language" from SHOOTING AN ELEPHANT AND OTHER ESSAYS by George Orwell by permission of Harcourt Brace Jovanovich, Inc.

leaves your audience with the impression that they have heard it before. Most of the paper churned out of our business, professional, and government offices has a doughy quality because of jargon.

Furnishing concrete expressions to replace jargon makes writing more difficult. As an engineer observed, "It's a lot like being told to write right-handed, after you have written left-handed all your life." Writing that plants a clear idea in the reader's mind demands hard work. Without it, though, the reader may never turn to page two.

Jargon can control the way you express yourself. If you feel that it dominates your writing, rather than your controlling it, follow these antidotes prescribed by Brooks and Warren in their book *Modern Rhetoric*:[2]

- use words that are as specific and concrete as possible.
- avoid stereotypes of all kinds—prefabricated phrasings that come to mind but may not represent precisely your own ideas and emotions . . . The rule of thumb would be: never shy away from an individual word merely because it is frequently used, but always be wary of frequently used *phrases*.
- use live words, remembering that . . . verbs are the most powerful words that we have.
- use simple sentences in normal order as the foundation of your writing.

These remedies imply that to rid business writing of jargon takes thought, time, and effort. However, a jargon-free report makes a favorable impression on your audience.

Positive Uses of Jargon

During a discussion of words among technical specialists from management, data processing, and engineering, the question of the exact meaning of jargon came up. A specialist in the design of computer programs suggested what might simultaneously offer a definition of the term and also a way to think about it. The specialist remarked:

What's jargon to you is terminology to me.

Considering the implications of that statement, jargon takes on a positive application, one that draws people to it the way moths are drawn to the porchlight. The use of special terms, jargon, from any discipline signals to the reader that the writer belongs to the select group of people who know the group's language. Knowing that, we should use jargon, but only enough to show that we belong.

Like spice in food, a little jargon in a report goes a long way. Too much and we spoil the message. The reader may begin to suspect, like Lady Macbeth, that we protest too much. Careful use of special terms that signal your membership in a group marks a positive use of jargon that adds credibility and power to your writing.

[2]Adapted from Brooks and Warren, *Modern Rhetoric*, 3d ed. Copyright 1970, Harcourt, Brace Jovanovich. Used with permission.

Use jargon:

- to signal that you belong to a technical or professional group
- to demonstrate your familiarity with specialized terms
- to communicate concisely with a group of specialists
- to acquaint a general audience with important and useful terms

Whenever jargon appears in a report, define it.

DESCRIBING OBJECTS, MECHANISMS, CONCEPTS IN REPORTS

Description orders space.

Few, if any, of the reports that you read or write consist entirely of description. It is nevertheless an important element of writing that adds credibility and clarity to your message by providing your reader with a mental picture. In most business and technical documents, illustrations provide a real picture, but a written description still has a place in reports.

Often a graph or a picture is not appropriate or physically possible. Here the written description works, and has the advantage of providing comparisons and associations that pictures cannot. Also the written description can direct your reader's thinking about an object through the careful selection of words that have the desired connotation. For instance, take this description from Poe's "The Fall of the House of Usher":

> During the whole of a dull, dark, and soundless day in the autumn of the year, when the clouds hung oppressively low in the heavens, I had been passing alone, on horseback, through a singularly dreary tract of country, and at length found myself, as the shades of the evening drew on, within view of the melancholy House of Usher. I know not how it was—but, with the first glimpse of the building, a sense of insufferable gloom pervaded my spirit. I say insufferable; for the feeling was unrelieved by any of that half-pleasurable, because poetic, sentiment with which the mind usually receives even the sternest natural images of the desolate or terrible. I looked upon the scene before me—upon the mere house, and the simple landscape features of the domain—upon the bleak walls—upon the vacant eye-like windows—upon a few rank sedges—and upon a few white trunks of decayed trees—with an utter depression of soul which I can compare to no earthly sensation more properly than to the after-dream of the reveller upon opium—the bitter lapse into every-day life—the hideous dropping off of the veil.

Compare that to this description in a weekly magazine real estate ad:

Elegantly Causal on 14 Acres
1760 Handsome Stone Classic

Impeccably Tended—Alive with Style, History, Tradition, Comfort—"Tranquility," Your European Countryside-Style Home overlooks: your own lake (fish/boat), a large landscaped pool with full cabaña graced alongside by a marvelous orchard, a superb large Barn, riding trail woods. Cozy to 2 original fireplaces in living room/eat-in kitchen, dining room, 4 bedrooms—3 + master suite. This property is in a Class by Itself. $385,000. Terms.

Both descriptions could possibly refer to the same physical place. The examples illustrate that written description evokes a mental and a physical picture as well.

Our discussion of description considers it not only as a report element, but also as a form of communication and a way of thinking about the physical world. After all, description focuses our attention on information that we gather through the senses. Since we are visual animals, description concentrates heavily on sight.

Three Kinds of Seeing and Description

To understand more fully the process of description, we can look to the visual arts for a way of thinking about the physical world that will aid us in writing. In *The Visual Arts as Human Experience,* Donald Weismann draws distinctions that help explain how people see the physical world.[3] Anyone who has witnessed a sporting event, a car accident, a play, or a movie knows we perceive differently. What you saw and what others saw reflects the many shades of difference in the way we look at the world.

Recognizing this, Weismann identifies three kinds of seeing:

- operational
- associational
- pure

Suppose that, on your way home from work or school, as you leave the building you come upon something in the doorway that partially blocks it. You are in a rush for a rather important dinner party, and you don't want to be late. You skirt the object, find your car in the lot, start it, and are on your way, giving the object no second thought.

That is an example of *operational* seeing. It allows you to function throughout the day. Most of us use this practical seeing every day, all day. It keeps us from bumping into each other with more regularity and frequency than we ordinarily do. However, it is the kind of seeing that rarely penetrates deeply into our consciousness.

Now suppose you leave the same building as before, but this time you have no pressing appointments, save the normal dinner at home. You walk out the door and notice it is blocked. But this time you stop to notice that what is in the doorway is an enormous ball at least six feet tall. You think about the absurdity of putting a ball that size in front of the door, and you remember the time you and your football team filled several practice balls with helium and you remember the expression on the quarterback's face when he threw the ball and it floated about 50 yards.

[3]Adapted from *The Visual Arts as Human Experience* by Donald L. Weismann. Englewood Cliffs, N.J.: Prentice-Hall, Inc., N.D. Used by permission of the publisher.

That kind is *associational*. Once you recognized the object as a ball, and put the label "ball" on it, immediately the thing became undifferentiated from all other balls. This allowed you to take the thing as the abstraction—ball—and associate it with your recollections and experiences with any ball, not necessarily the particular one that blocked the door.

Weismann labels the third kind of seeing *pure*. Using the same situation, suppose you again leave the building. Instead of being in a rush, this time you leave early and have no place to go. You have finished your work ahead of time and reward yourself with the afternoon off. As you approach the doorway the shape that blocks it stops you. Without putting a label on it you begin to experience its whiteness, apparent softness, and how the sunlight gently transforms into shadow at its edge. You walk around it, notice that it looks like fabric, and touch it. But the cold on your finger tips tells you it is stone. You see too the shape and its relation to the space that surrounds it, how it draws your eye along the curve.

As you look you begin to enjoy the shape for itself, not for what it is, or for why it is in the doorway, but for itself and the pure pleasure of the sensual experience.

A painter or sculptor must command your attention and focus your eye on shapes, colors, and textures. A writer, even a writer of technical and business documents, needs a little of the artist to help you observe the physical world about you. You can, then, better select information and communicate it.

Pure seeing can exist only if you observe the world around you freshly, wordlessly, so that you *see* ordinary things in a new way. Applied to writing, seeing freshly encourages an attitude that energizes your perceptions and helps transmit that enthusiasm to the audience.

Requirements for Description

Every description, whether casual, emotive, or technical, should demonstrate the following:

- the point of reference, either physical or emotional
- the list of physical characteristics
- an indication of what the item is like—similarities and differences when compared with other items.

The order of these requirements, of course, can be changed according to your purpose and the needs of the audience. However, you need to indicate each of them for the reader to get a clear mental picture of the item or concept under discussion.

Format for Description

Formal technical descriptions have three parts: an introduction, a breakdown of parts or sections, and an indication of the way the thing works or how it is used.

The introduction, like the introduction to any long piece, first indicates what the document is all about. Your introduction should answer these questions so the reader understands your motives:

- What is the purpose of the detailed description?
- What does the object look like?
- What major groups of parts does it have?

Like any good paragraph, the end of the introduction should lead into the next section, the list of parts.

The parts breakdown requires some strategy for order. No single pattern for ordering a description could apply to all complex items. So clearly let the reader know:

- the order that you are using
- the perspective, or physical point of view

The order and point of view are arbitrary. That is, the particular place from which the item is viewed and the part or section you choose to describe first, second, and so forth matters less than your conscious and clear selection of *a* point of view, and *a* pattern for arrangement.

For example, your company wants to buy a new manufacturing facility. In your report you describe the physical layout, even though you attach a photograph. In such a description, place the reader at a particular point. You choose that point. Only after your initial selection does the point become significant. Then you must adhere to that point of view throughout the description.

If you are Picasso, shuffling the visual point of reference is art. In reports, such shifts create confusion.

Once the point is selected, determine a simple pattern for the description. Both a pattern of presentation and a fixed reference point are necessary for a description to provide the reader with a clear mental picture.

- clockwise
- counterclockwise
- right to left
- left to right
- top to bottom
- bottom to top
- inside to out; outside to in
- large objects or parts to small; small to large
- sight to touch to hearing to smell
- bright colors to dark; dark to bright

Any of these patterns helps your reader to follow your description.

Because writing is presented linearly, one word at a time, a pattern is essential for description. We normally see the world all at once in a single burst. For your reader to understand the description, you must take the simultaneous sensations, break them down into separate parts, and arrange them to suggest simultaneity.

Looking at one word at a time, the reader arranges the information into a pattern that should, if we have written the description carefully, create a word picture. But since we have photographs, why do we need to write descriptions at all?

One reason, which has nothing to do with communication of facts really, is that the act of describing develops your powers of observation, so that you will operate more with the pure seeing. That power carries over into all areas of your writing. By becoming a more careful observer, you inject more concreteness and vividness into your writing, and it becomes more readable and informative.

The second is that a photograph can be interpreted in many ways. For instance, if we took the facility that your company is going to buy and included several shots of it showing weeds growing around the building and garbage in the street, the ordinary person might get a negative impression. Your description would focus on the dimensions and the possibility of converting some of the space for use by your organization. The photo would not really speak for itself. Your description would help the illustration, as much as the illustration would help your description. (See Chapter 11 for graphics and other visual aids.)

The physical description of a Double Locking Device that follows goes beyond the photo shown here (Fig 7–1). The description of the part makes the proper separation, treating each of the three sections in a separate paragraph. Notice that the photograph calls out the physical details, while the written version focuses on the ordering of the parts and on a description of how each part operates.

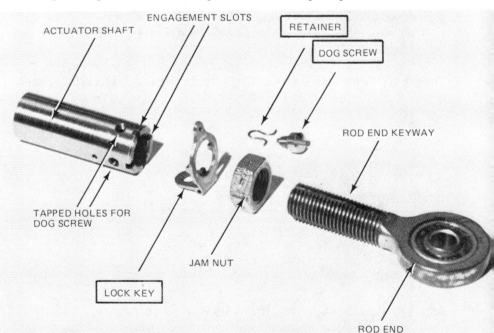

FIGURE 7–1   Rod End Double Locking Device (Used with permission from Grumman Aerospace Corporation)

2.1.1 Description of the DLD[1]

The patented Grumman DLD (Patent No. 4,274,754 and No. 4,232,978) consists of a three part integral assembly: a lock key, a retainer, and a dog screw (see photo). The lock key, an investment casting of 17–4 PH material, provides the primary means of preventing relative rotation of the adjustable rod end. It does this by engaging the slotted end of an actuator piston or bungee, as well as the longitudinal keyway in the threaded rod end. A lockwire hole in the lock key prevents rotation of the jam nut. A side shelf of the lock key acts as a guide for installation of the dog screw, thus providing a secondary means of preventing rod end rotation. In addition, the side shelf prevents the lock key from being installed backwards as the shelf will interfere with the jam nut during tightening. A hole is drilled in a portion of the lock key extended at the intersection of the side shelf and the jam nut interface to provide a hinge for the retainer. A limit stop is cast integrally at the hinge to prevent the dog screw from interfering with tightening of the jam nut.

The retainer is formed from music wire. It joins the dog screw to the lock key and guides the dog screw through the hold in the side shelf. The retainer snaps into a groove under the head of the dog screw. The retainer ends then are compressed and inserted into the hinge in the lock key. The dog screw and retainer are assembled to the locking device in a manner that provides a tamper-proof assembly.

The dog screw is a machine piece of 300 series stainless steel. The shank of the dog screw is threaded to allow engagement with any one of four holes tapped 90° apart in the outer diameter of the actuator cylinder. The end of the shank on the dog screw is undercut to a diameter and length sufficient to provide a secure engagement with the keyway in the threaded rod end. Bottoming of the dog screw in the rod end keyway not only provides the secondary lock, but by removing the play between rod end threads and piston rod, also prevents vibration. In addition, the length of the dog on the dog screw prevents possible damage to rod end threads, and does not impose a load on the side flange. This yields a secondary means of preventing rod end rotation. A lockwire hole in the dog screw, when lockwired to the jam nut, prevents the dog screw from rotating. See Appendix B for drawings of the three DLD components, as well as the assembly.

---

[1]Used with permission from Grumman Aerospace Corporation

The technical description of the Double Locking Device follows the format, complete with an introduction that covers the purpose, what it looks like, and the main divisions. The body of the description covers the parts and their use. The whole device is also covered.

This description incorporates the photograph that accompanies it. Knowing that the reader will get a physical picture and also line drawings, the writer concentrates more on the *purpose* of the device and its *use*.

Characteristic of technical descriptions, the comparison with other things that would help the reader see the device more clearly is missing. Description functions as an important element of reports, though as we have said a report will almost never be strictly descriptive. Nevertheless, the ability to describe things strengthens the writing of processes, instructions, and directions.

PUTTING INSTRUCTIONS IN WRITING

Directions and instructions essentially tell a story. They are narrative. When you tell someone how to do something, you tell them the story of that process, the narrative of it.

Narrative orders time.

With that in mind, you must arrange the sequence of events. For most directions and instructions, a step-by-step approach works well. Flow charts for computer programs and for the depiction of processes or the movement of materials offer another familiar device for organizing complicated actions. Whichever you select, the writing of directions requires you to know these items well enough to manipulate them easily:

- the steps in the process
- the equipment involved
- the materials used
- the purpose of the process
- the use of the result of the process

Format for Writing Directions

Directions respond to the same three-part format we use for description:

- an introduction
- step-by-step
- a conclusion

To that we can add two lists:

- equipment needed to do the job
- materials necessary for the project

Though this format resembles the write-up of a scientific experiment, business applications cover the same areas. But they adhere more loosely to formal patterns of arrangement.

Unlike the description of an object, the narrative expression of a process emphasizes what needs to be done, rather than what the physical characteristics are.

In the introduction, cover these items, but not necessarily in this order:

- What is the act?
- Who does it?
- For what purpose or goal?
- Steps?
- Why?

Much of your writing tells someone how to do something. The directions may be simple and not call for an elaborate format. On the other hand, complicated acts require you to decide which steps to include in the narrative.

CLASSIFYING INFORMATION

Another element of longer documents is the ordering of information into related groups. When we discussed definition, the act of defining exercised our ability to locate and identify a group that contained the word.

We all place items in groups. When we shop, for instance, we create categories—expensive, affordable, cheap. As humans we concentrate on placing the things and events of the world into compartments, groups, categories. Classification taps that natural tendency to order the chaos that the world presents to us. We create groups to simplify the world.

Classification allows us to group items based on shared features. In writing, we can classify by:

- listing the characteristics of the group
- listing acceptable additional attributes
- clearly showing the criteria

To demonstrate this process, ask a group of people to write a list of three nouns on a sheet of paper. One member of the group writes all the words on a flipchart or chalkboard. In a group discussion, everyone determines two groups for the words. Once the terms are separated into the two categories, each group is again divided, and so on until no further division is either possible or necessary.

This exercise illustrates how, through our natural desire for order and organization in our universe, we can find associations between words that are generated purely at random.

INTERPRETING FACTS

Interpretation provides meaning by clearly demonstrating relationships among facts. It presents facts, explains their meaning, and clearly articulates patterns the relationships might take.

In many business reports facts appear as graphs, charts, or drawings. Interpretation explains what the numbers mean. Since it requires a higher level of thought to see a pattern where apparently none exists, interpretation is more difficult for the writer. To offer a cut-and-dried way of interpreting facts would do you a disservice. Nevertheless, the scientific method, or inductive reasoning, offers the best way to seek a valid interpretation.

Inductive reasoning is the intellectual movement from the particular or the specific to the general. That movement is the basis of the scientific method and allows us to use facts to discover relationships which are not readily apparent. As practical application of the scientific method, inductive reasoning provides the writer with the means for interpretation, not a recipe.

To discover meaningful relationships try:

- the brainstorming techniques that we have mentioned
- placing yourself in the reader's shoes
- allowing yourself to think of the terms without preconceptions
- practicing the pure seeing used in description
- devising the "worst-case scenario"

In discovering relationships, remember that interpretation lets your reader know what facts mean.

Presenting Interpretation

Once you have discovered the meaning or relationship, you can present it by:

- providing the most meaningful details
- placing the interpretation prominently in the report
- selecting a format, such as a list, that focuses attention on the interpretation
- mentioning an authority
- putting the interpretation in the summary of background material
- demonstrating what it is like through a comparison and contrast with alternative interpretations

ARRANGING INFORMATION

Dramatic arrangement orchestrates the whole message and builds up to a climax. Movies, plays, and TV dramas use this approach to hold our interest throughout the presentation. For business and professional purposes, this method of ordering ideas may work against the reader's need for gathering facts quickly. The most important information might be placed anywhere in the report, forcing the reader to examine the whole document to get the relevant information.

On the other hand, dramatic arrangement strengthens unusual requests, and softens the impact of bad news. An analysis of the audience will help you decide whether you should come out first with bad news, or give reasons or reactions as a buildup to it.

Dramatic arrangement might appear graphically like this:

This simple diagram usually shows the plot development of a play. The action builds to a high point, or climax, then trails off to the final curtain. Under ordinary circumstances, professional reports do not lend themselves to this type of narrative arrangement. Nevertheless, if you can pull it off, it is effective. If you cannot, however, the writing could appear as if you are attempting to turn a monthly report into the great American novel. Like wearing a three-piece suit to the beach, the style does not fit the situation.

Arrangement through time, space, analysis, importance, or position applies successfully to business and professional reports.

Using Time to Arrange Information

In discussing the use of narrative to present instructions or directions, time functioned as the ordering principle. The natural perception of events in sequence provides a comfortable way to arrange information for a report.

Once we decide that time offers a clear principle for ordering, at least four methods provide the means:

- chronological
- sequential
- step-by-step
- dramatic

In each one, you manipulate and fashion events as the raw material to create a clear sequence.

Chronological arrangement is useful in a progress or status report. Briefly the thrust of this arrangement presents the major events in the evolution of a project in the time or historical sequence in which they occurred. Here we can think of time as calendar time, rather than clock time.

Sequential arrangement differs from chronological, and deals with the order of events unrelated to history. For instance, if you explain the way a bill of lading moves through your organization, the reader gets a clearer picture from a case sequence generally true for all cases, rather than a particular one, which follows the historical characteristic of time. Sequential arrangement lends itself to a discussion of general repetitive tasks. Assembly-lines embody the sequential concept of time events.

The *step-by-step arrangement,* used to create directions or describe a process, resembles the sequential. Both treat time as something divorced from history, isolated unto itself, but following a predetermined cycle of events, once the event has begun. Step-by-step arrangement, of course, applies appropriately to writing directions or instructions.

The *dramatic arrangement* which we have already discussed can be effective in carefully chosen situations.

Space as a Method of Arrangement

In addition to presenting a mental picture, description also provides a method of ordering information. We can use space as an organizational principle in reports. The point remains simple—the pattern you choose allows the reader to see the object as a pattern too. In constructing a mental picture, the pattern serves as a blueprint, sketch, or diagram. The pattern gives the reader a way to conceive of space. By providing a pattern, you make the reader's job easier. Of course, that makes your job more difficult.

Other uses of space as an organizational principle can be practical in a report. If, for instance, you must write a weekly report on several company groups that are scattered across the country, *geographical arrangement* provides order. Once you choose a pattern, stick to it. You can move from:

- east to west; west to east
- north to south; south to north
- big cities to small; to rural communities; coastal areas to mountains

Here, too, the particular pattern is not as important as having one in the first place. You can also use space to organize a report on activities in a factory. In reporting on the manufacture of a product, the arrangement roughly coincides with the manufacturing process. For instance, a quality check reveals that in making a toaster the actual chrome plating is half the amount called for in the specifications. The report on that must discuss the plating process and its impact on the entire assembly line. In doing that it might indicate that the two assembly lines will be slowed, or that you may have to shut one of them down, because the additional plating will now take 50% longer than it did before the discovery of the flaw.

The assembly line analogy helps to organize information, if you think of it as representing work flow. Usually a flow chart provides this for the audience. Your ability to create a flow chart represents your ability to arrange information in a useful, if not a logical order.

Flow charting provides the means to organize complicated events in either a physical space or a mental "space." It helps you to grasp and communicate them more clearly.

As a principle of organization, space is appropriate for many general situations, and is clearly understood by audiences that might have a broad range of education.

ANALYSIS USED FOR ORGANIZATION

Another method that can order the chaos of information is analysis. Put simply, analysis takes the object or event as a whole, and investigates it by looking at its parts. Often the parts are obvious, but many times they must be identified clearly in the discussion. As with the other approaches to organization, analysis takes several forms:

- comparison
- contrast
- reasoning, or cause and effect
- identifying and relating parts to the whole, and to one another
- separating and listing parts in order—alphabetical, numerical, functional, or occurrence

When we *separate, list, identify,* or *relate* parts we provide a pattern functionally similar to those for descriptions and processes.

Comparison

Analysis that involves comparison takes an object and provides a list of similarities. Anything can be compared with something else. Definition uses comparison to find similarities so that we may place the item in a larger group that shared several characteristics. In writing a comparison, first decide which and how many categories to consider. In deciding to build a new factory, your company may have two or as many as 10 possible sites in mind. Determining the factors for comparison clearly aids selection. You cannot, of course, put expensive factories in 10 places, so you must compare and decide on one. To write comparisons:

- choose the criteria
- select a manageable number of categories
- place them in a meaningful order
- indicate relative importance

Contrast

Instead of similarities, contrast shows differences. We usually think of these acts as the same. They are, but for the purpose of discussion we separate them. Contrast also requires selection of criteria, categories, a meaningful order, and relative importance.

Here is an example from the *Consumer Reports 1982 Buying Guide* which compares and contrasts the metals used for pots and pans.

METALS. The type of metal and its gauge (thickness) affect heat distribution, durability, ease of cleaning, and ease of handling. The Ratings depended primarily on how well the pans distributed heat. Poor heat distribution means scorched pancakes or ruined sauces. Copper distributes heat very well, but it's expensive, and the tested lines that had copper exteriors proved disappointing. . . .

A fairly thick aluminum pan provides much better heat conductivity than one of thin copper. Even the lightest aluminum pans we tested distributed heat well enough for routine cooking, but they were apt to warp or dent. . . .

Stainless steel, an alloy made with iron, nickel, and chromium, is very easy to clean. By itself, it has low heat conductivity, so stainless steel is nearly always used in combination with iron, copper, or aluminum. . . .

The lowest-rated pots and pans were those whose only metal is steel. . . .

Cast iron has low heat conductivity, but utensils made from it are thick, which helps to spread the heat. Cast-iron skillets and griddles provide a slow, steady heat that's fine for panbroiling. Since it's so heavy, though, cast iron isn't a good choice for everyday use. . . .

Reasoning

Use cause and effect to explain information and to supply reasons. For example, abuses by employees during their free time have forced your company to institute stricter guidelines. Tightened security is the effect, and some explanation of the cause usually accompanies any mention of effect. The following memo makes such an explanation.

TO: All Employees   Date: February 9, 19XX
FROM: Harold McPheters, Adm.   Subject: Side Door Entrance

The building entrance on the southside (near handicapped parking spaces) was broken on February 3, 19XX and has not been completely repaired to date. It has been the habit of many staff members to prop the door open when exiting the building to purchase coffee, etc., from the coffee truck; this improper use of the entrance has caused the door to break.

Effective immediately, the side entrance is to be used for employees to enter the building between 8 and 8:30 a.m. and to exit at the end of the work day only. When returning from lunch, field visits, etc., the front and rear entrances must be used for access to the building.

> I must remind you of the fire code regulation which pro-
> hibits the door from being left unlocked or propped open at
> any time.
>   I expect all employees to adhere to this policy. If this
> problem continues, further administrative action may be
> necessary, including removal of the coffee truck.

In reporting a cause and effect relationship, make the reader aware of what happened as well as why.

Importance as a Principle of Organization

When you arrange ideas in order of importance, clearly reveal the criteria used to determine the relative value or priority of the items you present. Let the reader know how to decide which item is more important than another. The order of presentation demonstrates the order of importance.

POSITION: THE UNWRITTEN MESSAGE OF ORDER

Business and professional writing, or course, places important information up front. This is called a direct presentation. Managers expect to find decisions and recommendations there, so use the first position in a report to satisfy their expectations. Where you put a fact in a document tells the reader as much about its importance as does the comment you provide. (See Chapters 8, 9, 10 for further discussion of the placement of information in a report.)

Don't assume that a person who reads 10 to 15 reports daily can spend hours pouring over your papers to discover the single important fact buried in the middle of your report. Clearly identify important data and present them as near the beginning of the paper as possible.

Practical reports provide first things first.

SUMMARY

- Definition of technical, uncommon, or esoteric terms provides clarity for your reader, and makes a report understandable.
- In reports we can define terms formally, informally, or through extended discussion and illustration. They appear in a report in context, in a parenthetical expression in the text, in a footnote, or in a glossary of terms.
- In business and professional situations jargon usually confuses the reader. However, limited uses of selected terms signals to the reader that you belong to the professional group.

- Description allows the reader to picture an object or a concept. It also provides you with a means of discovery. Writing description requires a point of view, a list of characteristics, and a suggestion of what the object is like.
- Giving instructions or directions is like writing a narrative. In doing so you provide the reader with a "story" of how to make or do something.
- Classifying items, putting them into related categories, helps a report to achieve order and clarity. It also enables the reader to see how items relate to one another.
- Interpretation communicates the relationships that exist among facts.
- The arrangement of the parts of a report demonstrates their importance and helps the reader see the main point by providing a pattern.
- Arrangement provides the reader with a pattern for the information, and demonstrates our knowledge and expertise concerning the material. It also helps both you and the reader understand the facts and ideas better, and represents a courtesy to your reader. Time, space, importance, drama, analysis, and position offer effective methods of arrangement.

EXERCISE 7–1

Make two lists of jargon. In one, place the words that signal your membership in a group of specialists. In the other, list the terms that are not useful or appropriate indicators. Show the two lists to a colleague and discuss your selections. Incorporate any additional items that result from your discussion.

EXERCISE 7–2

1. Write a definition of a term that you might come across in your profession. In doing so be sure to take the elements we have discussed into account, that is: the larger group to which it belongs, the differences, the common ground.
2. Suppose someone has asked you to prepare a brief presentation so that other employees will understand what your job is. Approach the presentation by thinking of it as an extended definition of your job title.
3. Take a term central to an idea that you know well. Use the context of a paragraph or more that you write on that subject to define it.
4. Create a glossary of terms that would be useful to an audience that is unfamiliar with the terms you compile. Make the glossary as clear and useful as you can.
5. Write an extended definition of one of the following terms, or of a term you select:

custodian	Keogh Plan	on-line
fiduciary	IRA	user-friendly
trustee	tax shelter	productivity

EXERCISE 7–3

1. Select a common object and describe it technically. Make sure that you cover all the criteria for a clear word picture and that your description recognizes the needs and expectations of your audience.
   Describe the same object, this time focusing on its more subjective qualities.
2. Clearly explain to a general reader how to perform one of the following tasks:

- operate a word processor
- handle an angry customer in a department store
- change the office morale from negative to positive
- sell a concept or idea to an unwilling buyer
- prepare a tax return
- request information from a company

If none of these appeal to you, select your own task.

3. Randomly select 25 nouns by pointing to words in a newspaper. Use only the first 25 nouns for this exercise. Create a classification for them. Write the criteria for each group.

4. Select an object that has no apparent use. Give that object to a small group and have them describe it for you. What was the result of this experiment? What does this tell you about the relationship between the audience and the object and the audience's needs in a description?

EXERCISE 7–4

1. Find a report that presents facts in a graph or in a table. Write the most complicated interpretation that you can imagine, based on the data in the graph or table. In another paragraph state clearly what you discovered about the meaning of unexplained data to your audience.

2. The child of a close friend knows that you have recently finished your university training and has asked your advice about what to concentrate on in college. Write a letter to the child, discussing the merits of three fields, and then select one as the best. For this you can assume that the child has not only the ability and skills, but also the willingness to learn.

3. (Advanced) Gather information from 50 people, 25 from the business and professional world, and 25 at random from the community. Ask them this simple question: "What is clear writing?" Based on this rough survey, write an interpretation of the information you compiled.

4. (Advanced) Select a topic you are interested in or know something about. Find an article in a daily newspaper, one in a magazine, and one in a professional journal on that topic. Based on the selections you have made, write an interpretation of the three audiences and the significant differences between them.

5. (Advanced) You have investigated existing replacements for obsolete office equipment. Submit your findings in writing. Support your choice of equipment by interpreting the factors that influenced your decision.

6. (Advanced) You commute to work each day. Survey the existing ways to get there, and determine the most efficient and least costly. Be sure to list clearly the criteria for judgment.

EXERCISE 7–5 (ADVANCED)

1. Select a group that you associate with at work, in the community, or at school. Analyze and interpret it.

2. Choose two cities that you are familiar with. Write a comparison of them. Make sure that you clearly present the criteria for your comparison. Also contrast the two cities.

3. Select two competing manufactured products. Analyze them and then decide which is best.

4. Tape a conversation, meeting, or telephone call. Arrange the topics of discussion to demonstrate the relative importance of each topic.

5. As a municipal manager, you have just been told that you must show a 10% reduction in your salary budget for the next period. A recent election overwhelmingly mandates the reduction. Assume that you have 10 employees, each making about the same yearly salary. All of these people have relative worth. Analyze where you will make the reduction. Or argue that reductions cannot occur without a significant, more than 10%, decline in the productivity of your section.

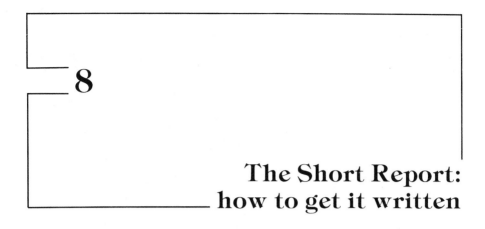

8

The Short Report: how to get it written

Information in the workplace must be communicated, assimilated, and acted upon quickly. The short report fills this need, however, short implies length only. The less space you have, the harder each word, phrase, sentence, paragraph, and page must work to put across your ideas. Writing a short report may take as much time—sometimes more—than writing a much longer account.

PUTTING IDEAS ON PAPER

Since that is the case, and since your time is limited, a clear approach to thinking about writing a longer project will help.

In the introduction to Part I we listed 10 steps to writing, not so much as a formula, but as essential parts of the process that you need to consider when writing. These steps become more important for a project that requires more than one sitting from gathering information to sending out the final papers. The 10 steps, like a map, indicate where you can go.

Because longer projects require more time, a slightly different approach to them is better. Because other demands at work, home, and play come between you and the longer project you are working on, you need to devise a strategy that accommodates daily pressures and your idiosyncrasies.

So let me offer an approach that you can use. (Checklist in Chapter 10 provides an overview of this strategy.) Or you can, like a good cook with a recipe, spice it up with your own personal touches. Whether you use this approach or change it, the method acquaints you with a strategy for longer written projects.

Getting Ready

Many people avoid writing by setting up "outs" for themselves. For instance, you are all ready to bang out a rough draft when you survey your desk and see no matches. The rest is familiar: the 10-minute search and the time wasted fall into what the psychologists call approach-avoidance. As the act of writing approaches, you fabricate an excuse not to.

The brainstorming techniques that we discussed in Chapter 1 should help you overcome delays and begin to think about writing. Here is a brief review of them:

- 10-minute "free writing," or directed to one topic
- forced associations with a key word
- questions for perceiving the abstract as concrete
- dialogue—placing yourself in the reader's shoes
- revision of a passage as absurdly as possible
- deadlining

Each of these provides a useful buffer between the rapid-fire pace of the office and the more considered one needed for writing.

Getting ready assumes you have all the materials you need—pens, paper, correction fluid—as well as a time and place to work.

Time.   Most important, though, is the need for you to clear time to write. Be realistic about the time you will need. Ten minutes of concentration doesn't happen without some preparation. The rituals that we have discussed provide the necessary transition from routine activity. The workplace can gobble up time, and the way you manage what you have reflects on the quality and promptness of your written reports. If you discipline your work habits, and in this way manage your time effectively, you methodically accomplish the reports.

To put time to work for you determine the part of your day best suited for writing. One vice-president of an investment firm responsible for commodities credit works best in the morning, the hours before the exchange opens and the phones begin to ring constantly. A particular hospital supervisor works best from 2:00 to 4:00 since the morning and early afternoon must be devoted to pressing problems from other shifts, patient rounds, and scheduled procedures.

No matter what time of the work day is best for you, set that hour or so aside for the generation of the report. To do this, inform your secretary and the switchboard that you cannot take calls or messages. Convince yourself that any decision can wait for the hour to be over. Doing this ensures that once you begin to concentrate, you will not be disturbed. I wish I could tell you that another way works, but as far as I have been able to discover, the act of writing is best performed in an atmosphere that fosters concentrated effort. That's why 10 minutes spent in this way can be as productive as two hours of effort punctuated by calls, messages, and other interruptions.

Once you have determined your best time, write at that time every day. Productivity in writing depends on routine and discipline. If you write just one 250-word page a day, at the end of a year you will have a substantial book of 365 pages. With that in mind, a week in which you produce 1,000 words a day, or approximately four pages, results in 20 pages. By contrast, you could write 20 pages in one day, but would you have anything left for the next? Most of us would not. In a business setting, you have a good productive day if you generate from 500 to 1,000 words. Writing at the same time every day provides the discipline needed for productive effort.

Space.   Just as time is important to writing, so is the need for space. If possible, write at the same place every day. If you happen to be in an office that clusters desks, or the space that you have been assigned is used for other purposes, try to devote space to the project you are working on. This will free you from the need to pile and unpile papers every time you write.

This represents an advantage for writing because many people organize concepts and information spatially. You may have notes, data, and graphs spread over one area of your table or desk, and while writing you "file" that information by remembering where on the desk you placed it. Such a practice can be difficult in a busy office, but in that way you can see, literally, the information that you have, and the relationships that exist between the facts.

Clearing time and space frees you to concentrate on the subject.

Concentrate on the Subject.   The ritual of writing helps you concentrate on the subject more quickly and easily. After all, your preliminary routine provides a smooth transition to the intensity that writing requires.

At the end of the ritual that you have devised you may find yourself staring at a blank sheet of paper. In that case, the brainstorming techniques can help you focus on the subject and also provide you with a method of discovery. This chapter and Chapter 9 present other techniques and provide additional discussion of brainstorming.

Remember the Purpose.   Early in the creation of any report answer the questions that help you determine the purpose. As you write, review those responses often.

You might type them up and pin them to a bulletin board so you can glance at them while working at your desk. This forces you to keep the goals of the report in mind every day. If we depend on memory, our perception of those goals changes. Time may discolor the paper a bit, but not the words on it; time colors recollections. Just read something you wrote some time ago to see if your present recollection of the event has changed. Throughout the writing, reviewing your original purpose and goals helps you maintain the proper focus on the material and also provides a method for injecting unity and coherence into your writing. We will discuss each of these elements of style in more detail later in this chapter.

Remember the Reader.   Careful audience analysis helps you maintain a sharp focus on the reader. Considering the reader also helps you maintain a consistent attitude toward the audience and material. Continuity of tone gives your reader the unwritten message that you:

- have considered their needs
- know your material
- are professional

Reminding yourself of the audience's needs and expectations during the writing of a report helps you to create a better rough draft, one that can be polished and edited faster, and with fewer major changes.

Discipline in Writing Reports

The strategy we have outlined provides a means to achieve the discipline necessary for longer pieces. It is a means to achieve discipline, not the discipline itself. Only you can obtain it by exercising the means when you write. Disciplined writing helps you meet the special characteristics of short reports, which are:

- harder to write
- precise presentations of ideas
- read only once
- read quickly

Writing the Short Report

Now, of course, we can employ the 10 steps of writing:

- brainstorm—ideas and graphics
- determine the purpose and the goals the report should achieve
- identify audience needs and expectations
- analyze information
- arrange data for maximum effect
- write a rough draft of the whole document
- allow time to cool off
- put conclusions, recommendations, or important ideas "up front"
- edit, revise, rearrange
- prepare a final draft, including graphics

The sixth item, write a draft of the entire document, becomes more important for reports than letters or memos. The initial draft of a short 3-page report might be twice as long. In creating the report, the rough draft written all at once can give the whole report coherence. Each part flows into the next.

Another advantage of this method of generating a rough draft is that you now have a clearer picture of the whole thing, as well as the portions that need more information and documentation. If you continue until you get to the end, the act of writing allows the "holes" in your information to surface. Writing is an act of communication and of contemplation. The effective writer does both.

Of course, writing a draft of a long report at one sitting might be impossible, considering that the final product might run 100 pages or more. With that in mind, a detailed outline serves as a rough draft that can be filled in later.

An outline can be formal, like the type you might see in a technical discussion of a piece of machinery. Or it might be nothing more than a few key words arranged randomly on a scrap of paper. Between these extremes lies an outline that you can use to aid you as you fill up pages for the report.

Remember, the outline serves as the underlying structure of your report, not the final version.

Getting It Written

> "Don't get it right, get it written.
> *Then* get it right."

At the risk of diluting the impact of Professor and critic John Thompson's advice to his students, let me point out that many business people fail to believe that good writing requires more than one draft. But in order to revise, you must have something to work with. Write it. Then revise. But make the two acts separate. If you do not, you will confuse the creative and critical processes essential to efficient writing.

Brainstorming helps to get words on paper. A practice that Ernest Hemingway used helps maintain momentum. He never ended a page with a period. That preserved the flow of thought from one page to the next during the act of creation. In a variation of this, begin the next section of the report immediately after finishing one. That practice starts the following section and also provides the reader with a transition or bridge between the two.

Another variation of this is: always stop a day of writing in the middle of a sentence, paragraph, or section that is going well. But why quit on a roll? So that the next day you will have a thread of that idea to pick up.

Cool Off.   Recognize that the creative and the critical processes of writing must be separate acts. Allow time to cool off between the writing and the revising of the draft.

Conclusions, Recommendations and Important Ideas.   Once you have written the whole report as a draft, look it over carefully to see the real conclusions that you can draw and also to see any recommendations that you overlooked beforehand. Put them at the beginning.

DETERMINING THE SUBJECT
OF THE REPORT

You, as the accounts coordinator, have resolved billing complaints from your California wholesalers. Now you must write a routine report on the changes that need to be made in the billing procedure. The obvious "subject" of the report came from the situation itself. But within the subject, the true subject waits to be discovered.

The "true" subject often involves a larger question. In this case that larger question might be that the existing computer equipment cannot hold together much longer. Like many information systems that were purchased several years ago, its programs continue to demand constant, major revision. The revision required now begins to cost more than the machine did in the first place. The present billing situation, you decide, reveals yet another symptom in the sick computer that you see no cure for.

But how do you go about discovering the hidden focus of the report? Many people say that the answer jumps out at them. They see the problem stripped of its trivial detail, and they get right to the heart of the matter. Barring a gift for such insight, most of us can cultivate our ability to make revealing connections, and in this way determine an underlying subject.

Brainstorming techniques also provide a method of discovery.

Focused Writing—Brainstorming
Technique 1

In Chapter 1 we discussed free writing. You can use that approach for writing a report. In this instance, however, focus your attention on the report itself, and then write for 10 minutes on that topic. Just as you did in the free writing, write for the entire time without pausing.

Forced
Association—Brainstorming
Technique 2

You will recall that this method helps generate ideas. It involves selecting a key word, and then forcing yourself to place two words next to it with the intention of producing a meaningful association.

Absurd Draft; Absurd
Revision—Brainstorming
Technique 3

If the first methods don't work, try writing about the idea absurdly. Take the topic and try to be as ridiculous as possible in your discussion. For instance, here is a cover letter for a proposal:

Dear Generous Company:

Subject: Proposal for MISSILE PROGRAM

Here is our proposal and we are certain you will find that it is
the best proposal you have ever received, particularly the
cost.

We are equally certain that you will feel our pricing is much
too low and our schedule is overly ambitious. But don't worry
about a thing, this is only a draft version.

Very truly yours,

Writing an absurd draft helps you place the message in the proper context and to determine the best tone for the communication. If you are in control of the situation, and if you need a fresh perception of the report, this technique may jar something loose, or even open some mental doors.

Questions—Brainstorming
Technique 4

Brainstorming not only allows you to discover ideas, but it also helps you to look at information again, more closely and with more understanding. The questions that we listed in Chapter 1 ask you to think of the abstract concept in concrete terms. Questions that force you to ask yourself what is the color of the company you work for, and what it eats and what animal it is, help you focus in on its qualities so that you can write about it more specifically.

Dialogue—Brainstorming
Technique 5

If the other techniques have proved less than satisfactory for the report that you are writing, you might try to write a dialogue. Place yourself in the reader's shoes and write a conversation between yourself and that reader. In generating that exchange, write as you would suspect both you and reader would respond. This technique helps you understand the nature of the audience you address. Writing a dialogue allows you to focus on the true subject by revealing to you a different perspective of the material.

Deadlining—Brainstorming
Technique 6

Another possible way to get words on paper simulates a deadline. Give yourself 10 minutes. For the first five, write the draft of a paragraph of something that you are currently working on. Take the next five minutes to revise the draft you just created, polishing and strengthening the original message.

This exercise demonstrates how to budget the time that you have. Writing usually has some explicit or implicit deadline. Keeping that in mind heightens your productivity and, if you are not used to writing, your anxiety. This method of generating ideas should allow you to establish the amount of time and energy you need to finish on time, as well as to acquaint you with the pressure of deadlines.

Storyboarding—Brainstorming Technique 7

Often business documents involve several writers, each responsible for a particular section. Storyboards, explained in detail in the chapter on proposals (Chapter 9), allow writers and interested reviewers quickly and efficiently to gain an overview of an entire project. Briefly, a storyboard provides a mechanism for presenting essential ideas on graphics and text so the writing team can create a coherent document.

This technique also provides a management tool for directors to keep track of several writers as they work on individual sections of the whole. For a short report, a single storyboard proves adequate; longer projects require at least one per section.

GOALS OF THE SHORT REPORT

During the brainstorming techniques and the preliminary investigations you sought the true subject of the report. The true subject may appear at first glance to be the same as the goal. But closer inspection suggests that the goals you have as a writer may not be the subject of the report at all.

So a clear articulation of the report's goals is in order. This makes the report more effective, readable, and useful for you and the reader. Separate the goals into at least two groups—your personal ones and the ones that you wish the report itself to achieve. Of course, your goals and that of the company may not be the same. Identifying this also aids writing, since such a situation forces you to present information tactfully, or may even require an attempt to change the attitudes of the people you write for.

To identify goals clearly, then, requires your realistic evaluation. For instance, don't suggest a total overhaul of the company's accounting system in a report on the cost of office supplies, even if the report has pointed out several defects in that procedure. Do suggest, however, that the problem is not limited to office supplies, and offer to provide a follow-up report that focuses on the larger problem, if requested. It might very well be that the person reading your report set up the present system and any suggested change would be seen as a direct challenge. In other words, keep in mind that a report has limited uses, to:

- inform or advise
- direct or instruct
- recommend or conclude

In a short report, try not to solve all the problems that you have identified at work. A realistic approach saves you from the appearance of too much ambition or too much arrogance.

Of course, the limits of time, personnel, and resources also determine the amount that can be forced into a short report. If you have only a week and you are doing all the work, including the final typing, that should suggest to you to keep it small. However, if you have a staff of 15, a typist, and a professional research assistant, then the report can present more thorough information and be broader in its approach.

Whatever you do in a report, be creative.

Business and the professions test you daily to perform routine tasks, but such performance often must go beyond the routine. Many qualities of good, clear writing result from a creative approach to the assignment.

MEETING THE READER'S NEEDS

In preparing a report of a single paragraph in the form of a memo, or several pages, list the needs of the reader. Be sure you:

- write down the needs clearly
- look at what they say they need
- reconcile the differences
- list similarities
- use the brainstorming techniques
- decide on an appropriate tone
- put yourself in the reader's shoes

LIMITING THE SCOPE

A short report, like a short story, must capture the audience's attention and get the needed information to the reader as quickly as possible. Information that is essential should appear in the body of the report. Place information you are not sure of in attachments, or not at all.

When limiting the scope, you can assume that:

- readers expect you to provide valid information
- they assume you can document conclusions
- if they want additional data, you can provide it
- you are hired to evaluate data and communicate it accurately
- you will not pull any punches

Limiting the report to the subject that you have and also limiting its physical length do not necessarily coincide. A short treatment of computer programming for a manager can run only a few pages. For a more specialized audience of data processing professionals, it may encompass several hundred pages. For an electronics engineer, the same discussion might fill several library shelves. The audience, the goals, the time, and the resources that you can manipulate, all are factors that influence the scope of your report.

Generally, the content of the short report should:

- clarify the goals and information for the reader
- contribute to the reader's understanding of the information
- relate directly to the subject, the true one and the stated one

To do this, write a single sentence, or a phrase, that clearly states the topic and goal of the short report. Put that on a card or piece of paper and pin it to the wall in front of you. Let it remind you, as you work, of the direction you are taking.

PLANNING AND ORGANIZING A SHORT REPORT

Planning and organizing the report require as much attention as the generation and polishing of ideas. Since most reports that you write are used for action—that is, someone will do something or think something because of it—the clearer the presentation, the more successful the report.

And clarity usually results from an uncluttered, easily followed organization plan. Any simple outline helps provide order for the information you present. Some companies have rigid report formats. You might further observe that, within a company, managers and executives have hobby horses. One vice-president may want the conclusions presented in only one sentence, and that sentence on one page at the beginning of the report. Your audience analysis should uncover such preferences. But usually short reports that achieve results adhere to these criteria:

- main idea up front
- appropriate ordering of other information
- use of graphic aids and attachments
- conclusions and directions labeled clearly; actions expressed directly.

The Central Idea Up Front

The first paragraph of the report must work hard, giving the reader a clear idea of the content that follows, and briefly informing the reader:

- why they received the report
- what decisions the writer made

- what information the writer uncovered
- what the reader must do or approve

Putting the essential information in the first paragraph may not conform to your sense of art, but it conforms to the practices of the workplace. (Fig. 8–1) To manage effectively, anyone in an organization who receives 20 or more reports each week needs to get information quickly, clearly, and concisely. That executive may want to read all the details that lead you to your conclusion, but may not have the time. Provide detail in the body of the report.

When you put *first things first* in a report, you give the reader a focus for attention, as well as a way to interpret the facts, statistics, and details.

Appropriate Arrangement of Other Elements

The strength of a report depends on your ability to select and arrange information appropriately. Like an artist who is able to suggest a face with only a few lines on a page, your information should work hard to suggest much more than the facts show in isolation. A carefully written report exceeds the sum of its parts.

Writing falls roughly into these four groups: exposition, argumentation, narrative, and description. No particular piece of writing belongs completely to any of the groups. Most writing mixes elements of two or more of them.

Four Types of Writing

Exposition.   This common type of writing is used to explain and inform. Many "how to" books and articles are expository in nature. Any set of directions is too. We use exposition to:

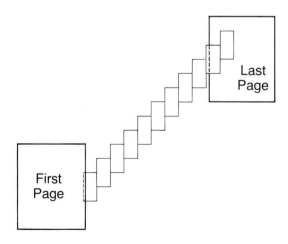

FIGURE 8–1   Audience's Relative Perception of Report Information

define	compare
identify	contrast
classify	explain
illustrate	clarify
analyze	

facts and ideas.

Argumentation.   Argumentation uses evidence and logic to convince or persuade the audience of the validity of an idea. We use argumentation to change a reader's

point of view,
attitudes,
feelings, or
ideas.

Recall exactly what convinced you of the truth of an idea or theory. Lead your reader along the same path.

Narration.   This type of writing presents events in sequence or chronological order. Works of fiction depend heavily on narration, and business and professional writing often require you to *tell* a story. Narrative provides the best way to communicate facts and matintain the audience's attention as well.

Description.   Description makes the reader aware of what the writer has perceived with the five senses and the mind.

Patterns of Development

Once you have selected the type of writing best suited to your material, you can develop your information in one of these ways:

Exposition

1. *Analysis.* Analysis begins with a condition or an object. It tries to identify each of the contributing factors or parts. Computers are programmed after a problem is analyzed.

2. *Synthesis.* Synthesis, on the other hand, begins with diverse elements and puts them together into a logical pattern of the whole.

3. *Simple to Complex.* Often a written message begins with simple situations and leads the reader through more complicated examples. This can be and is used in introducing a new piece of machinery to its operators.

Argumentation

4. *Inductive Development.* Induction takes details and specific examples and arrives at a general conclusion. The scientific method is inductive in nature.

5. *Deductive Development.* Deduction takes a general principle and applies it in order to discover a specific instance.

Narration

6. *Chronological Development.* A specific period of time—day, week, month, quarter, year—is selected. The report moves through that period.

7. *Topical Order.* Minutes of meetings or a report of a site visit can be developed in this way.

8. *Geographical Development.* Organizations with offices in many parts of the country use this method. The writer picks one, then moves on in a predetermined direction: for example, north to south and east to west, for all subsequent parts.

9. *Spatial Development.* This develops information by moving from one area to another within the organization—security, administration, central service, accounting.

10. *Directional Development.* Used to describe a process as it moves. It can be applied to documents, food, and blood as they move through a process or the institution.

11. *Order of Importance.*

12. *Conventional Order* follows a standard of practice such as ones approved by government agencies, the military, or a standards institute.

(left margin brace labels: Description for items 8–10; Convention for items 11–12)

Use of Graphic Aids

Graphics might not be practical for a short report because of the time required to produce them. However, two aids that add force to your report require no added time. One is the careful use of *headings,* and two, *white space.* Headings force your reader to see the pattern you have constructed for a body of information. In this way you direct perception of the facts. Headings also recognize that a large audience may want to focus on one group of details over another. The headings act the way a newspaper headline does, grabbing the reader's attention and stimulating interest in the information that follows.

White space, of course, recognizes that the use of generous margins and spatial breaks for headings, lists of items, and paragraph breaks make the report pleasing to the eye of the reader, and easy to read.

Other graphic aids include:

- pictures
- charts
- graphs
- drawings
- maps
- representations
- mock-ups

These focus the reader's attention. Good graphics allow the reader to gather information more quickly and efficiently. (See Chapter 11 for more detailed discussion of graphics in reports and for several examples.) Generally, any graphic aid should displace its own weight in text. In other words, a single-page illustration should replace 300 or more words of explanation.

Relevant Attachments

Some readers of your report will certainly be interested in the details. Provide that information in attachments. Tables, calculations, data, and tangential notes can follow the text of the report. Carefully selected attachments make the difference between a report that informs the reader appropriately and one that suggests to the reader that you have no idea what relevant means. Choose only the details that reinforce the true subject of the report. Leave the rest in a file folder.

Clearly Labeled Conclusions, Recommendations, Findings, and Actions

Though the introduction provided the important information in the beginning of the report, a section appropriately labeled draws the reader's attention to the items that you consider important.

Also, short reports often request action. A project that runs behind schedule might require additional employees. To do that you must say that hiring is needed, or that people must be channeled from one project to another. In either case, the action to be taken should be clearly stated and labeled. When you provide directions for the reader, anticipate their questions. In the report answer: Who? What? When? Where? Why?

Be sure that the words conclusion, recommendation, action, or findings in your heading reflect the content.

RECOGNIZING UNITY, COHERENCE, EMPHASIS

Unity—Placement of Details

A report, or a memo or a letter, for that matter, has unity if its elements relate clearly to the whole message. Either an item belongs in the report or it does not. In some cases, parts that we would not expect to fit, do so comfortably. As Brooks and Warren point out in *Modern Rhetoric*, "We recognize unity. We do not impose it."

To test for unity:

- determine the true subject
- distinguish relevant from irrelevant information
- keep minor topics minor by limiting the space devoted to them

Each bit of information that you gather for the report becomes special to you. Like packing for a trip, however, you must decide what you need to take with you and what to leave behind. So, too, information must be presented selectively.

In the body of the report place only:

- dramatic or illustrative details
- details that represent trends or generalities
- cases that are particularly graphic
- essential instances

Other details belong:

- in the appendix
- in charts or tables

And still other details belong in your file and are:

- only mentioned in the report
- unmentioned in the report.

Coherence

While talking with people on the job and listening to them attentively, they might suddenly mention their farm and then the weather. Then they return to the subject. In conversation digressions such as these allow your listener to rest. Asides often help to win the sympathy of listeners by adding a human touch. But in writing, tangential chatter suggests that you may not be in control of your material enough to order it.

Coherence is that quality of a report that describes the order and continuity of its elements. In our discussion of the patterns of arrangement, we have provided methods to achieve order. Logic, time, space, and emotion furnish order and provide the means to a coherent report. Our subjective perceptions of objects and occurences, however, may not prove as useful in writing business and professional documents. Nevertheless, it represents a legitimate way to provide order.

Achieving Coherence.   As you drive from your home to work you expect that the road you take runs continuously, using bridges or other structures to overcome any natural obstacles. When writing, you should provide your reader with bridges to follow your thoughts. Don't leave your reader stranded on the wrong side of the river. When you make the whole report flow smoothly, it has coherence, and the connections you provide form the basis for it. Planning, clear paragraph structure, transitional words and phrases, unambiguous reference of pronouns, and the repetition of key words and phrases help build a coherent report. Transitional words and phrases, and sentence and paragraph structure, which we discussed in Chapter 4, give the impression that the elements of the report belong together in the order in which they are presented. They help promote a clear understanding and a useful interpretation of facts and ideas.

Emphasis

In addition to unity and coherence, any piece of writing must exhibit appropriate emphasis. It must indicate clearly which fact or idea is most important. Some ways to achieve emphasis are:

Flat Statement.   A declarative statement provides emphasis. However, the statement must be supported by the facts and discussion in the rest of the document.

Position.   First and last are the most emphatic positions. Business, journalistic, and institutional writing put the most important information first. Building to a conclusion, though, is a more dramatic approach. It is, nevertheless, impractical for most daily writing.

Proportion.   The most important part of the subject gets the fullest treatment.

Other Devices.   You can achieve emphasis through the repetition of a key word, phrase, idea, or sentence. Overuse of repetition can be mechanical and dull, so use it with caution.

A short isolated paragraph is emphatic. It draws the reader's attention, but be sure that such a paragraph is worth the spotlight. Brooks and Warren observe that "certain devices of emphasis are worse than useless." And what could be worse than useless? So beware of RANDOM UNDERLINING, *use of italics,* exclamation points (!!!!!), and CAPITAL LETTERS. Those devices might be fine to sell used cars or stereo equipment, but they contain too much hype for the ordinary reader. Too much mechanical emphasis has the breathless tone of an excited, excitable adolescent.

Let what you have to say stand on its own merits.

ENDING A SHORT REPORT

We have suggested that you effectively present important information in a short report up front. These are usually findings, results, or conclusions. You could end a report with the conclusions in a section labeled "conclusions." But in a short report, you might offend the reader's intelligence if you merely repeat the same information a page or two later. In spite of that, every piece of writing requires some indication that the discussion is over, especially a report that has arranged ideas in descending order of importance.

It may seem awkward to just stop, though that is a common and acceptable way to end. However, just stopping neglects the human desire for a sense of closure in writing. Since writing is a manipulated presentation of reality, and since the writer controls information and presents it as orderly as possible, readers expect some signal that the message is over. Of course, the last page itself does that, as does "THE END" at the end of a movie. But in both cases the audience may not be entirely convinced that the presentation is complete, so you must draw the matter to a close. Here is an example from a report on a change in the way bills were presented:

> In summary, the recommendation outlined in this report will satisfy our California customers' needs, maintain proper billing controls, and will not substantially change the present billing system.

This example demonstrates the function of the short report ending by:

- restating the most important information in a new, but recognizable, way
- providing the reader with a sense of closure

Don't underestimate the necessity for a sense of closure. If the report has it, the reader feels that you know the subject, have explored the available material, and can make a valid decision. The closure in a report ties up the loose ends for the reader.

Without some ending, the reader could assume that you did not budget your time well enough to think things through, or worse, that you did not know how to end it.

Here is another example. It is the last sentence of a report on a conversation one or two senior managers had with an outside consultant. The report concerned recommendations. The writer sent it to the division director and the other manager.

> I think that the above recommendations merit some real thought and discussion by the three of us. Shall we?

This ending draws one portion of the evaluation process to a close, and provides a transition to the next. Its tone is simultaneously cordial and candid. The qualities of candor and respect suggest the attitude of the writer to the material and to the audience addressed.

REVISION AND EDITING: ACCEPTING YOUR UGLY CHILD

Once the draft of the report appears complete, begin to rework the material you have produced to make it meet the reader's need for direct and clearly presented information. To do this you must accept the notion that your child is ugly and could stand some improvement.

As we said, revision is a critical process and should be a separate act from the creative generation of the whole draft of the document. But to know how to improve the document, you must know where to look and what to look for. Here is a list of items that you can use to flag your writing for improvements. The instances reveal candidates for revision.

- locate the verb
- locate pronouns
- locate prepositions
- determine the voice: locate any "by" in the report

These spots should help you decide if the writing

- uses strong verbs
- uses active voice verbs
- provides transitional words
- selects concrete expressions

and if it avoids:

- jargon that distracts from clarity
- strings of prepositions, such as, "as to whether"
- strings of helping verbs, like "will be made to think"
- the use of nouns instead of verbs
- strings of infinities, for example, "to be used to fit," for "fits"

In your writing, read the draft through completely at least once to get an idea of the whole piece. Go through it again, picking out the flags. Then write those undesirable constructions out of the draft. (See Table 10–1, Flags for Rewrite: A Checklist for Editing.)

INTRODUCING THE REPORT: BEGINNINGS, INTRODUCTIONS, SUMMARIES

To start your report, you usually write a paragraph and label it INTRODUCTION. That approach is fine as long as you know that you will rewrite the introduction at least once. An introduction that gets your paper read:

- captures the reader's attention
- suggests the substance of the report
- launches the reader into the rest of the document.

To meet these criteria, of course, you must know what follows. Experienced writers create an introduction last.

If writing were a science, then we could provide a formula for beginners. Since it is not, consider Aristotle's definition of a beginning—that part which is preceded by nothing and followed by something. His observation is so simple and obvious, yet so profound, that it almost seems absurd to mention it. But realizing that the report has nothing before and something after, helps you as a writer to act on that knowledge, bringing home the point that the reader does not know what you know unless you provide the information.

Thinking about the beginning in this way provides an appropriate focus for writing the first paragraph or section.

Methods for beginning fall roughly into two groups:

- traditional
- unusual

Traditional Openings

Under traditional openings we can list these familiar headings:

Summary   presents the main ideas of the report, acts as an abstract and presents the entire report in a condensed form, or lists results.

Scope   sets the limits of the report to let the reader know the extent of the information used in its preparation. A carefully worded title can also provide such information.

Purpose   states clearly and concisely why the report was written.

Background   provides a foundation of other work in the field upon which the report is built.

Problem   expresses a clear statement of the problem to provide a common understanding of actions or events.

Point of View   functions as a variation of the statement of the problem, but focuses on the writer's perception of the material.

Quotation   a popular and overworked method of starting a report. Before you reach for *Bartlett's Familiar Quotations* to find out what Shakespeare had to say about extending credit, consider the potential for pretention. A quotation often sounds ridiculous in a report.

Definition or Classification   allows you to explain a key term for the reader. Since this, too, is a common device, make it a bit different if you use it in your report.

Conclusions, Recommendations, or Actions   have become traditional approaches to the beginning of a report. Business professionals expect them up front.

Of course, the short report can mix these approaches, providing more than one in the initial paragraph, and others in the title or subject line. Any of them launches the reader into the report.

Unusual Openings

Other ways to begin don't fit into traditional groups. Among the more unusual ways to start, avoid these two:

Jokes   are difficult to control in writing; more likely than not the joke will be misinterpreted. Avoid them in writing; reserve them for working lunches, conferences, meetings, and social situations.

Belittling the Opposition   may suggest to the reader that you have nothing positive to say about your company. Instead, present the other side as a contrast to your own, providing the reader with a more positive impression of you.

Not all unusual beginnings are negative. The examples that follow can inject freshness into what might be an ordinary situation.

Forecast or Hypothesis generates interest in the topic. Most people are naturally curious about the future, and this technique uses information about the growth of a market, for example, projections on the use of computers, to capture the reader's attention.

Questions stimulate the reader's curiousity to read on for the answers. For example: "What would happen to XYZ Corp.'s debt service if HRW & Co. merged with us?" Be sure the answer to the question is in your report. A carefully considered question provides an easy, but potent beginning.

Illustrations, Examples, and Cases, carefully selected, can arouse interest. Selecting any of these means that you must build up to a conclusion. Weigh the trade-off of not putting first things first and decide whether these are suitable beginnings or not.

Details. The presentation of a dramatic and carefully selected detail from the report hooks the reader, suggests material to follow, and launches the rest of the report. If the facts back it up, say, "ABC's method of materials management makes manufacturing 27% more efficient." Of course, this strategy works only if the facts bear it out.

Interesting Findings. Information that is new, different, or exciting can also provide a fine opening. Make sure that you state the information, though, rather than asserting that it is interesting. Instead of writing: "It is interesting to note that only 17% of American families fall into the traditional husband-breadwinner/wife-homemaker category," why not let the information suggest its own interest? "Since only 17% of American families fall into the traditional husband-as-breadwinner and wife-as-homemaker category, we have studied this impact on the marketing of our household products."

Comparisons and Contrasts can introduce new ideas, information, or suggestions. For instance, "Automating the Bronco Division of Trinetco increased weekly output by 26% after the third month, while at the Able Division output declined 6.5%."

The last three approaches assume that after writing the draft of the entire report, you select information that allows the reader to focus on the pertinent ideas.

Of course the creation of a report always places a deadline on the information. Following the advice we gave earlier in this chapter, leave yourself enough time at the end of the writing to rethink and rewrite the introduction so that you can incorporate facts and ideas that generate interest, as well as understanding.

Spend more time and effort on the introduction of the report than on any other portion because professionals read the first few sentences of anything that crosses their desk. Hook your reader with the beginning, and deliver what you promise in the rest of the report.

SUMMARY

- Employ the 10 steps essential to writing the way you might use a map to travel from one city to another. When preparing to write, clear time and space, collect essential materials, and establish a comfortable routine. Begin writing with brainstorming techniques designed to get words on paper.

- Short reports require a realistic determination of the purpose, honest analysis of the reader's needs and expectations, and efficient limitation of the scope. Reports require a disciplined approach to writing because they are read quickly and they must present ideas clearly.

- To meet a deadline, get the rough draft written in one sitting. Then revise vigorously to get it right.

- Short reports usually put first things first: all important conclusions, recommendations, and actions up front. Plan and organize a short report by (1) placing the central idea in the beginning, (2) ordering the rest of the parts, (3) using graphic aids, (4) providing clearly labeled conclusions and recommendations, and (5) collecting data in attachments. Short reports possess the unity, coherence, and emphasis related to the purpose and audience.

- Revise the draft of the report by being sure the final version uses strong verbs, the active voice, and quantitative expressions.

- Though the introduction comes first, it is the last part that you write. Be sure it captures the reader's attention, suggests the substance of the report, and launches the reader into the rest of the report.

EXERCISE 8–1

Assume that you are a company Vice President. Your organization wants either to buy or to invest in a small company. Present a short report explaining your decision concerning which option to take. For this exercise you may select an actual company or create a composite company as the example. Use plausible facts and figures that you associate with the fictional company.

EXERCISE 8–2

Submit a status report of an ongoing project. You may use your own career for this.

EXERCISE 8–3

Your company Blakely, Blanck, and Gillmond Associates maintains a small division in rural Maine to manufacture nacelles, the covers for jet engines. The last pieces they shipped to your Florida assembly plant had minor damage from a truck accident outside Memphis. The investigation report mentioned a crack in the nacelle, which could be related to the accident, a result of the manufacture of the part, the packing of the part, or a design flaw. The part was returned to Maine for analysis. You go to find out what happened and find the operation there plagued with management and employee problems ranging from the rapid deterioration of the physical plant to open hostility between the employees and foremen. You notice that mechanics have neglected preventive maintenance of the machine tools, including the big extrusion press. In addition to a report on the nacelles, submit a report of your other findings and suggest solutions to the problems you identified.

EXERCISE 8–4

(Advanced) Investigate a company that offers a service, one that might suggest an area of expansion for your company. Write a report of your findings.

EXERCISE 8–5

(Advanced) Read the following report carefully. Write comments in which you:

- determine its purpose
- identify the audience
- express any conclusions
- comment on its organization

Look over the comments you have made. Suggest solutions to any problems or weaknesses you have identified.

Write a version of the report that incorporates the solutions you identified, and also demonstrates the qualities of an effective and organized short report.

```
Masters & Company

Date: October 31, 19XX
To:   Christine A. Watson
Re:   SANE Integrated Information System


     On October 30, 19XX, I, along with Gil Paul and Judith
Kasile, attended a seminar and demonstration of the SANE
Computer Systems.
     SANE 1450 Integrated Information System (IIS) combines
their word processing system and a virtual storage computer
system. The advantage of this system allows one CRT terminal
to be used for both word processing and data processing sys-
tems. This, naturally, would be an advantage only if, in
fact, Masters & Co. decides to go into a word processing
system.
     We spent about an hour with a SANE representative who de-
monstrated the data processing system to us. This system is
SANE's answer to HBE's W20.
     In my opinion as a user, the HBE W20 system is much simpler to
use if, in fact, the user decides to extract various pieces of
data from the tables set up on the system rather than "paging"
through the whole table. This, of course, is a prime use of
this type of on-line statistical information system. It
appears that the SANE system uses more complicated steps to
re-sort tables, put two tables together, or extract pieces of
information from the tables, using specific parameters for
the information needed.
     In W20 the user can place his parameters into each column of
the table format, allowing the user to see what he is doing.
```

All these parameters can be put in at once on the same screen
format. The SANE system requires the parameters to be put in
on two or three different screens which contain information
about the table parameters in word format, not report format.
I found this to be somewhat confusing compared to the W20
system.
  I am enclosing some brochures for your perusal.

                                        Paul D. Gold
PDG:b
Enclosure(s)

EXERCISE 8–6

(Advanced) Three short reports follow. Choose one and evaluate its strengths and
weaknesses, using Scorecard 1 in Chapter 5 on memos.
  After you have completed your analysis of the sample, rewrite the report. The revision
should strengthen the weaknesses that you have identified.

Example A

Through careful study of data gathered it is the recommendation of this department to purchase said equipment
1234, from the ABC company.
  On January 1, 19XX, this office received a requisition from
Mr. Jones, manager of purchasing for the XYZ company. The
requisition was for an investigation of the testing procedures of the ABC company, specifically to evaluate if said
equipment 1234 was adequately manufactured.

Investigation Report

Monday, January 4, 19XX

Upon arrival at the ABC company began close study of drawings, and specifications of said equipment 1234. Drawings,
and specifications were acquired from front office. Inspected said equipment 1234 to prescribed measurements.
Variations were as follows:

  1. Coupling (as found)   19.998"
     Coupling (specification)   19.999"
     Variation   − .001"

  2. Journal (as found)   6.999"
     Journal (specification   6.999"
     Variation   1.000

3. Resistance fit (as found)   0.499"
   Resistance fit (specification)   10.497"
   Variation   +.002"

4. Bolt Holes (as found)   1.000"
   Bolt Holes (specification)   1.000"
   Variation   .000"

5. Overall Length (as found)   74.250"
   Overall Length (specification)   74.250"
   Variation   .000"

6. Diameter (as found)   27.000"
   Diameter (specificiation)   27.000"
   Variation   .000"

Note: All variations within tolerance of ± .003"

Tuesday, January 5, 19XX.

Began nondestructive testing procedures.

    1. Magna Flux Test: Covered all external parts with flux solution, and cleaned. Magnetised said equipment and checked with ultraviolet light, and visually inspected parts. Found minimal signs of pitting and no cracks.
    2. X-ray Test: Using cobolt system, x-rayed said equipment to check for possible cracks internally. Found no signs of cracks.

Wednesday, January 6, 19XX

    Said equipment was hooked up to simulator for stress test. Running at 115% there were no signs of wear. However, there was a rise in bearing temperature, which should be expected.

Example B

SPRAY SYSTEM OPERATIONAL PROBLEMS

Problem Overview

    Andy Peoples and Phillip Nickelson from SOIL Associates conducted an extensive survey of the spray irrigation at the MEGA-CONGLOM pulp mill from May 26, 19XX, through May 28, 19XX. During the visit, all phases of the spray irrigation system's operation were reviewed, including: the aerated lagoon; the sprinkler system; the groundwater monitoring wells; the mill's laboratory procedures; and characteristics of the wastewater from the mill to the lagoon and from the lagoon to the spray fields.

The basic problem with the system is that in their present condition, the spray fields cannot accommodate the original design wastewater application rates. Reduced application of wastewater is necessary to prevent surface runoff from the fields. As a result, the system storage capacity is nearing its limit and the system is operating such that effluent flow to the mill essentially equals flow out to the fields. Under normal operation, at this time of the year, the system should be operating to reduce volume stored at a rate which is greater than the influent flow to the lagoon.

Causes

Several factors contribute to the runoff problem, and these factors will be discussed throughout this report. Briefly, the operational problems associated with the disposal system could have resulted from:

(1) The sprinklers may be located on ground of excessive slope, or may spray wastewater into forested areas, causing water to run off the fields before it is able to infiltrate the soil.

(2) The cover crop may not have been correctly established, and/or the irrigation system not adequately managed, as required for correct operation of a wastewater spray irrigation system.

(3) The soil characteristics may not have been correctly assessed or are not suitably compatible to the wastewater being applied.

(4) The wastewater's characteristics may not be suitable for land disposal.

It should additionally be noted that the system has experienced several piping system breakdowns. These breakdowns have forced sections of the spray fields to be closed until replacement materials could be obtained and installed. The closing of fields has decreased the ability to spray and has increased the volume in the storage lagoons.

Evaluation & Recommendations,

The following sections of the report review the causes listed above, evaluate their impact on the system's problems, and make recommendations for solving the problems. The recommendations will be based on providing long- and short-term solutions.

The long-term solutions are to provide a system which can be correctly operated. The short-term solution is to enable the mill to continue proper disposal of its wastewater throughout the year, while instituting the long-term remedies. Disposal of the existing excess volume in the aerated lagoon must also be incorporated into the short-term solution.

In addition to the problems and causes detailed above, several additional areas are to be reviewed in this report. These include: localized soil erosion due to sprinkler bleedoff; localized erosion due to the underground pipeline's drain valve operation; review of the mill's laboratory procedures and priority pollutant analysis; review of the system's groundwater monitoring wells; and review of the system's equalization lagoon.

Example C

Dear Professor Goodman:

This career progress report is submitted in response to your Final Exam question one.

It is my career objective to become an executive administrator in a major corporation. I am particularly interested in the expanding role of computers and the impact of telecommunications on offices in the future.

Recently developed "programmerless" computer languages, based on simple English commands, will decrease management's dependence on operating technicians. Inexpensive telecommunications equipment has made it feasible to replace the desktop calculator with a computer terminal. Administrators have immediate access to decision-making information and can communicate rapidly throughout the corporation. The role of the secretary is changing. Word processing equipment is becoming commonplace. This equipment, once emphasized as a labor saving device for the typist, today more closely resembles a computer than a typewriter.

The administrator of the "office of the future," frequently called "the paperless office," will be presented with unique challenges. Personnel management and communication skills will be as important as technical knowledge in doing an effective job.

Early in my career as a secretary, I was introduced to word processing equipment. Later, as an office manager, I planned the installation of word processors and of centralized dictation system for a regional sales force. In my current position as administrative supervisor for an executive recruiting firm I am involved in planning and implementing a computerized filing and coding system for resumes. While these are valuable learning experiences, in order for me to achieve my goals, a formal education and a business degree are mandatory.

Objective #1

Complete B.S. in Business Administration. Anticipated date of completion is June 19XX, based on current level of credit points.

Objective #2

Employment by a major corporation with a progressive
management training program by January 19XX.

Educational Objectives

I will continue attending New City University, con-
centrating my course work on computer science, particularly systems analysis. Secondary emphasis will be
placed on personnel administration courses.

For the next several years I may seek other job changes to
gain more experience. However, the demands of my academic
schedule, as well as my lack of academic credentials, pre-
clude rapid career advancement for a few more years.

Sincerely,

Susan Brown

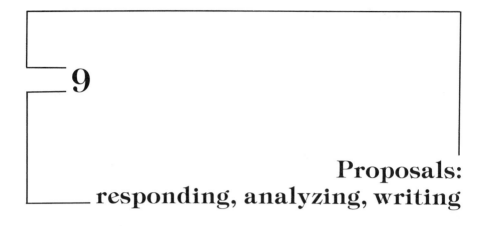

9

Proposals:
responding, analyzing, writing

The proposal may be the most unique and, at the same time, the most challenging form of professional and business writing. A proposal can be as short as a letter to confirm an agreement, or as long as a multivolume work on the design of a future spacecraft. No single pattern describes them adequately, but some general observations about them are helpful. An effective formal proposal stresses the organized presentation of technical solutions to specific problems, teamed up with a convincing management plan, and at a realistic price. These three goals account for the major divisions:

- *Technical*—the offer to solve the problem, including detailed description of the approach, possible plans and designs.
- *Management*—the explanation of how the project is run, who manages it, and when—according to a schedule—each task is accomplished.
- *Cost*—the detailed analysis of the amount of money each portion of the project requires.

The major difference between writing a proposal and other business papers is the way it is created. Most proposals require a group effort. If you are not accustomed to writing as part of a group, you must adjust your habits and attitudes to work effectively with others. Some large companies assemble experienced writers to produce proposals when a need arises. Others maintain a permanent group to manage proposals from a center expressly set up for that purpose. Such organizations recognize that the *ad hoc* approach can actually end up costing more time, money, and effort. Not only can preparation time and money be wasted, but the company's image can be severely tarnished by a poorly written proposal.

RESPONSIVENESS

Cost, contrary to popular belief, is not always the determining factor in the award of a contract. The proposal that offers a *responsive* solution to the problem often wins over a lower bid that is not exactly what the customer wanted. If responsiveness in a proposal leads to success, how can that be incorporated into the writing? First, you must know what you are responding to. In the jargon of proposals, you will soon meet an RFP—a Request for Proposal. Sometimes the RFP includes an SOW, or Statement of Work. The SOW usually lists specific technical requirements for the project. For example, if the city of Denver wants a new multi-purpose truck, it will detail the weight, color, pay-load, fuel efficiency, maintenance, and hundreds of other details in the RFP and the SOW.

Where can you find out about these projects that are open for bid? State and local governments advertise the project in local newspapers, usually on the business pages or in the public notice sections. In the ad is an address to write for the RFP. Also, the federal government publishes its needs in the *Commerce Business Daily*. These notices indicate that substantial effort and documentation must often go into a proposal. In other words, when a team gets an RFP, the work is only beginning. They must read the RFP thoroughly first, decide how best to solve the problem, select an appropriate presentation for the solution, write the proposal, and submit it before the deadline. The deadline is not easily extended since one exception allows all bidders more time. Of course, that delays a project. Though an extension can occur, count on submitting your proposal on time.

Read the Request for Proposal

Before writing the proposal, first read the RFP. This sounds so obvious that it is almost insulting. However, it is important advice when we think of an RFP as a set of instructions. Most of us have the habit of throwing instructions away or misinterpreting them, based on our prejudices. But ignoring the RFP, according to government evaluators, almost guarantees that your proposal will be rejected. The people selected to write the proposal are often tempted to ignore the directions because they feel they know the project and the needs of the agency that issued the request. In many cases they have worked closely with that agency before, and even helped them articulate the problem presented in the RFP.

Nevertheless, no matter how familiar you are with the project, you must read the Request for Proposal carefully and be prepared to give the customer what he asks for, not what you think he needs. Close reading of the RFP represents the first step in writing a proposal that responds effectively. A winning proposal is always responsive. If the specifications call for a lead pencil and your solution offers a word processor, you have not been responsive.

However, if you can clearly demonstrate that what is specified cannot solve the problem adequately, most government RFPs state that an alternate proposal can also be submitted. A detailed discussion of exceptions is responsive because it recognizes, rather than ignores, the problem.

The impulse to disregard directions may be a result of their chaotic presentation. Requests for Proposal often appear as if the agency collected forms and stapled them together with little thought and with even less attention to obvious contradictions and paradoxes. Whether the RFP is a jumble or not, sloppiness from the client never condones sloppiness and unresponsiveness in your proposal.

Outline the Response

After reading the Request for Proposal carefully, the proposal team leader prepares an outline that covers *all* the points mentioned there, even if they do not apply. You must acknowledge each item since your proposal will be evaluated against the RFP. If you leave something out, the evaluator assumes you forgot to include it, not that you felt the item did not apply. This gets you a zero for that item. Since covering each point of the RFP is fundamental to a responsive proposal, an outline acts as a basic tool and helps you touch all the bases. You can outline a proposal in at least two ways—the RFP Outline and the Topic Outline.

The RFP outline.   As its name suggests, The RFP Outline follows the order and numbering system found in the Request for Proposal. If, for example, the RFP calls Section 3.4.1 Deliverables, and includes that as part of the management portion, the RFP Outline follows that arrangement even though the item might make more sense at another place.

Why, you might ask, would any thinking person want to compound the poor organization of a badly written RFP by mirroring its presentation in the proposal? The answer is simple. Evaluators use the RFP as a guide in going through the material in each submission. The RFP Outline organizes information so the evaluator can find it easily, check it off, and award points. Large projects designate some evaluators to check only for parts covered and not covered. Others judge the quality of each part.

For any project, an RFP Outline makes the presentation of information easy for the evaluator to follow. The more organized the RFP, the easier it is to arrange the proposal logically.

The topic outline.   The second possible approach is the Topic Outline. As the better alternative, the Topic Outline resembles the way we organize and present a report, arranging information in one or a combination of ways that are detailed in Chapters 7 and 10:

- narrative
- chronological
- sequential
- general to specific (deductive)
- specific to general (inductive)
- step-by-step
- analytical
- spatial

The Topic Outline provides a more coherent structure than the RFP Outline.

The drawback of this approach is not so obvious. Time is its major disadvantage. Writers require more time to develop a response when using the Topic Outline than with the RFP method. Proposal leaders must decide whether to trade the time it takes to develop the topic approach for a potential sacrifice in the quality of the end product. Often the period from receipt of the RFP to its deadline is less than a month. Considering typing, editing, printing, and delivery, a week or two of concentrated writing is common. These almost ridiculous deadlines frequently force trade-off decisions that favor speed over quality.

Because so much must be written, the task may demand several or many writers working as a team. Until the day they sit in the same room with a copy of the RFP in front of them, they may never have met one another. Remembering that the RFP statement of requirements may be disorganized and that time is extremely limited, are we talking about a piece of writing that can be accomplished at all, even without considering the quality of the job?

Though it seems impossible to anyone who has never been involved with a proposal, many groups perform the impossible daily. In part, their attitude accounts for their success. They approach proposal writing the way they would the manufacture of a special product. In other words, they transfer management skills to the act of writing. Without such coordinated intellectual effort, the production of a five- or six-hundred page technical proposal with detailed drawings, graphs, charts, and illustrations could not be accomplished before the deadline.

The outline, in effect, provides a management plan for the proposal project. A carefully organized outline also furnishes the superstructure on which the professionals, artists, and craftsmen hang the fruits of their specialties. Coordinated and organized proposal efforts:

- cover all the points in the RFP
- support the main theme of the proposal with appropriate detail
- meet the objective of the RFP
- consider the specifications of the SOW
- provide graphic and artistic material that allows the evaluator to picture the concepts, data, and plans of the proposal clearly and effectively
- tie the company experience and expertise to the solution
- convince the evaluator that you know what you are doing

According to Herman Holtz's *Government Contracts,*[1] a proposal is "a sales tool." It certainly is, but that is not all it is. In fact, many professional proposal makers warn that over-attention to packaging loses points for you in the evaluation. The proposal that looks more like a brochure, with little factual content and lots of glossy pictures, sends the evaluator the wrong message. Most of them are down-to-earth people who can see through the glossy package. They look for a no-nonsense solution

---

[1]New York: Plenum Press 1979.

to the problem. Most of them are familiar with the way it could be solved, and they will not easily fall for the hype that we associate with a Madison Avenue ad campaign.

A proposal must sell. But, as Proposal Operations Director at Grumman Aerospace Corporation Bill Tebo observes in briefing proposal writers: to sell, it must also have the "ring of truth." That quality convinces the reader that you know what you are doing. Clear, organized writing demonstrates your professionalism as well. A reader convinced of your expertise and your ability to get the job done is sold if you have proposed a workable scheme.

Parts of a Proposal

Each proposal presents a unique design, service, or object. Because of this individuality, the organization of a particular proposal is also unique. However, some general points common to proposals help make the information easier for the evaluator to understand. Typical proposals have Technical, Management, and Cost Volumes. The list of parts which follows represents possible categories arranged in an understandable order:

- Letter of Transmittal
- Executive Summary
- Inside cover: "Response Index"—a cross-reference guide that locates the proposal responses to the Request for Proposal and the Statement of Work (RFP/SOW).
- Title Page
- Table of Contents
- List of Illustrations
- Introduction. Since this is the section of the proposal that everyone reads, the introduction should be written most carefully.

A good proposal Introduction should touch on these items:

1. demonstrate your understanding of the RFP by stating generally the customer's problem
2. explain why you are best qualified to solve it by mentioning related experience
3. briefly present the major points to be explained
4. indicate the method or plan used in the project
5. mention the limits placed on the discussion or project
6. indicate the essential points
7. say something about the organization of the proposal itself.

Write the last two to stimulate your reader's interest by including the most important item in the proposal. In that way you launch the reader into the technical discussion that follows and achieve the primary goal of the introduction, to "hook" the reader. Then continue with:

8. Summary (often included with the introduction)
9. Specific understanding of the requirement or problem

- Technical approach
- Discussion Sections should cover these items:

1. Statement of alternatives, comparison, and selection of the best solution to the problem
2. Presentation of facts and information in response to the RFP and the proposed solution
3. Indication, often by context, of the significant facts
4. Demonstration of the important relationships between facts and ideas
5. Analysis and interpretation of facts
6. Use of properly placed, relevant diagrams, pictures, drawings, schedules, and other illustrations

Some typical subheadings in the discussion sections:

Program Approach	Test Plan and Procedures
Design Description	Test Objectives
Design Characteristics	Test Set-up
Materials and Processes	Test Sample Description
Test and Evaluation	Risk and Tradeoff Analyses

The discussion sections can also include the following areas that are indicated in the RFP/SOW:

Project Organization and Staff (résumés of key people)	Schedules
Tasks, Subtasks	Organization
Management	Management Controls
Quality Control	Cost and Schedule Controls
Deliverables	Subcontracts and Purchasing Procedures

- Experience and Qualifications Sections include:

1. The "Company Résumé"
2. Corporate Structure and Management
3. Related Corporate Experience (past or current)
4. Facilities
5. Résumés (including contractors and consultants)

- References
- Appendices
- Inside Back Cover: Glossary of Key Terms

SOLICITED AND
UNSOLICITED PROPOSALS

So far we have discussed solicited proposals. These, of course, follow the solicitation or RFP. The public request suggests that the agency has a problem that their permanent staff cannot solve. The request invites companies to submit a written proposal in which they discuss the specifications and offer a solution to the problem.

An *unsolicited proposal,* on the other hand, is not the result of a public call for bids. Instead, it must first convince an agency or a client of the need for the product or service detailed in the offerings, and secondly that what is proposed is valid and worth the money. Usually a proposer knows the client is mildly receptive to an idea after informally suggesting it, even before launching the project. Successful unsolicited proposals often lead to a sole-source procurement. That means that the client awards a contract on the basis of the proposal, rather than competitive bidding.

THE STORYBOARD—A
DYNAMIC MANAGEMENT
AND BRAINSTORMING TOOL

Whether solicited or not, a proposal will ultimately emerge from the work of more than one writer, in some cases several hundred. The leader or manager of the effort needs an efficient and effective way to handle the paper and control the work. To do this the aerospace industry modified storyboards from their use in the creation of movies and television programs, and applied them to the management of complicated technical proposals.

Storyboards provide a sensible control mechanism for writing and organizing a coherent response to complicated technical projects that have come to mark the world of business, industry, and government. Storyboards offer a technique for the group production of detailed technical documents within a limited time.

Storyboards also represent a variation of an organizational method that you are familiar with—the use of index cards in writing a research paper. By contrast, a storyboard is larger and accommodates more detailed information concerning the origin of the ideas and their placement in the proposal itself. Figure 9–1 is an example of a blank storyboard planning sheet developed and used by Grumman Aerospace Corporation.

The completed storyboard effectively helps to coordinate a project involving many professional people. The blanks at the top call for information that allows other members of the team to identify the part to be analyzed on that sheet, where the part fits into the whole, and how long that part should be. When complete, the blanks also indicate when the part was written, who wrote it, and how to contact the writer if someone has a question about it.

The graph that covers almost the entire right-half of the planning sheet encourages the writer to sketch illustrations that might accompany the written statement.

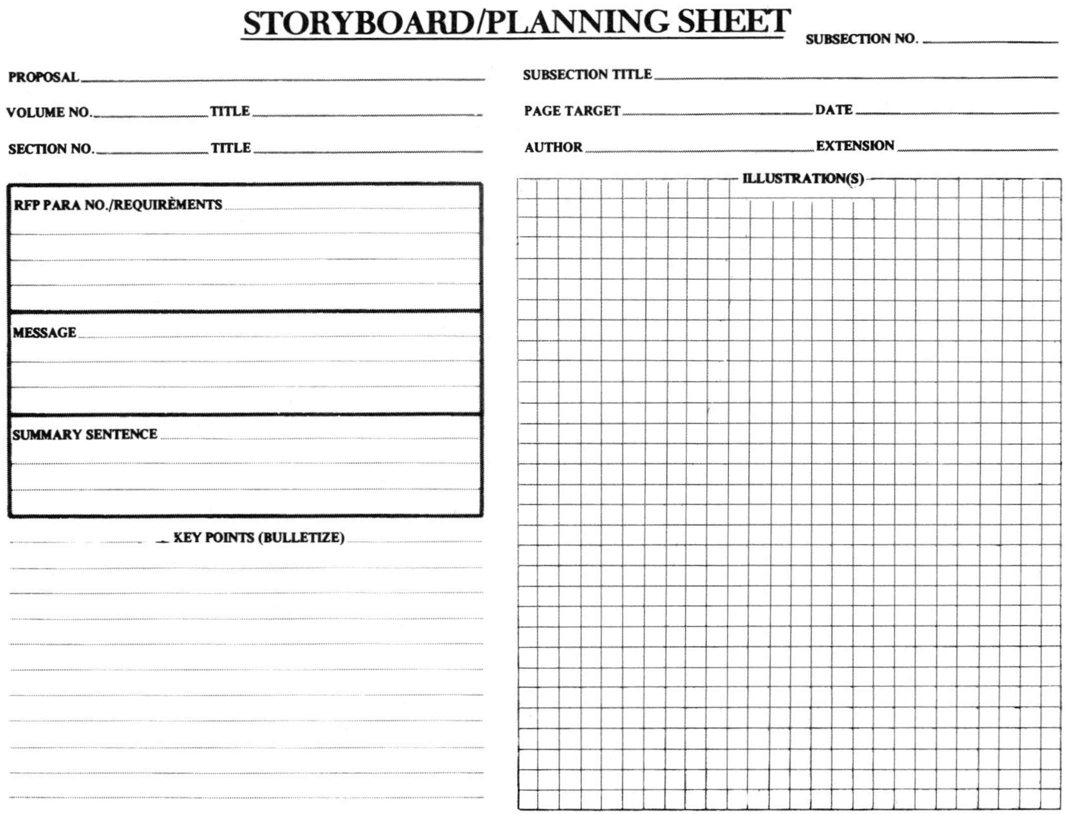

STORYBOARD/PLANNING SHEET

SUBSECTION NO. _____

PROPOSAL _____

VOLUME NO. _____ TITLE _____

SECTION NO. _____ TITLE _____

SUBSECTION TITLE _____

PAGE TARGET _____ DATE _____

AUTHOR _____ EXTENSION _____

ILLUSTRATION(S)

RFP PARA NO./REQUIREMENTS

MESSAGE

SUMMARY SENTENCE

KEY POINTS (BULLETIZE)

FIGURE 9–1 Storyboard Planning Sheet (Used with permission from Grumman Aerospace Corporation)

Since a proposal must sell the idea, and because it also deals with a concept for a device or service that exists in the minds of the developers at the organization, illustrations play an essential role in telling the story to the person who may buy the idea. The illustrations may be drawings, graphs, maps, and photographs. (See the discussion of graphic aids in Chapter 11.)

In a proposal these illustrations help to clarify information. They provide a visual oasis for the reader, communicating information without pages of textual explanation. A good graphic displaces at least its own weight in text, if not more. A poor illustration can confuse a reader more than none at all.

The text is also important. The highlighted square beside the graph provides for three important considerations:

The part of the RFP covered.   This information helps leaders and managers to determine whether the proposal *responds* to every item in the RFP. It also provides other project members with a written understanding of the requirements stipulated in the RFP.

The message.   The limited space for this item suggests that a clear, direct statement is required. Previously we discussed audience analysis and ways to determine the purpose of a message. Think of this message as the essential expression of purpose of that section of the proposal. It must clearly inform everyone what that section of the proposal should do for the whole document. Determining this message helps writers to focus not only the concerns of that part, but of the whole as well.

The summary sentence.   This statement articulates the purpose, but for use inside the proposal itself. The wording of this sentence may be quite different from the message because it is a statement constructed to present the idea coherently and to meet the needs and expectations of the audience.

The completed square with these three pieces of information alerts others working on the project to the content of that part. In this way others can see, in writing, what someone intended to communicate, and then compare that with what actually was written. Doing this aids the coordination of separate sections and that promotes continuity. As we have emphasized, most proposals are group efforts, and that makes their production as much an exercise in management as one in writing.

Proposal leaders decide which illustrations to include, which to leave out, and what information belongs in each section of the proposal. Storyboards offer an efficient method for this critical review. An individual or a group can evaluate the entire proposal early in its development by spreading out completed sheets on walls or on display boards. Sheets arranged this way furnish an overview of the whole project. Group members then comment on essential messages, key points, focus, organization, depth of detail, and approach in an effort to achieve a coherent and unified presentation. During such a session ideas surface for meeting the requirements and for better identification of the strengths of the solution.

In this sense, storyboards act as yet another brainstorming technique that a group of writers and managers can use. In addition to the people actually involved with the production of the proposal, storyboards inform other members of the company who are interested or responsible for parts of the project. During a conference, with individual storyboard sheets pinned to the walls around the room, even people not acquainted with the details and the progress of the project can quickly get an idea of its scope. In such a session, participants comment on the project. A secretary or stenographer might take down these statements. Later the proposal team can review them, incorporating useful ideas, and discarding others.

Reviewing a large proposal this way can uncover wrong approaches, fruitless efforts, and blind alleys. The storyboard also helps reveal the true messages of the project, and leads to valid approaches to the solution of the problem.

RESPONDING TO AN RFP

Now that we have considered the reading of an RFP, outlining, and storyboarding, we should look at an example of one that you might receive from the government. The solicitations have many parts in common, so agencies try to standardize their requests as much as possible and seek better ways to present their needs to the companies who write the proposals.

However, some RFPs are a collection of forms and boilerplate paragraphs and the RFP extract shown here is an example. Items relevant to the project are indicated by an "X." Following the order and organization of the RFP would result in a proposal that had few qualities of good writing. It would have little coherence, combining personnel and facilities with detailed analysis of the problem.

Given that assessment, the proposal should be approached with a Topic Outline. Doing this requires the evaluator to search the proposal for sections that correspond to the RFP. To avoid possible misunderstanding of the response, provide a Response Index to locate each specific item easily from the RFP in the body of the proposal. Usually the Response Index appears on the inside front cover so the evaluator can refer to it while reading and scoring the proposal.

A Response Index is included here which coincides with the RFP extract on the rating system for PCB or Printed Circuit Board testing. This index, of course, is a clearer presentation of items as they appear in the RFP. In addition, it shows that the writers chose to rearrange the information and present what they considered to be the important technical information first. Past technical experience forms a single volume consisting of the completed synopses of relevant contracts.

The first sections of the next volume consolidate the true purpose of the proposal—the technical approach to a test and rating system for printed circuit board testing—from five sections scattered in the RFP. The writers consolidated the arrangement of requirements in the RFP and presented two clear topical sections.

Instead of the 12 or more separate sections that would emerge by following the RFP, a Topic Outline combines related items, and presents them in seven sections.

ADVANCED APPLICATION OF THE PCB TESTABILITY DESIGN AND RATING SYSTEM

38. TECHNICAL PROPOSAL/GENERAL. A technical proposal is the most important consideration in the award of a contract; it should be specific and complete. Dollar values should be removed so that preliminary evaluation can be made without regard to cost. The proposal should contain an outline of the proposed lines of investigation, method of approach to problem, any recommended changes to the statement of work, the phases or steps into which this project might logically be divided, estimated time required to complete each phase or step, and any other information considered pertinent to the problem. The proposal should not merely offer to conduct an investigation in accordance with the statement of work but should outline the actual investigation proposed as specifically as possible. The proposal shall also identify any areas of uncertainty together with specific proposals for their resolution.

39. TECHNICAL PROPOSAL/SPECIFIC CONTENT. Technical proposals shall at a minimum contain information in paragraphs marked "X."

X   a. General experience and background of offeror on similar projects. Available specifications, photographs, technical descriptions may be submitted to support the proposal.

X   b. Statement of available plant, equipment, and test facilities proposed for use on this project, including any Government-owned facilities. industrial equipment or special tooling intended to be used in the performance of the proposed contract; the value thereof, identification of the Government contract under which acquired, rental provisions and other relevant information.

X   c. Specific statement of additional plant, equipment. and test facilities required for this project.

X   d. Names of persons to be assigned for direct work on the project and as direct technical supervisors, plus: (1) Their experience on similar projects or equipment. (2) Their general qualifications, including education and specific accomplishments. (3) Percent of total time each will be available for this effort for the duration estimated by the offeror.

X   e. Statement of additional engineering personnel required for full employment, subcontract or consultation, and source from which they will be obtained. Statement of assurance that proposed additional personnel will be available for work on this contract as specified in the proposal. Alternate personnel sources should be listed if assurance of availability cannot be stated.

X   f. Outline of basic difficulties of the problem and general approach toward solving it.

X   g. Principles which may be applied in the solution of the problem and an evaluation of the various methods considered, with justification for that selected.

SECTION L

39. TECHNICAL PROPOSAL/SPECIFIC CONTENT (continued)

 h. Complete detailed statement of solution, including preliminary design layout, sketches, and other information indicating configuration; also functions of components.

X i. Specific statement of interpretation, deviations, and exceptions to the solicitation.

X j. Period of performance proposed in months by phases or steps, if applicable, and delivery schedule of all items.

X k. Hourly time estimates by labor classes.

X l. Degree of success expected and major difficulties anticipated.

X m. Specific concurrence in requirements for reports.

X n. Extent of subcontracting anticipated, and preliminary information on the availability of subcontract sources.

X ð. PREDETERMINATION OF RIGHTS IN TECHNICAL DATA (1976 JUL)

 (a) The offeror is requested to identify in his proposal which of the below listed data (including data to be furnished in whole or in part by a subcontractor) when delivered, he intends to identify as limited rights data in accordance with paragraph (b) of the "Rights in Technical Data and Computer Software" clause of this Solicitation. This identification need not be made as to data which relate to standard commercial items which are manufactured by more than one source of supply.

 (b) Limited rights data may be identified as such, pursuant to (a) above only if it pertains to items, components or processes developed at private expense. Nevertheless, it cannot be so identified if it comes within paragraph (b)(1) of the "Rights in Technical Data and Computer Software" clause. At the request of the Contracting Officer or his representative, the offeror agrees to furnish clear and convincing evidence that the data which will be so identified comes within the definition of limited rights data.

 (c) The listing of a data item in paragraph (a) above does not mean that the Government considers such item to come within the definition of limited rights data.

X p. Synopses of relevant past performance on comparable Government contracts shall be provided utilizing the format set forth below. Limit submissions to not more than five (5) relevant contracts. Air Development Center contracts are preferred. Submit only unclassified data. Consolidate all synopses into a separate volume of the technical proposal with each synopsis a separate entity itself. Relevant past performance means quality of work, essentially comparable to the instant acquisition, completed under and in accordance with a contract. It includes but is not necessarily limited to work in the same or similar acquisition phase or

SECTION L

39. TECHNICAL PROPOSAL/SPECIFIC CONTENT (continued)

category, for the same or similar item, for the same or similar scope,
performed by the same company/division profit center, and in a time period
reasonably recent to the instant acquisition. Past performance shall not
be a prerequisite to submitting a proposal or receiving a contract; however,
see paragrph one of Section M. Offerors with no relevant past performance
must so indicate in their proposal. Each synopsis of performance on a past
contract shall address the contractor's technical performance as it relates
to each ranked evaluation factor in Section M, paragraph one. For example,
if Soundness of Approach is a ranked factor for this acquisition, discuss
past performance as it relates to Soundness of Approach on each of the cited
past contracts. The synopsis should also outline the cost and schedule
performance for each identified contract. Note that the contracting office
reserves the right to review contracts they consider representative of
relevant and recent past performance, although not volunteered by the
offeror. Such contracts will be made known to and discussed with the offeror.

SYNOPSIS

Contract Nr_____

Title_____

Date of Award_____

Contract Type_____

Contract Price_____ (Negotiated)

Contract Price_____ (Actual)

Reason for Price Increase (if applicable)_____

_____

Acquisition Activity (Complete Address, Symbol & Phone Number)_____

_____

_____

ACO (Complete Address, Symbol & Phone Number)_____

_____

_____

PCO (Complete Address, Symbol & Phone Number)_____

_____

_____

RESPONSE INDEX CROSS-REFFERENCES PROPOSAL AND RFP SECTIONS

This proposal for the study of Advanced Application of the PCB Testability Design & Rating System consists of three volumes:

- Volume 1a - Technical Proposal
- Volume 1b - Technical Proposal (Past Performance)
- Volume 2 - Cost Proposal

The following cross-reference index facilitates evaluation of this proposal by identifying where technical responses to the RFP and SOW requirements may be found:

RFP SECTION L 39	Subject	Response Section/Subsection
a	Experience and Background	5
b	Available Facilities	6
c	Additional Facilities	6
d	Resumes of Key Personnel	7
e	Additional Personnel	7
f	Problem Difficulties and General Approach	2
g	Principles Regarding Problem Solution	2
i	Interpretation, Deviations and Exceptions	2.5
j	Period of Performance and Delivery Schedule	3.2
k	Hourly Time Estimates by Labor Classes	3.2
l	Success/Difficulties Anticipated	2.2 thru 2.4
m	Concurrence in Requirements	2.5
n	Subcontracting Extent and Availability	2.5
o	Rights in Technical Data	2.5
p	Synopsis of Past Performance	Tech Vol 1b
SOW		
4.1.1	Survey of Recent Systems	4.1.1
4.1.1.1	Intermediate Level Maintenance Assessment	4.1.1.1
4.1.1.2	Organizational Level Maintenance Assessment	4.1.1.2
4.1.1.3	Assessment of Overall Testability	4.1.1.3
4.1.2	Preparation of Backup Reference Materials	4.1.2
4.1.3	Verification/Validation	4.1.3

The Table of Contents shows how a Topic Outline can achieve a more orderly presentation for the technical proposal volume. In conjunction with the Response Index on the inside cover, an evaluator can clearly determine what points are emphasized and which element of the RFP is involved, while reading and scoring the proposal.

The sample outline presents an understandable organization of the response to the problem, and it follows the suggestions for preparation most often given in RFPs.

CONTENT OF PROPOSAL SECTIONS

Though a typical proposal does not exist, some general statements can illuminate effective ways to approach the writing of the technical, management, and cost sections. A closer look at the elements of each should clarify your perception of the whole proposal. And an analysis of these sections should enhance your understanding of how to produce a clear, organized document that sells.

Technical

When a customer provides them, guidelines for proposal preparation strongly suggest that the technical part should be detailed, specific, and complete because it is a description and justification of the solution to a problem. This part answers *what*. The discussion of the solution should:

- present ideas simply and economically
- state expertise and past performance concisely
- relate practically to the solution
- organize information coherently and clearly

Customer's directions further suggest that evaluators consider mere repetition of specifications, or of the details that appear in the request, unresponsive.

The proposal should:

- specify the chosen approach to the problem
- outline the method of investigation
- recommend and justify any and all requested changes in the specifications
- divide the project into logical steps or phases
- estimate the time required for each step
- furnish other information important to the solution

Such guides emphasize that the technical section must not merely parrot the information that has been provided, but outline and discuss the actual project.

To be specific and factual about something that does not yet exist might be the most difficult part of writing a proposal. To do that, you must think as if the thing

TABLE OF CONTENTS DEVELOPED FROM TOPIC OUTLINE

Section		*Page*
INTRODUCTION		1
1	CONTRACTOR'S UNDERSTANDING OF THE PROBLEM	1–1
	1.1 The Testability Problem	1–1
	1.2 Importance of the Problem	1–1
	1.3 Developing Successful Testability Methods	1–2
2	TECHNICAL APPROACH	2–1
	2.1 Selection of Suitable Subsystems	2–1
	2.2 Basic Testability Rating System	2–2
	2.3 Analysis for Built-in-Test (BIT) Effectiveness	2–2
	2.4 Areas of Technical Risk	2–4
	2.5 Other Considerations	2–4
3	ORGANIZATION	3–1
	3.1 Project Organization	3–1
	3.2 Schedule	3–2
4	PROPOSED STATEMENT OF WORK	4–1
	4.1 Tasks	4–1
	4.1.1 Survey of Recent Systems	4–2
	4.1.2 Preparation of Backup Reference Material	4–4
	4.1.3 Verification/Validation	4–4
5	RELATED CORPORATE EXPERIENCE	5.1
	5.1 Automatic Test Equipment	5–1
	5.1.1 AN/USM-429 CAT-111D	5–2
	5.1.2 Other CAT Family Products	5–3
	5.2 ATE Software	5–4
	5.2.1 Atlas	5–5
	5.3 Computer Test Program Generation	5–5
	5.4 Electronics/Avionics Circuit Design & Analysis	5–6
	5.4.1 Digital Word Generator	5–6
	5.4.2 Multiple Matrix Switch	5–7
	5.4.3 Colorgraphics Terminal	5–7
	5.4.4 Bus Monitor Interface Unit	5–7
	5.4.5 Weapons Release System Test Set AN/AWM-67(V)	5–8
	5.5 Electronic Circuit/System Modeling/Simulation	5–8
	5.6 Related Studies/Programs	5–9
	5.6.1 Optimum System Testability Using (BIT) Built-In-Test	5–9
	5.6.2 Objective (PCB) Circuit Testability Evaluation System	5–9
	5.7 Contract Performance	5–10
6	FACILTIES: COMPUTER RESOURCES & SOFTWARE RESOURCES	6–1
	6.1 Management & Engineering Facilties	6–1
	6.2 Computer Facilities and Equipment	6–1
	6.3 Technical Information Center	6–2
7	RÉSUMÉS OF KEY PERSONNEL	7–1
Appendices		
A	PCB Testability-Evaluation & Scoring System	A–1
B	Subsystem Testability Assessment Procedure	B–1
C	Directives, Standards Instructions & Handbooks	C–1

Used with permission from Grumman Aerospace Corporation

were in front of you as you write. This not only helps you see the problem and your solution more clearly, but you can also convince the evaluator that you have really thought through the solution to the last detail. A detailed discussion inspires the reader's confidence in your ability to get the job done right, on time, and within reasonable cost.

Additional suggestions for the general content of a one-volume proposal often request a specific format. Here is a brief sample outline:

1. Table of Contents
2. List of Tables and Drawings
3. Short Introduction and Summary
4. Technical Discussion of Approaches. This major section of the technical proposal provides as much detail as possible and contains as a minimum:
   (a) Specific Statement of the Problem
      (1) describe the elements of the problem
      (2) point out critical elements
      (3) identify solutions never before attempted
   (b) Problem attack
      (1) complete and detailed approach to the solution of the problem—analysis, experiments, tests.
      (2) describe or discuss the principles and techniques used as the basis for the solution
   (c) Specific statement of any interpretations, deviations, and exceptions to the RFP.
   (d) Alternate technical approaches presented
5. Program Organization (A Management Section)
   (a) Relationship of the program to the overall company structure
      (1) General experience and background on similar projects
      (2) Estimate of subcontracting
      (3) Itemized list of equipment and materials purchased for the project
      (4) Substantiated purpose and need for travel
6. Personnel Qualifications (A Management Section)
   (a) Identify specific people assigned to the project
      (1) List the key personnel and related experience on similar projects
      (2) Estimate of the hours each person is to work on the project
   (b) Identify specific personnel—subcontractors, consultants
   (c) Provide resumes for all people listed.
7. Facilities and Equipment (A Management Section)
   (a) List of the plant, lab, computer, equipment, and test facilities available for the project
   (b) Additional facilities required for the project, with an indication of their importance to the project, and possible alternates or substitutions
8. Program Schedules (A Management Section)
   (a) The duration of the project by weeks or months,
   (b) The steps or phases, the time of delivery of items, and the times of submission of reports are included in this section of the proposal
9. Supporting Data and Other Information
   (a) Reference material used to prepare the proposal
   (b) List of contracts, past and present, that relate to the project or a field close to it.
   (c) Other pertinent information

This format offers a guideline for the coherent presentation of a solution to a problem. The organization might seem odd to you at first, but remember that proposals are not essays; neither are they reports. They are offerings to solve problems, which also act as sales tools.

Finally, many RFPs repeat a 1969 directive that proposals be prepared simply and clearly. The regulation forbids the "unnecessarily elaborate" proposal. In concrete terms, the proposal should not really look as if it were a magazine or an annual report, printed on glossy paper with numerous color photos. It should look professional, however. Its intended purpose is clearly to inform the customer of the solution to the problem articulated in the RFP.

Management

The management section, like the technical portion, must also eliminate any statement of cost. It presents *how* you will accomplish the solution that you presented in the technical section. With this in mind, the management section will follow the arrangement of the technical section. It may ease the evaluator's job considerably if you also use the same numbering system. That way you explain what you plan to do, and then how you plant to do it.

The main points of the management discussion appropriate to all proposals include:

- project organization
- cost control
- quality control
- flow of work
- staffing
- control of scheduling
- contingency planning

Often the management plan requires an estimate of labor and time, and an articulate expression of plans to react quickly to changes. If the project is large and requires additional personnel, provide a plan for recruiting, too.

If the project requires, you may need to separate technical management from the general management and administration of the project. Since this section describes a service, give it concrete expression and provide documentation on the plan. To do this, explain clearly company policies and practices that apply to the project. If necessary, provide copies of procedure manuals and guidelines. At least, mention them briefly, and in a few cogent phrases summarize portions that apply specifically to the project.

Provide forms or documents that further demonstrate the responsible technical and administrative approach to the solution of the problem, mentioning past performance on similar projects. Furnish other pertinent and relevant information, such as a subcontractor's financial statement.

Cost

For government contracts, submit cost proposals as a separate, detached document. That allows the technical proposal to be evaluated separately.

Like the management section, the cost section should follow the organizational system that appears in the other sections. This allows the whole proposal to exhibit continuity. Since the technical part provides the *what,* and the management section the *how,* the cost part tells the evaluator *how much.*

Escalating costs of goods and services focus more attention on this section of the proposal. It is too confusing to dive into the quagmire of cost-plus, fixed-fee, and other types of contracts. It is better to mention only the type of information to include in presenting cost.

The forms and documents that make up this section address three criteria for cost—reasonableness, realism, and completeness:

- Reasonableness results when sufficient information enables the evaluator to judge whether or not the method of estimating the cost is acceptable.
- Realism results when the cost and the scope of the estimate are compatible and a rationale is provided.
- Completeness in the cost proposal is the result of thorough consideration of every item and task.

In addition, recent Air Force efforts to standardize the presentation of estimates suggest that the cost proposal have four parts:

1. Introduction
   (a) Table of Contents
   (b) Overview—a general description of the scope of the program, limitations, and qualifications with mention of any deviations
   (c) Index—(1) a correlation matrix showing RFP paragraph and the response in the cost/price proposal
   (2) a listing identifying specific proposal paragraph references for supporting data not included in the cost volume, but in others
   (d) Summary—an explanation of how the cost was calculated so the evaluator can get an idea of how calculations were made
   (e) Changes in estimating or accounting practices
   (f) Significant risks in the project
2. Supporting data for appropriate forms, and data applicable to other standardized forms.
3. Work breakdown structure (WBS) and other forms
4. Other information, such as inflation rate summary, past cost performance, liability of the customer, cost related to the design.

To furnish information on the cost of material, labor, overhead, and other expenses, use forms and supporting documents. You also include the costs of such items as:

- special tests
- special equipment
- travel
- consultants' fees
- computer use
- the cost of money
- cost of facilities
- profit or fee
- G & A rate (rate of general and administrative costs).

Although the proposal does not usually call for such detail, it is good practice to be prepared to explain and justify the figures for each item.

Even though the cost proposal is the section that is closest to a contract, it too is a sales tool. Until the award of a contract, cost is still negotiable. The cost portion usually represents the work of accountants, lawyers, and administrators, some of whom will run the project if the proposal wins a contract. Understandably, they want to ensure that the company makes money on the project, more so than the technical professional who focuses primarily on the solution of the problem. Both groups of professionals must realize that a proposal is not a contract, but an effort to win a contract.

To win, the solution must be valid, the management competent, and the price reasonable and justifiable. But reasonable price is not always the lowest. The misconception that low bids win, masks the more frequent emphasis on superior quality in the technical solution. Fierce competition for contracts invites speculation on the way proposals are judged. Rather than review common beliefs, we will look at what evaluators tell us they expect to see in a proposal.

EVALUATION OF PROPOSALS

The proposal offers a unique opportunity to analyze audience needs and expectations. For instance, many calls for bids, like the Section M of a military RFP, detail the evaluation factors in writing. They explain what should be part of the document, and how the final proposal is judged. Readers rarely articulate their desires in print. Formal proposals allow us to gather a clearer perception of what should be accomplished in a piece of writing.

Even with written statements from the reader, the analysis of the audience is less than scientific. An experienced writer feels intuitively that the words, sentences, clauses, paragraphs, headings, and illustrations are just right for the topic and the reader even though some rules may be bent slightly, and some ideas may be expressed in an offbeat way.

Whether you have the touch or not, proposal writing requires a firm sense of audience. Even though you may know the persons you write to, and they have told you what they want, that sense always carries with it an element of fiction. In other words, no matter how much you know about the needs and expectations of the evaluators, and no matter how much they tell you, they cannot possibly tell the whole of their desires. You must, then, weigh each piece of information, select its place in the presentation, and judge its effect.

You can consider the evaluation criteria provided in an RFP, not as commandments in stone, but as clues to an evaluators' desires. For example, look at the sample of Proposal Evaluation Criteria that accompanies the RFP on printed circuit board testing.

Notice that here, too, the information is general and could be applied to numerous proposals. The relevant items are checked. By looking at the criteria we see that in this particular project the evaluators consider past performance on similar projects important. The directions in Section L, part 39p, reinforce this since they stipulate a separate volume containing the synopses of related contracts. Because the project is a study, the evaluators want to know if the bidder can do the job. The "track record" shows how well the bidder produced, and helps predict future performance.

Evaluators say they will give equal weight to understanding of the problem, soundness of approach, compliance with the requirements, and technical factors. And, according to the criteria, cost "will not be a controlling factor," but will be of major importance. The evaluators penalize an unrealistically low or high price, which should alert the writers of the cost section. Notice that the study allows for only a total of 1,500 engineering hours. As one professional observed, that is a small project for a formal bidding procedure, but quickly added that it appeared to be a preliminary study leading to others.

By looking closely at evaluation criteria in this example, the evaluator's strong interest in the cost emerged. It also showed, however, that the study required more time than specified in the request. Analysis of the reader's expectations suggested that the government wanted several companies to work on the initial stages of the project. After that they would issue another request and choose the most promising. Research projects often operate that way because it helps an agency avoid being locked-in to one contractor or one approach.

The proposal that was submitted in response to the RFP that we have used for our example was prepared in less than a week. To commit more time and manpower would have cost more than the amount of the contract. The decision to bid in the first place was based on the potential of a larger project to follow. Of course, management decides to bid, or to "no-bid," a contract.

But once money and people are committed to a project, the quality of the writing should be high, no matter how small the contract. Whether you win or lose, a well-written document reflects positively on your company and pays off in a reputation for quality. Howeve, with the pressures of immediate deadlines, the quality of the writing often suffers. In spite of limited time, produce the best content and employ the cleanest writing style you can.

SAMPLE OF PROPOSAL EVALUATION CRITERIA

EVALUATION FACTORS FOR AWARD   SECTION M

1. For the purpose of making an award under this solicitation, the technical
factors rated below will be considered. The relative importance of the
factors is indicated by the number in the space opposite each. Number one
indicates the most important; number two; the second most important; three -
third, etc. If two or more technical factors are rated of equal importance,
the same number will be assigned to each equal factor.

_____1_____   Relevant Past Performance (see definition in paragraph 39.p.
               of Section L). Each of the other applicable factors listed
               below will be measured against the offeror's relevant past
               performance. Offerors with no relevant past performance
               shall not be scored under this factor. This factor will
               include cost and schedule performance on relevant contracts.

1. Technical performance and the degree to which successful attainment of
study objectives were realized.

2. The extent to which the study task and overall program schedules were
met.

3. The adequacy of the management of financial and manpower resources.

_____2_____   Understanding of Problem

1. The offeror's apparent degree of understanding of the Statement of Work.

2. The offeror's apparent knowledge of fault diagnostics and testability
design techniques, electronic circuit design, microelectronic design,
partitioning and packaging, and advanced instrumentation techniques that
include Automatic Test Program Generation (ATPG).

3. The offeror's knowledge of electronic systems/equipments design in
order to determine the effects of partitioning and module design
configuration and their effects on testability analysis.

4. The offeror's familiarity of the Air Force maintenance concepts to
include awareness of the potential problems associated with defining
testability design requirements and constraints with respect to
measurement requirements, built-in-test (BIT) requirements, interface
requirements, and automatic test equipment (ATE) requirements at the
various levels of maintenance.

_____2_____   Soundness of Approach

1. The approach, its practicality and how well it will satisfy the basic
objectives of the Statement of Work.

Section M (continued)

2. The details included which describe how the requirements will be met;
the type and depth of the surveys to be performed; the configuration;
functions and complexities of the PCB/modules/subsystems to be investigated
and selected as candidates; and the sources and types of systems/equipments
to be surveyed.

3. The methods proposed to be used in developing and applying the
testability rating system; evaluating the results; and the types of analyses
to be performed in determining the effectiveness of the rating system at the
various indenture levels of systems/equipments.

_____2_____ Compliance with Requirements

1. The detail provided as to how the requirements will be met.

2. The compliance of the proposal with the requirements of the Statement
of Work.

3. Exceptions to the Statement of Work requirements, their merit and effect
on the overall study.

_____ Ease of Maintenance (Hardware Buys Only)

_____2_____ Special Technical Factors as listed below:

1. The offeror's experience in designing for Testability as it relates to
Air   Electronic systems/equipments.

2. Extent of background in the development and application of design and
cost trade-off techniques to optimize testability design and cost-effective
requirements.

3. The test modules chosen, their complexities, data availability.

2. The cost or price proposed may not be the controlling factor in contractor
selection, but will be considered to be of major importance in contractor
selection. The degree of cost realism will be evaluated, and a proposal
which is otherwise acceptable may be penalized to the degree that the
proposed cost/price is unrealistically low or high. However, the above
specified technical factors will, as a group be of greater importance than
cost or price.

3. Qualifications based on offeror's data will also be assessed. This will be
a judgement assessment, determined from technical data the offeror submits
in response to the solicitation. Such assessments are limited solely to
technical aspects confirming whether a source has the technical capability
to perform the required project successfully. These assessments consider
specific experience, technical organization, special technical equipment

and facilities, and any other factor mentioned below. These considerations are rated but never scored, and will be taken into consideration by the award determining official. In any case, these factors will be of a lesser importance than either the technical or cost factors in paragraph 1 or 2 above.

4. In addition to being evaluated specifically as set forth in Section M, paragraph one above, past performance will also be evaluated generally. This latter evaluation covers aspects of past performance considered by Air   which are not provided with an offeror's proposal. Past performance under this paragraph will be rated, but not scored. Further, past performance as a general consideration will be of lesser importance than the factors considered in paragraphs 1, 2, or 3 above.

5. Item(s) listed below is/are requested on a term basis at the level of effort indicated:

    Item 0001 only - 1,500 <u>Engineering</u> Hours

6. If this solicitation indicates the level of effort desired for specific items, proposals furnished in response to this Request shall be on the basis of that level of effort using only direct hours. <u>The Government reserves the right to reject proposals where the level of direct effort in any designated category, or in toto, is less than the specified level</u>. However, if the offeror or quoter believes some other level of effort is more appropriate to economical and successful completion of the work desired by the Government, he is invited to submit an alternate proposal <u>in addition to a basic proposal</u> for the requested level of effort <u>and adequately justify his conclusions</u>. If the Contracting Officer determines such other level of effort is in the interests of the Government or if a Contractor proposes to complete work requirements, he further reserves the right to enter into a contract on any basis which is the most advantageous to the Government.

7. Competitive advantage arising from the use of Government production and research property shall be eliminated by use of an evaluation factor established in accordance with DAR 13-502, except when the principal contracting officer determines that the use of an evaluation factor would not affect the choice of contractors.

8. The amount of any royalty payable under the royalty sharing provisions of a previously accepted value engineering change proposal incorporated in the solicitation will be considered in evaluating offers when the value engineering change is one of two or more acceptable alternatives under the solicitation.

9. Any alternate proposal (see para 19 of Section L) submitted in response to this solicitation will be evaluated in accordance with this Section M, using the criteria specified in paragraph one above.

AUDIENCE ANALYSIS

Close reading of the evaluation criteria found in the RFP provides a method to further acquaint you with your audience. You can also ask yourself questions to get a better feel for the people who will eventually read your proposal. You can modify the questions we furnished on audience analysis to accommodate the uniqueness of proposals. To present your ideas and information positively, answer the following:

- Who is the audience?
- What does the Request for Proposal (RFP) tell you about them?
- What unwritten message does the Request for Proposal present?
- What clues to the reader's needs and expectations exist "between the lines" of the RFP?
- What are the reader's needs and expectations concerning the project?
- What level of expertise does the reader have with this field, or with this topic?
- Is the reader business or technically oriented?
- What misconceptions does the reader have about the project?
- What type of information does the reader need?
- If the audience has several readers with different needs, what is the common ground?
- Would this proposal find its way to anyone else? Who?
- What is your attitude toward your audience?
- What do you want your writing to make your reader think, feel, do, or know?
- What does the reader think about you and your organization?
- What are your reader's hobby horses—pet projects, ideas, prejudices?
- How much convincing will it take for you to get your reader to think, feel, do, or know what you intend?
- What tone will work most effectively with this audience?
- What marketing intelligence on the project exists?
- What internal problems in the office soliciting proposals might influence the project?
- From the preproposal briefing, what information is available on the competition, the contracting officer, the technical people?

EXPECTATIONS AND CRITICISMS

Generally, evaluators expect a proposal to show:

- Clear understanding of the problem or need presented in the RFP
- Concrete technical solution of that problem
- Compliance with the specifications or requirements, or a valid explanation of any deviations; assurance of quality
- Acknowledgment of other possible solutions, and understanding of their strengths and weaknesses (what the neighbors are doing)
- Responsible and concrete management plan for the project, including:

> qualified staff
> adequate facilities
> believable work schedules
> responsive reporting schedules

- Realistic cost; financial credibility
- Clear presentation

The list of expectations implies the following evaluators' criticisms of proposals:

- Failure to understand the problem and requirements spelled out in the Request for Proposal or the Statement of Work
- Overall vagueness in response to the Request for Proposal
- Writing to impress the reader with obscure vocabulary and ornate language
- Oversimplified technical explanations; too little test information
- Misinterpreted specifications; deviations from the specifications explained inadequately

Careful writing will help you meet evaluators' expectations, and avoid their criticisms. The exercises that accompanied the earlier chapters should have strengthened your ability to write a proposal that is concrete, detailed, and clear.

DEBRIEFINGS

Even with several attempts to find out what a reader wants in a proposal, you can never know exactly why one proposal wins over others, but you can get close. Regulations require the government to meet with unsuccessful bidders after a contract is awarded to answer questons and provide explanations. This is called a *debriefing*. Characteristically, ground rules for these sessions sometimes forbid taping the discussion, and agency officials familiar with the project answer written questions orally during the meeting. After the debriefing, the company representative prepares a memo outlining the content for distribution inside the organization. Such memos can reveal strengths in the proposal to incorporate into the next one, and weaknesses to correct in following efforts.

During the debriefing, government officials seldom reveal details concerning their judgments, because regulations forbid direct discussion of competing bids, but also because of possible challenges to their decisions which could delay the project. On the other hand, they may be refreshingly candid in their responses in an effort to encourage future bids.

For example, a debriefing following the award of a scientific experiment for the space program yielded some useful general information directly applicable to writing. From the discussion at the debriefing, the company representative inferred that the contract went to the competitor for other than technical or cost factors.

Though the government assured the company that the proposal was technically sound and that the price quoted was well within reason, the comments actually given concerned the proposal directly, not the other factors we mentioned. Some of those comments may help the preparation of your proposal since they generally concern the presentation of ideas:

- the proposal did not use page allocations well; not concise and to the point
- though a full-page design drawing was included, it did not have the detail required for the evaluator's complete understanding
- supporting details—test data, simulations, and mock-ups—make a proposal more believable

From these and other statements and responses that surfaced at the debriefing, the company representative concluded in his memo that:

- the technical approach in some areas did not come across well enough
- details and supporting information were not deep enough
- format and page allocation were not as efficient as they could be
- the presentation did not provide enough back-up material—test information, mock-ups
- the company did not know the customer well enough concerning specific technical, management, and proposal aspects.

More careful writing could have proved useful for the identification and eventual repair of the problems of this proposal. An examination of an early draft could have revealed the need for more detail early enough to include it in the project. The irony in this example is that the information existed, but did not come out strongly in the writing. Among other things, this debriefing demonstrates that a reader may not know exactly what a piece of writing needs, but can easily determine that something has been left out.

Though no one can scientifically predict the impact of a proposal on an evaluator, you should nevertheless write as directly and clearly as possible. When an investigation of the reader leaves important questions unresolved, think of the reader as someone capable of understanding complicated, even difficult technical or philosophical concepts that are presented in an organized, uncluttered way. Writers bring news to other people, so assume that what you have to say is news. Even if the reader knows the general outline of your idea, assume that they do not know your specific approach.

Details, specifics, clarity.

It boils down to that.

THE LOSING PROPOSAL

Louis Juillard's article "Proposal Evaluation in the Aerospace Industry,"[2] concludes with negative instructions for writing a proposal. Stating that everyone has tips on

[2]Louis Juillard, "Proposal Evaluation in the Aerospace Industry," in *Proposals and Their Preparation* (Washington, D.C.: Society for Technical Communication, 1971), p. 26.

how to win a contract, he lists seven items that guarantee that a proposal will be a loser:

1. Forget about the Request for Proposal. Quote what you think the customer needs rather than what he asked for. Consider that all data which he requested and all those instructions are for other bidders—the ones he's not as well acquainted with as he is with you.
2. Don't be specific. Don't follow the crowd with a simple reply. Show him that you are industrious by submitting a few alternate approaches. Cover all the angles and let him make the choice from your offerings. Hope that he doesn't realize that you couldn't decide on the approach . . .
3. Camouflage your approach if you haven't done your homework or if you can't find a single message to drive home throughout your proposal.
4. Show him how learned you are with invisible, intangible, and symbolic language . . . Keep his attention for the longest possible time by making him go back and reread each paragraph to figure out the meaning. . . .
5. . . . Stuff (the proposal) with all kinds of preprinted documents and irrelevant data and hope the customer will be impressed with the weight of the package.
6. Promise that you can do anything . . . but don't bother to tell him how you intend to do it.
7. Offer again and again to negotiate a reduced price if he finds your original submittal too high.

THE WINNER

A proposal must:

- Tell a story
- Respond to the Request for Proposal
- Sell

Its presentation should be general, yet specific enough to satisfy the technical requirements. Use the outline and storyboards to:

- Cover *all* points in the Request for Proposal, without glossing over any point
- Support the message and meet the criteria and objectives
- Tie expertise and experience to the solution
- Relate to the topic
- Select appropriate photos, graphs, drawings, maps

In writing the proposal:

- respond to the Request for Proposal
- put yourself in the evaluator's place
- build on the outline and storyboards

- drop the seeds of the information to follow
- use active voice
- use personal pronouns
- be concise and economical with words and phrases
- provide enough detail, but not a flood
- supply a clear summary to make it easy for the evaluator
- Stop.

SUMMARY

- An effective formal proposal emphasizes the organized presentation of: a solution to a specific problem, a convincing management plan, a realistic price.
- Successful proposals respond to the request for proposal or RFP. First, read the RFP. Then prepare a response, using either a topic or an RFP outline.
- The storyboard provides an efficient brainstorming and control mechanism for writing and organizing a coherent response to a complicated problem.
- A Response Index on the inside cover of the proposal provides easy cross-reference between the RFP and the proposal.
- The content of the technical section should be detailed, specific, and complete because it describes and justifies the solution to a problem.
- The management section presents the approach to the solution. It discusses project and corporate organization, cost, and contingency planning.
- An appropriate cost section addresses three criteria: reasonableness, realism, completeness.
- Many RFPs include evaluation criteria. Close attention to them provides a further approach to understanding and analyzing the audience's needs. Evaluators generally expect a clear presentation of ideas and detailed information. With government contracts, unsuccessful bidders have the opportunity to ask questions at a debriefing.
- A losing proposal ignores the customer's needs.
- A winner covers all points in the RFP, builds on the outline and storyboard, uses strong verbs and economical language, and sells.

SUGGESTED PROJECTS

1. Contact an organization and ask what kind of proposals they write, if any. Find out as much as you can about how they approach the task. Write a short report concerning the information you uncovered.

2. Contact an agency of the government and ask if they have guidelines on the submission of proposals. Carefully examine what they send. Write an evaluation of their material, keeping in mind that this is really a first step in the audience analysis. Be sure to include any items that give you insight into the preparation of a proposal for that agency.

3. Prepare a proposal in which you respond to the following RFP:

For some time the paper clip has been the main method of putting papers together. Design a suitable and improved version of the paper clip that will be more effective, cheaper, and more durable than the metal ones now in use.

4. Submit a five-page technical proposal for a program designed to solve the following problem:

Unproductive time at the end of the working day—workers stand around, waiting to leave, for 10 to 15 minutes before time to go, especially on Fridays.

5. Three outlines for proposals appear in this chapter: (1) the section on proposal parts, (2) the table of contents from the sample proposal, and (3) the suggested format found in preparation guidelines. Compare the three approaches and discuss their shared structure. Draw conclusions about uses of the proposal from your comparison. Be sure to see the document from both the writer's and the reader's points of view.

10

The Long Report:
research and documentation

The long report presents facts, research findings and ideas. It depends more heavily on formal structure than does a short report or proposal, because its form represents a modification of scientific and scholarly papers. Although it is not so obvious a sales device as a proposal, every report "sells" ideas, concepts, conclusions, and recommendations. Since a long project demands more time and concentration, a disciplined approach helps you write a quality report and complete it within the deadline. The checklist on the following page outlines the writing strategy for reports discussed in Chapter 8. Instead of repeating the methods for putting ideas on paper, a brief review of brainstorming, audience analysis, report elements, short reports, and proposals should help you prepare for a long project.

PRELIMINARY
CONSIDERATIONS

Starting a long writing project is like swimming in the ocean. Some people run right in and get wet, ignoring the chilly water. Others stand for what seems hours, inching their way into the water. Decide which type you are. Sooner or later you have to make the plunge. You might as well jump right in. In writing a long report consider your own attitude toward a long project, the approach you take to beginning, how you budget time, the parts your report could have, and when to begin and end writing each draft.

The creation of a long report that achieves desired results takes several days. If you are a deadliner, in the habit of putting everything off to the last minute, you should reassess your approach because reports written hastily often fail to deliver the information that readers need. Once you convince yourself that the report will take some time, think of it as a continuing project that requires part of your daily concentration for several days, and almost all of your time on the last day.

```
┌─────────────────────────────────────────────────────────────────────┐
│                    GETTING IT WRITTEN:                                │
│                                                                       │
│          A Review of the Strategy for Long Projects                   │
│                          Checklist                                    │
│   ● Get ready to write                                                │
│                                                                       │
│      Clear time. Find out your best time for writing, and write       │
│      then every day.                                                  │
│      Clear space. Write at the same place every day.                  │
│      Avoid interruptions: no calls, no appointments, close the door   │
│      if you have one.                                                  │
│      Set a regular pattern or routine.                                │
│      Make sure you have everything you need: coffee, paper, cigarettes.│
│      Concentrate on your subject                                      │
│                                                                       │
│   ● Remember your purpose.                                            │
│   ● Remember your audience's needs and expectations.                  │
│   ● Think about what you really want to say.                          │
│   ● Prepare a map of your message as your own guide.                  │
│                                                                       │
│      Use BRAINSTORMING techniques; free writing, forced associations, │
│      questions to get ideas on paper, dialogues, deadlining, absurd   │
│      revision.                                                         │
│                                                                       │
│   ● GET IT WRITTEN.                                                   │
│   ● Then get it right. Revise, edit, add to, and delete from what you │
│      have. Keep your purpose and the audience in mind.                │
│   ● Always stop in the middle of a sentence, paragraph, or section    │
│      that is going well.                                              │
│   ● Write the introductory pages after you have completed the whole   │
│      document.                                                         │
└─────────────────────────────────────────────────────────────────────┘
```

Budgeting time for the report may prove difficult, since most people need little if any excuse to avoid writing. Think of the writing time needed for the report as you would a conference, meeting, or appointment with a client. You must attend the writing appointment you have with your report. If you have singled out a certain hour each day for writing, stick to it every day. Postpone other activities until that hour and that writing is over. Try to precede that hour with tasks such as reading mail or writing routine memos and letters. This should help you make the adjustment to writing.

A long project requires time and concentration. Often you will think about the project throughout the day. Put your thoughts on paper. Write all the time when working on a long project. It helps you to organize your thoughts and to further understand the problem you must discuss. Writing in this way also presents a form of discovery. Often you make connections you had not considered before, and you might see a pattern to your ideas and possible leads to further information that had not occurred to you previously. To take advantage of the thoughts you have, and also to anticipate the inevitable draft of the report, carry some blank 3 × 5 note cards with you and jot down the ideas you have as you think of them. Start now if you are not in this habit. What you thought was a blockbuster idea will most likely disappear in the shuffle of daily activity if you wait to write it down. You may not subsequently use the idea, but putting your thoughts on paper sharpens the focus and your understanding of the report. While writing notes and ideas, you can also consider where in the report they might belong by thinking of the parts a report can have. The Outline of Long Report Elements illustrates this method.

Some of the material that you write in the preliminary stages will most likely be thrown away. Nevertheless, it is an essential step in creating a quality report.

OUTLINE OF LONG REPORT ELEMENTS

   I.   Title
  II.   Table of Contents
 III.   Table of Illustrations
  IV.   Foreword or Preface
   V.   Summary (Conclusions and Recommendations can be placed here.)
  VI.   Introduction
     a. Mention of relevant history or background
     b. Indication of the limits placed on the subject
     c. Statement of the method of research
     d. Presentation of any special definitions
     e. Outline of what to expect
     f. Mention of the plan of the report
 VII.   Discussion
     a. Presentation of facts and information according to the appropriate form of discourse
     b. Indication of which facts are significant
     c. Demonstration of important relationships between facts and ideas
     d. Analysis and interpretation of facts
     e. Use of properly placed, relevant charts, tables, maps, graphs, and pictures
VIII.   Conclusions and Recommendations
     a. Placement indicates the order of importance
     b. Facts and ideas in the body of the report substantiate statements
  IX.   Documentation
     a. Footnotes
     b. Bibliography
   X.   Appendices
     a. Exhibits
     b. Attachments
     c. Photographs
     d. Oversize drawings
     e. Models

SELECTING A MANAGEABLE SUBJECT

In considering the subject of a report, you still have choices, even though someone else determined the general topic. The focus, purpose, and organization of the report require both your attention and manipulation.

Focus

Conventional methods of focusing the subject involve limiting the scope of the discussion. You can place boundaries on: (1) time—January–July, 19XX; (2) space—Mining Operations in Alaska; and (3) the depth of detail—Chemical Analysis of Core Samples. Or you can use all three: "Chemical Analysis of Core Samples in Alaska Mining Operations from January–July 19XX." Brainstorming techniques help you pare down the concerns of the report in the initial stages of writing.

Functional analysis. In addition to brainstorming, a functional analysis can also help focus your report. A product of engineering, functional analysis essentially demands a clear determination of how a mechanism or system functions. The expression of the function consists of a list of phrases beginning with verbs. If, for instance, we were to analyze a paper clip, we might generate a list like this:

- binds papers together
- forms a temporary bond
- costs little to manufacture
- is reusable

During the analysis you decide which functions are primary and which are secondary. Primary functions are basic to the operation; secondary are supplementary. Applied to writing, we have a representational way to determine if each topic in a report pulls its own weight.

Purpose

Value analysis provides a method of determining the purpose to add to those discussed earlier in this book. Simply, value analysis takes the functions revealed by a functional analysis and considers the primary and the secondary ones with the intention of eliminating unnecessary items to reduce the cost of the report. A very brief look at new automobiles reveals the results of value analysis applied to the interior of the car. Many metal parts are now made of plastics, replacing ashtrays, dashboards, and instrument panels with more serviceable and lighter materials, at lower cost.

When writing a report, list the primary and secondary functions of each goal that you have identified. Decide which items on the list can be eliminated or replaced by better, more efficient ones.

Organization

The organizational principles of time, space, and importance apply to long reports. Since the document is much longer than a letter or memo, a clear pattern of organization helps the reader understand the material. Several valid approaches make the task of organizing material in a report less time consuming. The Outline of Long Report Elements is one graphic representation or map for organizing. This and the following examples can provide a structure for building your report. No single example is ideal, but the guides presented here should accommodate a broad range of report material and situations.

The Inverted Pyramid (Figure 10–1), as discussed in Chapter 8 on short reports, represents a common and effective pattern for presenting information in business reports, since it loads important information up front. It also provides a useful structure for general audience discussions similar to those found in newspapers and magazines. Memos written with this pattern usually achieve results.

The Flying Pyramid (Fig. 10–2), a representation used in the Environmental Protection Agency's employee writing manual, illustrates a pattern for the formal presentation of research findings or investigations. It follows a structure that responds well to information gathered by experiment or research. Such an approach recognizes the need for any discussion of data to have a serviceable beginning and ending.

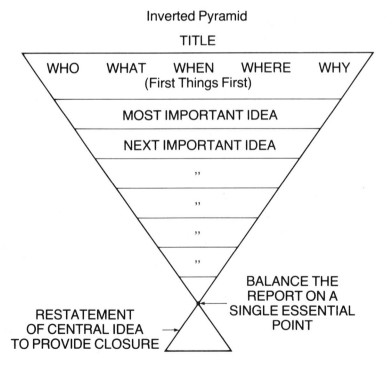

FIGURE 10–1   The Inverted Pyramid Structure for Memos, Short Reports, and News Stories

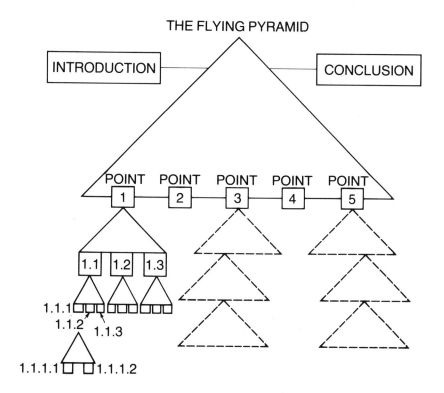

FIGURE 10–2   The Flying Pyramid: EPA Structure for Formal Presentation of
Research or Investigations

The Keyhole[1] (Fig. 10–3) uses an indirect approach you might be familiar with, building to the climax. Though, as we have pointed out, such dramatic organization of facts makes for entertaining and delightful reading, the professional person needs important information keyed to the beginning of the document. Compared with business reports that start fast and trail off, this pattern depicts a slow build up. Follow an *indirect* approach if you wish to:

- bring bad news
- present a new idea
- change an old attitude

The *Report Organizational Flowchart* (Fig. 10–4) recognizes the linear nature of writing. It also recognizes that reports of various lengths have parts that serve similar functions. It furnishes a gentle reminder that the sections of a report must be joined with transitional elements that provide an exit from one portion of the

[1]An adaptation of a figure from THE PRACTICAL STYLIST, Fifth Edition, by Sheridan Baker. Copyright © 1981 by Harper & Row Publishers, Inc. Reprinted by permission of the publisher.

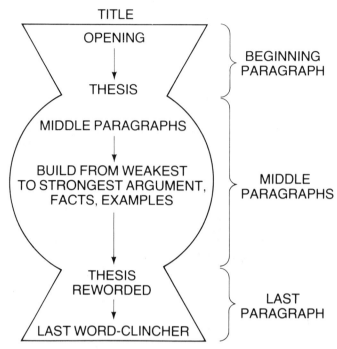

BAKER'S KEYHOLE

TITLE

OPENING

↓

THESIS

} BEGINNING PARAGRAPH

MIDDLE PARAGRAPHS

↓

BUILD FROM WEAKEST
TO STRONGEST ARGUMENT,
FACTS, EXAMPLES

↓

} MIDDLE PARAGRAPHS

THESIS REWORDED

↓

LAST WORD-CLINCHER

} LAST PARAGRAPH

FIGURE 10–3   The Keyhole Structure: An Indirect Approach that Builds to a
Strong Conclusion (An adaptation of a figure on page 32 from THE PRACTICAL
STYLIST, Fifth Edition, by Sheridan Baker. Copyright © 1981 by Harper & Row
Publishers, Inc. Reprinted by permission of the publisher.

discussion and an entrance into another. The discussion of paragraphs in Chapter 4
suggested that every paragraph have an entrance, a tour, and an exit. Sections of a
report need the same elements, providing the reader a way to think about the
information in a context of the whole report.

These four depictions of report organization are by no means the only ones that
exist or, for that matter, work. Nevertheless, they do provide useful maps for plotting
a course for the material you must present in a long report.

Once you have analyzed the information and determined your approach to a
manageable subject, you must develop the idea.

DEVELOPING IDEAS: LOGIC
AND ARGUMENT

Thinking about the longer project, focusing the subject, and selecting an organizational structure are not separate steps in the process of writing a long report. They
occur simultaneously as you develop an idea. Time, space, and analysis offer
methods of development. Analysis includes the use of cause and effect, comparison

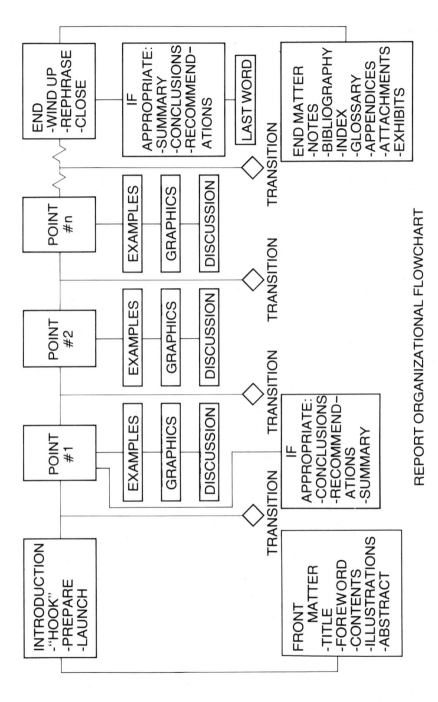

REPORT ORGANIZATIONAL FLOWCHART

FIGURE 10—4   An Example of Flowchart Structure for Organizing a Report

and contrast, flowcharting, and numerical and alphabetical ordering to develop ideas for coherent written presentation. In addition, the devices of inductive and deductive reasoning, analogy, and argument serve to develop ideas clearly.

Inductive Reasoning

Inductive reasoning enables us to make general statements based on particular examples and evidence. Because of this, it provides a method of explanation. It also furnishes a strategy for discovery, allowing the opportunity for an inductive leap to occur.

Since this type of reasoning draws conclusions from evidence, we can ask questions about the evidence and the conclusions to test the argument.

- Is the evidence reliable? In part at least, the reliability of evidence rests on the strength of the source. We base most of our general statements on observation, experiment, and reading. We then must determine if what we have seen is representative, what results of our experiments are valid, and whether what we have read comes from an informed and objective person.
- Is the evidence adequate? A reasonable number of samples that are representative provide adequate evidence.
- If more than one inference is justified, is the simplest selected? Always select the inference that you want to stand out for your audience. Usually, the simplest has the most beauty.
- Have the exceptions to the generalizations destroyed its worth? When the exceptions to the old rules grow too numerous, abandon the old notions and salvage what continues to maintain validity.
- Does the cause explain the effect convincingly? Often more than one cause exists, and an explanation must take the possibilities into account.

Inductive reasoning works well if your audience needs a great deal of explanation, especially of a new or unusual solution to a routine problem.

Deductive Reasoning

Deductive reasoning works in the opposite direction from induction, drawing inferences from general statements. It uses generalizations that are valid in one case and applies them to a related one. Deduction, then, has limited value for prediction.

Deduction usually moves from the general to the specific, but it can also move from one general statement to another. The general statement must come before the conclusion. In formal argumentation the information is reduced to its essential premises and conclusion—a syllogism. In writing, the *categorical* syllogism often compresses the major and minor premises into the conclusion. The "if . . . then" structure of the *conditional* syllogism resembles closely what occurs when a writer strives to establish a point. And the *alternative* syllogism, with its "either . . . or" structure, excludes one premise from another, though in some instances "either . . . or" implies "perhaps both."

Reasoning by Analogy

Because it establishes no logical conclusion, analogy presents the weakest form of argument. The writer states a conclusion, but the reader does not have to accept it simply because two contexts that may involve similar ideas are brought together. The two may have significant differences as well.

Nevertheless, analogy often makes your audience see your point more clearly by providing illustration. If the association between the things or ideas is close enough, your audience may accept your conclusion in spite of the absence of necessary logical connections. In using analogy, the lack of connections may mislead the audience, so be careful to examine the choice of words.

If it is not pushed to prove a point, the comparison and contrast of analogy can be effective and extremely illuminating for your audience.

Common Fallacies

Think of a logical fallacy as a phoney argument, valuable only because some people take it for the genuine article. Fallacies can be formal or informal.

Formal

• Begging the question—when the conclusion essentially presents nothing more than a rephrasing of the initial assumption, we "beg the question." The statements are circular, getting nowhere in proving the point, because they assume to be true what should be proved.

• Confusion of *all* and *some*—You may know this as the fallacy of the undistributed middle term.

• *Post hoc, ergo propter hoc*—This means literally, "After this, therefore, because of this." In its erroneous cause-and-effect reasoning, one thing happens and another follows. The first is inferred as the cause of the second. As the story of Chicken Little illustrates, however, one thing that happens before another need not necessarily be its cause. You can easily reduce a *post hoc* fallacy to absurdity. Reducing an argument to absurdity can also be a fallacy if it only ridicules without clarifying the position it mocks.

Informal

• Confusing the argument—Using ambiguous language or distorting evidence are the main methods for confusing the argument.

• Evading the issue—Ignoring the question by introducing irrelevant issues represents a common fallacy in every day reasoning. Anyone who ignores the question usually attempts to avoid controversial or unpleasant issues.

• Substituting emotion for evidence—A substitution of emotion for evidence appeals to feelings. It pushes reason aside, and often clouds the issue. The most common emotional argument attacks the individual instead of the issues. Another appeals to the emotions and prejudices of people.

Patterns of Argument

Argument implies an opposite position. A writer can argue in three ways:

- Take an affirmative position and present arguments.
- Give additional evidence that counters the arguments of the opponent. In the absence of the opponent's position, he can anticipate the arguments.
- Discredit the opponent's position by pointing out fallacies.

While preparing for argument, find a common ground. Try next to anticipate the arguments of the other side. Our discussion of audience analysis can help you do this more effectively. Argument requires objectivity; so avoid loaded language, accusations, and unsupported generalizations. Argument forces a writer to:

- accumulate strong, adequate, and verifiable evidence
- define terms
- arrange information
- anticipate the arguments of the other side
- detect fallacies
- use a clear, unconfused order of presentation
- select and use precise and appropriate language

ACCUMULATING STRONG INFORMATION

The long report, of course, demands a lot of discipline, time, and digging for information. Most of you are familiar with the writing of a research paper. The same abilities and skills apply to a business or professional report. The difference might be that sources of information may not appear in places you are accustomed to. Nevertheless, approach the gathering of information with discipline, organization, and curiosity.

The act of research is, after all, playing detective. Everyone wants to know the answer, so rely on your natural tendency to ask questions and seek answers. If anything works in gathering information it is that you not quit looking until you have found the answer to the questions you ask.

Along with the cards that you carry to jot down ideas, also keep a separate set of cards to record the sources that you used. This practice anticipates the last thing that you will do in a report, the documentation through notes and a bibliography.

Using Standard Research Materials

In spite of major advances in the way we store information and go about retrieving it, the library still offers the printed word as its major research tool. That means books and periodicals. Many of you have written a research paper, using the library, and those skills apply to the workplace. Libraries, of course, organize information. Approach the library with this thought in mind and you will avoid some of the frustration and wasted effort that accompany thinking of the act of research differently. In the end, nothing, not even computer-generated lists of books, replaces contemplative and disciplined sifting through information to find the facts and ideas that you will need to write your report.

Here are several groups of reference materials that you will find useful in gathering information for your report. The selection of books is, of course, eclectic. It is far from complete, since a mere list of available reference sources fills several volumes. This list provides examples of the types of books useful for finding facts, opinions, and other information.

General Guides and Bibliographies.   Books that provide lists of other books usually group information into several categories. The authors of such lists explain in the introduction to their works how they have chosen to arrange and present the books and their locations. Often in such lists, the editor or author provides a brief description, or annotation, of the contents of the book. Annotations and a close reading of the title inform you about a book's relevance to your topic.

> Androit, John L., *Guide to U.S. Government Publications*. McLean, VA: Documents Index (revised periodically).
>
> *Books in Print*. New York: R. R. Bowker Co. (annual).
>
> *Business and Economics Books and Serials in Print*. New York: R. R. Bowker Co. (annual).
>
> Daniells, Lorna M., *Business Reference Sources*. Cambridge, MA: Harvard Business School, 1979.
>
> Sheehy, Eugene, compiler. *Guide to Reference Books,* 9th ed. Chicago: American Library Association, 1976.
>
> Walford, A. J., ed. *Walford's Guide to Reference Material,* 4th ed. London: The Library Association, 1980
>
> Wasserman, Paul, Charlotte Georgi, and James Way, Eds., *Encyclopedia of Business Information Sources: A Detailed Listing of Primary Subjects of Interest to Management Personnel with a Record of Sourcebooks, Periodicals, Organizations, Directories, Handbooks, Bibliographies, On-Line Data Bases and Other Sources on Each Topic,* 4th ed. Detroit: Gale Research Co., 1980.
> Vol. 1, Science and Technology.
> Vol. 2, Social and Historical Science, Philosophy and Religion.
> Vol. 3, General, Language and Literature, the Arts.

Indexes and Abstracts.   Since no one could possibly read all the information that appears in print on a particular subject, you need a way to selectively screen publications. Indexes provide the means to make your search for books and articles much easier. In much the same way, library card catalogs guide you to publications by listing items by author, title, and subject; at least, most indexes will list information by title and subject. As with other reference sources, the introduction to an index should explain its organization and list the publications surveyed. Some, like *The New York Times Index,* provide a brief summary of important articles, often saving you the trouble of finding on microfilm a story that proved less than useful to your purpose.

Abstracts list publications, but much more. In professional journals, articles often begin with a brief summary paragraph that provides you with the essence of the article—an abstract. A book of abstracts reproduces these paragraphs that appear in the journals and books. By reading a particular abstract, you can decide if the full article is pertinent or not. Some sources are:

Applied Science and Technology Index. New York: The H. W. Wilson Company (monthly except July, with periodic cumulations).

Business Periodicals Index, New York: The H. W. Wilson Company (monthly except July, with periodic cumulations).

Dissertation Abstracts
Section A: Humanities and Social Sciences (monthly)
Section B: The Sciences (monthly)
Section C: European Abstracts (quarterly)
    Masters Abstracts (quarterly)

F & S Index United States. Cleveland: Predicasts (weekly, cumulations monthly, quarterly, annually.)

Index of Economic Articles in Journals and Collective Volumes. American Economics Association. (annual).

The New York Times Index. (semi-monthly; quarterly cumulation, last issue an annual).

NTIS National Technical Information Service.

Personnel Management Abstracts. Ann Arbor: University of Michigan. (quarterly).

Psychological Abstracts. Washington, D.C.: American Psychological Assn. (monthly).

Public Affairs Information Service Bulletin (P.A.I.S.). New York. (weekly).

Reader's Guide to Periodical Literature. New York: The H. W. Wilson Company. (semi-monthly, with periodic cumulations).

Social Science Index. New York: The H. W. Wilson Company. (quarterly, annual cumulation).

Sociological Abstracts. San Diego. (5 per year).

The Wall Street Journal Index. New York: Dow Jones Books. (monthly annual cumulation).

White, Jane, and Patty Campbell. *Abstracts of Studies in Business Communication 1900–1970.* Urbana, Ill.: The American Business Communication Association, n.d.

Electronic Data Bases.   The library is no stranger to the computer; many schools of library science have incorporated the appropriate information on sciences of the computer. In the library, computers aid in finding information quickly and accurately.

Lists of publications, and sometimes abstracts, are available from these sources. Use these electronic lists the way you would use an abstract. Look for the publications that each covers, and be especially careful about the dates included. Most of the computer data services are current. For instance, the Dow Jones News Retrieval data base goes back only 90 days. So if you look there for something six months old and can't find it, don't be surprised. Other lists constantly add information going backward, but entering published indexes into the computer memory is a slow process. Use electronic data bases, but do so intelligently. The computer will neither write the report for you, nor will it evaluate the relevance of the article. The following is a list of some useful online databases available by subscription from publishers or computer information services:

Dow Jones News/Retrieval.

Energyline (data base version of *Energy Information Abstracts*).

INFORM (Management and administrative abstracts of articles from about 1971).

ERIC (education information).

LEXIS (Legal articles).

Management Contents (from 1975, similar to INFORM).

MEDLARS (medical literature).

The New York Times Information Bank (abstracts of all articles in *The New York Times* from 1969, and from 1972 of articles from 50 other newspapers and periodicals).

The New York Times Index.

Psychological Abstracts.

PTS (Predicasts Terminal System).

Sociological Abstracts.

Dictionaries.   In addition to a standard college dictionary, hundreds of specialized dictionaries on limited topics can often provide you with in-depth information rapidly and efficiently. If you are not sure that a dictionary on your topic exists, check for one in the subject listing in *Books in Print,* and in the card catalog in the library. Here are several examples of special dictionaries:

Ammer and Ammer, *Dictionary of Business and Economics.* New York: Free Press, 1977.

Johannsen and Page, *International Dictionary of Management.* Boston: Houghton Mifflin, 1975.

Kohler, E. *A Dictionary for Accountants.* Englewood Cliffs, NJ: Prentice-Hall, 1975.

The McGraw-Hill Dictionary of Modern Economics.

Tuer, David, compiler, *The Gulf Publishing Company Dictionary of Business and Science.*

Directories.   Business and professional groups and associations often list member organizations and individuals in a directory. Some of these lists offer no more information than a phone book. Others can serve as a guide to services, information, and products. Here are a few examples of the many listings that are available:

Ayer Directory of Publications. Philadelphia: Ayer Press (annual).

Consultants and Consulting Organizations Directory. Detroit: Gale Research Co. (biennial).

Directory of Business and Financial Services, 7th Ed. New York: Special Libraries Association.

Directory of Directories, 2nd ed. Detroit: Gale Research Co., 1982.

Dun and Bradstreet, Inc., *Million Dollar Directory* (annual).

Dun's Census of American Business (annual). Organized by Standard Industrial Classification (SIC) number.

Encyclopedia of Associations, 16th ed.
Vol. 1, National Organizations of the U.S.
Vol. 2, Geographic and Executive Index.
Vol. 3, New Associations and Projects.

The Foundation Directory, New York: Foundation Center, 1983.

Jane's Major Companies of Europe. New York: Franklin Watts (annual).

National Directory of Newsletters and Reporting Services, 2nd ed.

National Trade and Professional Associations of the United States, and Labor Unions. Washington, D.C.: Columbia Books (annual).

Rand McNalley Commercial Atlas and Marketing Guide. Chicago: Rand McNally & Co. (annual).

Standard and Poor's Register of Corporate Directors and Executives. (annual).

Standard Periodical Directory 1981–1982. Detroit: Gale Research Co., 1982.

Standard Rate and Data Service. Skokie, IL: (monthly).

Thomas Register of American Manufacturers. New York: (annual).

United States Government Manual, Washington, D.C.: General Services Administration (annual).

Who Owns Whom: North American Edition, London (annual).

Financial Newspapers and Periodicals

Advertising Age (weekly)
Asian Wall Street Journal. Hong Kong: (5 per week).
Barron's (weekly).
Business Week
Computerworld (weekly)
Dunn's Review (monthly)
Financial Times, London and Frankfort (daily)
Forbes (biweekly)
Fortune (monthly)
Industry Week
The New York Times (daily)
Publisher's Weekly
The Wall Street Journal (weekdays)

Information from Industry Associations and Trade Groups

Almost every profession or industry has at least one national group or association. Usually that group publishes an annual book of facts and statistics on the association's activities. Although the group might not seem to be an unbiased source of statistics and data, the best interest of the group is served by an association that upholds industry and professional standards. Look at the publications individually and judge them on their own merits. Often the most demanding critics are the professionals themselves. These few random examples should introduce you to the range of information published by trade groups:

> American Paper Institute, *Statistics of Paper and Paperboard*. New York, 1964, and annual supplements.
>
> *Department Store and Specialty Store Merchandising and Operating Results*. New York: National Retail Merchants Association (annual).
>
> *Electronic Market Data Book*. Washington, D.C.: Electronics Industry Association (annual).
>
> *Insurance Facts*, New York: Insurance Institute (annual).
>
> *Prescription Drug Industry Fact Book*. Washington, D.C.: Pharmaceuticals Manufacturers Association (frequently updated).
>
> *Textile Organon*. New York: Textile Economics Bureau (monthly).

Biographical Information

> Dun and Bradstreet, Inc., *Dun and Bradstreet Reference Book of Corporate Management* (annual).
>
> *Who's Who in America*. Chicago: Marquis Who's Who (bienniel).
>
> *Who's Who in Finance and Industry*. Chicago: Marquis Who's Who (bienniel).

Government Publications and Documents

A fine source of information and publications is the United States Government. No matter what is made, or what service is required, the government in one of its bureaus or agencies more than likely has used the service and has written about it. In addition, the government gathers much of the statistical information useful to business and professional people. They disseminate the data through books and periodicals.

The information gathered and published for the United States by the Government Printing Office, except for administrative papers and classified documents, is available to anyone. To make access easy, Congress has designated certain libraries as Government Depository Libraries. More than 1,300 of them across the country receive the publications, which range from the Federal Register, the report of congressional action, to pamphlets by the Department of Agriculture on growing vegetables at home.

Depository libraries are roughly located two in each congressional district. In addition, other libraries of government bureaus and agencies have the publications, as do the service academies, land grant colleges, and law school libraries. Check the nearest library to find the closest repository for government documents.

A list of government publications fills a book itself. For a particular title look into *The Guide to U. S. Government Publications* mentioned above. Here are some published government sources.

U.S. Bureau of the Census, *Statistical Abstract of the United States* (annual).
————, *Census of Manufacturers*. 1982 (every 5 years)
U.S. Bureau of Labor Statistics, *Monthly Labor Review*.
U.S. Council of Economic Advisors, *Economic Indicators*.
U.S. Department of Commerce, *Survey of Current Business* (monthly).
————, *U.S. Industrial Outlook* (annual).
U.S. Government Printing Office, *Monthly Catalog of Government Publications*.
U.S. Internal Revenue Service, *Statistics of Income* (annual).
   Corporate Income Tax Returns
   Business Income Tax Returns
   Individual Income Tax Returns

Company Reports

You can also gather much information about a company from its own published reports. Although nothing requires a firm to do so, most issue an annual report that outlines generally the year's activity. Many annual reports look very impressive, with excellent photography and printing. Of course, these reports are as much for public relations as they are for information. Other papers open to the public can provide more solid figures, for example, a company prospectus, proxy statements, Securities and Exchange Commission 10Q and 10K reports. You can get SEC reports by requesting a copy from the company. If you can't wait, many libraries have the 10K reports on file. Sometimes the file is quite current, but not always. Reports are usually issued after tax returns are filed in late spring. Ask your librarian about such company reports and other information published by the company you are interested in.

Other Sources

In addition to the research library, you can gather information from the special libraries of professional associations, organizations, and societies. Groups such as the Public Relations Society of America provide a library for their members. These collections naturally concentrate on books and materials directly related to the profession, and usually restrict access to members. But you can write to the librarian and request permission to use their collection, explaining why you need to see the books they have.

If you have been unable to secure the information that you need by combing standard sources, you might write a letter to the organization or an expert in the field. Avoid doing this unless you have a valid reason, have exhausted your search, and have limited your question to one that can be answered briefly. Writing an expert under those conditions demonstrates not only your understanding of the material, but also your respect for the time and energy that it takes to respond to questions.

ASSESSING THE MATERIAL
YOU GATHER

No matter how much experience you have in research, you must still condense the facts and information for the report. To do that effectively, you need to determine the relative worth, or the quality, of the information that you have uncovered. Some simple items to look for in assessing your sources should provide direction for judgment. Credibility provides the best characteristic of useful data. To establish the credibility of a work:

- Note the author. Is the person known in the field? Does the writer have a track record?
- Note the publisher. Is the company reputable? What kind of related material do they publish?

Once you have a credible reference, check it for usefulness by reading the introduction carefully. If the source is a book, check for a bibliography, notes, and index. Books with these elements make use efficient, and a good bibliography directs you to other material.

The few extra minutes that you take to evaluate the credibility of any source of information will pay off in the quality and credibility of your own report.

PRIMARY SOURCES OF
INFORMATION

Often the scope of a report requires original information that you must gather. Primary sources include company files and records, observation, experimentation, and interrogation. Though original information makes your report more timely and specific to the topic, it also makes production more difficult and costly. The more information you gather from primary sources, the more time you will need to evaluate and arrange it for your reader's comprehension.

Company Files and Records

Looking back through documents in company files can give you insight into a project that you are working on. Such a search provides a historical background for your discussion, as well as some facts and figures for comparison. If, for example, you looked at the number of employees in the company 10 years ago, and compared the personnel breakdown with similar current ones, you might see a trend. Your digging provided facts that supported the conclusions of the report, so an effort to get relevant information makes the report more believable.

Observation

Most of us like to look at what someone else is doing. If we use any natural powers of observation, we can often get ideas and information to solve problems or make suggestions. For instance, a bank noticed that on Thursdays, the payday for two hospitals each only a block from the branch office, the crowds lined up in front of teller windows. It was not unusual for customers who happened to choose a "slow" line to become angry, creating tension, which occasionally resulted in shouting matches between guards and those in line. The branch manager noticed this and replaced the separate lines with one line that snaked through the bank, allowing everyone to wait on the next available teller. No one needed to feel that they did not wait their turn. Such a solution was the result of observation.

Experimentation

Securing information through experiments presents the most difficult and time-consuming method. The discussion of experimental design, set up, execution, and evaluation is lengthy and complex. But briefly the scientific method applies. Try to perform an experiment that can yield verifiable results. In other words, set up an experiment using a control group and changing only one variable. For instance, if you wanted to see the effect of introducing word processors in your typing pool, you would have two groups in similar offices of the same size, workload, personnel, experience, and so on. You would try to make the two groups as alike as possible. The only item that you would change would be that one group got word processors while the others continued to work as they had in the past. Any differences in morale, performance, absenteeism, and so forth, could then be attributed to the introduction of the word processor.

Interrogation

Another method of obtaining information is to ask questions. You can do this in writing, on the telephone, or in person through interviews.

Letters written to an individual often elicit important and helpful responses. In writing a letter in which you ask for information, follow our previous suggestions. That is, address your question to a specific person, keep it short, and keep it very specific. You are more likely to get a positive response to your inquiry if you:

- provide detailed information of the subject—if you have them, file numbers or names, dates, persons involved. This lets the person know what you want and helps them find the information.
- ask for specific, rather than general information. Most companies are busy places and they will not spend very much time with a broad question like: "How did you become so good at making printed circuits?" But they might answer, "How much research and development commitment in manpower and resources did you initially make in perfecting a production method for your printed circuits for television sets?"
- let the persons you are questioning know that you have searched elsewhere and they are the logical source for the answer. This shows that you have done your homework and that you do not expect them to do your work.

If you have many people to question, you might try an *interview*. You can interview in person or on the telephone. For a *personal interview*, set up the questioning in advance, either in writing or by telephone. In either case, explain who you are, what you want, why an interview is necessary, when you are available, and how long you think the questioning will take. Prepare your questions in advance, and if you plan to tape the session, be sure that you ask them if it is all right to do so. Offer to send them a copy of the tape or a transcript of the conversation. You might, however, want to take notes instead. Taking notes makes the process of interviewing easier because the atmosphere is more flexible and relaxed. Here, too, you might offer to send a summary of your interview, asking if they would like to add their comments. This is a courtesy and a way of building good will. How should you create the questions? Similar to the way you would in creating a questionnaire.

The telephone interview offers an alternative to the personal interview. It has the advantage of speed. The telephone provides an effective tool for interviewing, but be aware of its limitations. Be sure, too, that you have drawn up your questions in advance, and that the responses you seek are relatively short. Unlike a face-to-face interview, the telephone is a poor medium for discussion. Initially make sure that you identify yourself clearly and your reason for calling. Make the questions brief and clearly understandable, give people time to answer, and thank them after the interview.

A *questionnaire*—a printed form addressed to a selected group with questions intended to gather information or perform a survey—is obviously cheaper to administer and can canvas more people in less time than telephone or personal interviews. Its disadvantages include the quality and quantity of the response. Often the segment of people who respond to surveys hold strong beliefs that can bias your conclusions. Also, if you mail a survey, you cannot force responses. If possible administer a questionnaire in a room in which you have assembled the group you wish to question.

Though in business and industry many companies offer survey services and will design and administer them for you, you should at least be aware of the rudiments of constructing a questionnaire since it is a useful source of primary information.

Make a questionnaire brief and simple to fill out. Since you essentially ask someone to give you their time, try to:

- provide clear instructions
- make questions unambiguous
- ask for specific, not general, information
- make the format easy to understand
- use tact in requesting personnel information
- arrange questions systematically

Instructions on an effective questionnaire are clear and easy for a first-time user to follow. If you are like most people, you don't read directions, you scan them. Keep that in mind. Also, it aids the person taking the survey if you provide instructions in outline form. Instead of this:

"Read the statement of confidentiality at the end of the questionnaire before you provide any answers. Print or type your answers. Complete all questions and statements. Indicate "no," "none," or "not applicable" where appropriate. Fill out, sign, and return to the division indicated on the back. If you need additional room for comment, use the back or additional paper."

Try this:

1. Before you begin, read the statement of confidentiality at the end of the questionnaire.
2. PRINT or TYPE your responses. Complete all questions. Indicate "no" or "not applicable" where appropriate.
3. After completion of the questions, sign the sheet and return it to the division indicated on the back.
4. Use the back or an additional sheet of paper if you need more room for comments.

When constructing questions, ask for specific information. For example:

Did you like the AlphaBeta scheduling chart?   _____Yes   _____No

Rather than,

How did you like the AlphaBeta chart?

Avoid loading questions; make them specific and clear.

At least five types of questions are possible in a questionnaire. The content, the group questioned, and your use of the answers should dictate which type you choose.

1. yes/no. Ask a straightforward question that can be answered with a yes or no. An ambiguous question results in indeterminate responses because it confuses the reader. For example, avoid a question like:

Do you use a blade or an electric razor?   _____Yes   _____No

Try this instead:

Do you shave with a blade?   _____Yes _____No
                     electric razor?   _____Yes _____No
                     not at all   _____Yes _____No

Such an alternative allows you to interpret a response better.

2. multiple choice. Some questions have several possible responses. In an effort to group them you can provide a list of choices for the reader. For instance, to determine preferences for vacations you might ask:

How would you prefer your two-week vacation?

_____   two weeks together
_____   split the two

_____ one week of consecutive days, the other of separate days
_____ all vacation time split into single days spread out over the year
_____ separate days restricted to Tuesdays, Wednesdays, and Thursdays
_____ no restrictions on separate days

Such questions allow better analysis of answers and, in this case, preferences.

3.   check response boxes.   This type of question is a variation of the multiple choice and provides boxes next to information that can be checked off. This alternative makes the questionnaire easier to fill out. For instance, when you ask for simple information such as marital status, instead of providing a blank, try this:

Marital status   ☐ Single
                  ☐ Married
                  ☐ Divorced
                  ☐ Separated
                  ☐ Widow/Widower
                  ☐ Single, head of household

Such an arrangement is clearer and easier to follow than:

Marital Status   Single_____   Married_____   Divorced_____
___   Separated_____   Widow/Widower_____   Single, head of
household_____

4.   scaled response.   When you want to find the degree of feeling, you can use a scaled response, or a continuum. For instance,

	Very Satisfied					Very Dis- satisfied			
	10	9	8	7	6	5	4	3	2	1
Overall rating of the meeting	└──┴──┴──┴──┴──┴──┴──┴──┴──┴──┘									
. Adminstration										
(a) registration procedure	└──┴──┴──┴──┴──┴──┴──┴──┴──┴──┘									
(b)room seating/atmosphere	└──┴──┴──┴──┴──┴──┴──┴──┴──┴──┘									
(c)meals	└──┴──┴──┴──┴──┴──┴──┴──┴──┴──┘									

or

How useful was the AcuCount Charts tax table?
very useful   average   useless
+   0   −
└────┴────┴────┴────┴────┴────┴────┴────┴────┴────┘
+5   +4   +3   +2   +1   0   -1   -2   -3   -4   -5

5.   short comment.   Often several questions invite comments from the person filling out the questionnaire. Even with these general solicitations, ask for specific information without slanting the question. Avoid vague questions that are a guaranteed invitation to broad but almost useless answers. Rather than a question such as:

What did you like best about Wash Daze detergent?

ask instead:

What results do you expect from a laundry detergent?

You can even solicit constructive solutions to continuing problems by inviting comment:

How would you change the current method of assigning vacations?

Word questions carefully when requesting comment, so that you get a relevant response.

Attitude and Order.   Often you might need to collect information that some people consider personal. A tactful approach usually draws out the answers you need. For instance, instead of asking:

How much money did you make in 19XX before taxes?

try:

Income for 19XX was   _____   less than $10,000
                        _____ $10,000 to $20,000
                        _____ $20,000 to $30,000
                        _____ $30,000 to $40,000
                        _____ $40,000 to $50,000
                        _____   more than $50,000

This approach indirectly asks for personal information. It exhibits tact, and is more likely to result in responsive answers than a more brusque approach. Also, put questions in logical sequence. Don't ask, "How many miles do you drive in a year?" and then follow with, "Do you have a driver's license?" If you put questions out of logical sequence, you might receive confused answers.

If you mail question forms, be sure that you include an addressed, post-paid envelope to encourage the return of the completed questionnaire.

GRAPHICS

The next chapter details the use, impact, and types of graphic aids that help you make your report stronger. However, two aids—*headings* and *white space*—make a report ·visually pleasing by providing breaks on the page. Think of headings and subheadings as similar to the headlines in newspapers. Carefully placed and worded headings provide a brief running outline that allows the reader to follow the report more easily. Though you may have considered headings an obvious device, they are considered essential elements in the writing of a quality business or professional report. White space is nothing more than a visually pleasing arrangement of type on the page. That includes generous margins and a sufficient amount of space between paragraphs and sections.

REVISING AND EDITING THE DRAFT

Once you have generated an idea through brainstorming, developed it, gathered enough information, and written a draft of the report, you then must go over and polish the draft. Few writers include all they want to say in the the first draft. Expect to go over your draft at least once for editing, and one more time for proofreading. Editing your own material is difficult, but necessary for clear communication with the reader. To edit efficiently, begin to think critically about what you have written. See the material from your reader's point of view. To do this, recall this essential step in the writing process:

Allow time to cool off.

That time gives you the opportunity to shift mental gears and begin to judge the material you have written.

Revision requires the critical effort of reseeing the material that you write. When you revise keep in mind that audience's needs and expectations, your goals, and the clarity of expression.

Ways to Edit and Revise

Putting yourself in the reader's shoes helps you to be less defensive about what you have created. No writer easily discards a sentence or a whole group of paragraphs. So when editing material, a review of your audience analysis will prepare your thinking for the reevaluation of your material. For this kind of revision, you can give the material to someone else to read. If a question about your meaning arises, don't look for excuses or grammatical rules. Listen to comments carefully. A reader usually finds portions of your material that are unclear. Rather than defend a sentence in question, revise it to make the point clearer.

To be able to edit that way, you let go of the understandable feeling that the draft material cannot be changed. After all, a draft is preliminary in its essence. It invites change. An outside reader illuminates sticky spots, but you must determine whether to rewrite.

Putting yourself in the reader's shoes is a good approach to editing and revising drafts. All editing techniques will work better for you if you keep in mind that writing is always an act that assumes some reader.

Flagged constructions identify places in your writing for improvement. This technique directs your attention to specific types and groups of words, and suggests what to look for when you find them. Four useful locations to "flag" for your editorial attention are verbs, voice of the verbs, pronouns, and prepositions. Though these four items are by no means the only locations that can be singled out for improvements, they do give you very useful hints for your rewriting and help to identify some of the characteristics of weak sentences and paragraphs. Chapter 2 discussed word choice, and the exercises furnished several examples that required you to find and rewrite sentences that had weak verbs, abstract constructions, and so forth. In those exercises, you practiced writing a concise version of wordy constructions. To edit your own material, first identify a sentence that could be revised, and then rework it. Table 10–1 lists "flags" that indicate that a sentence needs revision. The flags and suggestions should provide enough tools to make the initial revision that improves the quality, as well as the clarity of your own writing.

Look again at the list in Table 10–1, and notice that its emphasis for revision is on the verb. The heavy stress on verbs in this editing guideline is not accidental. It results from a principle that the verb forms the basis of strong writing. By focusing your attention on verbs while revising and editing, you identify rapidly and efficiently sentences that pull their weight, and ones that do not. Revise the ones that contain a message, and discard the ones that do not. Flaccid writing works against your main purpose, and suggests your lack of control over the material.

Another editing technique is *cut-and-paste*. In a longer paper or report, certainly you may come across information that belongs in a section produced previously. Do you retype the whole thing in revising? No. To save time and effort, take out a pair of scissors and cut the paper, inserting the additional material in the appropriate spot. If the material is a page or more, place it in the draft at the proper spot and merely number the added pages with letters. For example, to add three pages that logically follow page 17, mark the additional material 17a, 17b, 17c.

Cutting and pasting achieves a logical arrangement without retyping. Often a block of material can be moved from one portion to another easily, resulting in greater clarity for the audience. When you use this technique, check the end of the passage immediately before the inserted material and the first passage of the inserted material to make sure the transition between the two is smooth. Do the same at the end and with the passage that follows. Like driving from one town to the other, make the reader's road smooth, and provide bridges to connect one part with another.

Peter Elbow's Advice.   In his book, *Writing with Power*, Peter Elbow says, "Learn when not to revise." He makes this suggestion to keep you from getting stuck in the revision of material, cutting the life out of it. He also reminds you to continue generating new material. In other words, use brainstorming techniques while you are revising.

Table 10–1 Flags for Rewrite: A Checklist for Editing

Flag	Remedy
"is" "there is . . . " "there are . . . "	Use a strong active verb w/o "is" Find a verb, and eliminate "there" if possible
"is + (verb)-ed by" (passive voice construction)	Use the active voice of the verb.
string of helping verbs; for example, "will be made to think"	Select a single powerful verb.
string of prepositions; for example, "as to whether"	Use a single preposition or a strong verb.
string of infinitives; for example, "to be used to fit"	Use a single well-chosen verb.
-ing, -ed verb endings	Check to see if the verb is indeed a verb or, if used as the name of something, a more concise alternative exists.
Jargon	Find a plain English equivalent.
Abstract expression; for example "salary enhancement"	Use a concrete equivalent.
Poor connections between sentences	Provide transitional words, such as "and," "however."
Pronouns	Be sure that the word a pronoun replaces is clear to the reader.

Samples of Revision. Figures 10–5, 10–6 and 10–7 are passages from Chapter 12 on the use of computers and writing. Compare the first draft (Fig. 10–5) before editing, with the second version (Fig. 10–7). Figures 10–5, 10–6, and 10–7 demonstrate the improvements you can make by editing and revising your own text. Of course, you could continue to revise the two pages of the example, but effective revising requires that you know when to begin and when to end. Revise, but don't cut the life out of a piece of writing.

The efforts to automate the office with electronic mail, and electronic meetings, memos, and reports all sent and received on screens that are part of expanded word processing are not always the answer. The machines have forced managers and executives who are mostly talkers and not typers to sit in front of a screen and type a memo onto it. Most of them do not want to "look like a jerk," as The Wall Street Journal put it in a June 24,1980 article. Most consider the computer rigid. They feel they could work on the mail, reports, rescheduling, or memo writing in a far less restricted way than the machine calls for. Some top executives simply turn their display terminals to the wall and pull the plug, choosing not to type memos and messages on the machine that they were used to dictating to their secretaries.

From the point of view of this book, it would be a miscalculation to overlook the applications of the processes of writing to the workplace through the new technology. It does require a change in the work habits of managers, mainly because writing requires much more thought and organization than simple speaking. When you speak you appear to make sense, but writing often brings back mercilessly the absurdity of an idea. An uneasy feeling that the written message sounds dumb and therefore makes a writer look foolish, or even to appear incompetent, provides the major barrier to the use of the new technology. Also it is a technology that is based on a written culture, rather than on an oral one. The use of the radio, TV, and movies suggests that the orality of our times makes the ordinary person less of a reader and less of a writer as a result. Such folks are now expected to look professional on paper or on screen. Often the fear of writing sent them into the sciences and

FIGURE 10–5a Chapter Draft before Revision

technical occupations in the first place. These are the same
folks , like yourself, who took the science and mat courses not
because they did not have some interest in sciences but because
they really did not like the idea of writing long term papers or
writing essay exams. True/false and multiple choice tests
exercise thinking that works against the ability to write. In
order to communicate clearly and directly, writing requires that
you be so familiar with your information that you can shape it,
manipulate its parts, and make make its message understandable to
your audience.

FIGURE 10–5b (cont'd)

BARRIERS TO AUTOMATED WRITING

The Efforts to automate the office with electronic mail and
electronic meetings, memos, and reports all sent and received on
video terminals create, as well as solve, problems. screens that are part of expanded word processing are not always
the answer. The change the machines have forces managers and executives who
are mostly talkers and not typers to sit in front of a screen and
type a memo onto it. Most of them do not want to "look like a
jerk," as The Wall Street Journal put it in a June 24,1980
article. Most consider the computer rigid, and They feel they
could work on the mail, reports, rescheduling, or memo writing in
a far less restricted way than the machine calls for. Some top
executives simply turn their display terminals to the wall and
pull the plug, choosing not to type memos and messages on the Stet
machine that they were used to dictating to their secretaries.

FIGURE 10–6a Revisions to the Draft

From the point of view of this book, it would be a
miscalculation to overlook the applications of the processes of
writing to the workplace through the new technology. It does
~~Sending~~ Written messages on the computer
require̲s̲ a change in the ~~the~~ work habits ~~of managers, mainly~~ because
writing requires ~~much~~ more thought and organization than ~~simple~~
speaking. When you speak you appear to make sense, but writing
often brings back mercilessly the absurdity of an idea. An
uneasy feeling that the written message sounds dumb and therefore
makes a writer look foolish, or even to appear incompetent,
provides the major barrier to the use of ~~writing with~~ the computer. ~~new technology~~
Also it is a technology that is based on a written culture,
rather than on an oral one. ~~The use of the~~ Increasing dependence on radio, TV, and movies
for information and entertainment,
suggests that ~~the orality of our times makes~~ the ordinary person
less of a reader and less of a writer as a result. Such folks
the VDT.
are now expected to look professional on paper or ~~on screen.~~
Often their fear of writing sent them into ~~the sciences~~ business and
people
technical occupations in the first place. These ~~are the same~~
~~folks,~~ like yourself, ~~who~~ took the science and math courses not
because they did not have some interest in sciences, but because
they really did not like the idea of writing long term papers or
~~writing~~ essay exams. They found more memo, letter, and report
~~True/false and multiple choice tests.~~
writing on the job than they expected.
~~exercise thinking that works against the ability to write. In~~
~~order to communicate clearly and directly, writing requires that~~
to
~~you be so familiar with your information that you can share it,~~
~~manipulate its parts, and make make its message understandable to~~
~~your audience.~~
Clearly,
~~It seems clearly that~~ anxiety associated with an automated
from the belief partially fear of
office ~~might be the results of psychological resistence to the~~
being replaced by a machine, as well as fear of writing
~~notion that people are somehow the tool of the machine rather~~

FIGURE 10–6b (cont'd)

228

FIGURE 10–6c   (cont'd)

BARRIERS TO AUTOMATED WRITING

Efforts to automate the office with electronic mail, meetings, memos, and reports sent and received on video terminals create, as well as solve, problems. The change forces managers and executives who are talkers and not typers to sit in front of a screen and type a memo onto it. Most of them do not want to "look like a jerk," as The Wall Street Journal put it in a June 24,1980 article. Most consider the computer rigid, and they feel they could work on the mail, reports, rescheduling, or memo writing in a far less restricted way than the machine calls for. Some executives simply turn their display terminals to the wall and pull the plug, choosing not to type memos and messages on the machine that they were used to dictating to their secretaries.

Sending written messages on the computer requires a change in work habits because writing requires more thought and organization than speaking. When you speak you appear to make

FIGURE 10–7a   Revised Draft Incorporating Changes

sense, but writing often brings back mercilessly the absurdity of an idea. An uneasy feeling that the written message sounds dumb and therefore makes a writer look foolish, or even to appear incompetent, provides the major barrier to the use of the computer. Also it is a technology that is based on a written culture, rather than on an oral one. Increased dependence on radio and TV for information suggests that the ordinary person is less of a reader and less of a writer as a result. Such folks are now expected to look professional on paper or the VDT. Often their fear of writing sent them into business and technical occupations in the first place. These people, like yourself, took the science and math courses not because they did not have some interest in sciences, but because they really did not like the idea of writing long term papers or essay exams. They found more memo, letter, and report writing on the job than they had desired or anticipated.

Clearly, anxiety associated with an automated office results partially from fear of being replaced by the machine, added to the fear of writing. After all, the communications features of some computers allow messages to be sent all over the world, bypassing jammed telephone lines and shortcutting the unpredictable mails.

With this in mind, the act of writing becomes even more important, since many messages and letters can be written directly on the computer and transmitted to other terminals.

Whether with a computer or not, writing demands more than an ability to spell correctly and write grammatically. It is an

FIGURE 10–7b (cont'd)

230

```
outward  reflection  of  thought,  and  must  shape  ideas  for  the
reader.  When  we  speak,  we  digress.  We  respond  to  questions,
facial  expressions,  body  movements,  and  sounds  that  are  not
words.  In  writing  we  do  not  have  the  chance  to  interact  with  our
audience  as  we  would  in  conversation.  Writing  pressures  us  to  be
```

FIGURE 10–7c   (cont'd)

DOCUMENTATION

Documentation can be to your report like the tar baby was to Brer Rabbit. Whatever system you choose, remember to *be consistent*. In a business report, documentation should provide reference to authority. But references or notes should not interfere with the reading of the text. They should also avoid any question of plagiarism. The content of references should support the text, and information necessary to understanding should be part of the text and not a footnote.

Place references:

- at the end of the report in endnotes
- at the bottom of the page in footnotes
- in the text itself in parentheses

Handbooks on the format of notes and bibliographies provide you with accepted methods of citing authority. *The MLA Handbook for Writers of Research Papers, Theses, and Dissertations,*[2] a readable and accurate guide that has wide professional approval, suggests on page 119:

> The conventions of documentation are a means to an end: to lend authority and credibility to your work and to enable the reader to locate sources with ease. Provide a note only where there is a reason.

Rather than discuss specific formats which, as we have said, can vary widely depending on the style manual you use, a general discussion of the rationale behind

---

[2]Reprinted by permission of the Modern Language Association from *The MLA Handbook for Writers of Research Papers, Theses and Dissertations,* 1977, by Joseph Gibaldi and Walter S. Achtart.

documentation should be more helpful and be considered in addition to the technical and format data in the style guides.

Generally, many notes[3] could distract the reader. If possible, incorporate references to the source of your information in the discussion itself. This makes your material more readable. When you use notes, make sure they follow in numerical order, starting with 1. Place the number *after* all punctuation marks, and slightly above the line. The number should follow material that you quote, paraphrase, summarize, or refer to.

Endnotes.   At the end of the report, begin on a fresh page titled Notes, Endnotes, or References. List them in numerical order the way they occur in the text. Gathering your documentation on sheets at the end of an article makes the preparation and typing easier. Also, since productivity in a business setting is at a premium, it makes good sense to gather these notes in one place and save preparation time. Publishers and scholars now have largely abandoned the once common practice of placing all documentation at the bottom of each page in footnotes; a footnote then appears at the bottom of the page on which the information is referred to. Endnotes and Notes appear after the text.

General Information Included in Notes.   For documentation, three broad categories usually are used: author, title, and publication information. In notes the *name* of the author appears in normal order. The full title appears next, with each major word capitalized. If the work is an edition, translation, or compilation, indicate that following the title.

The publication information includes the place of publication, the publisher's name, and the date of publication. For books, the copyright year is the date. You find that on the reverse side (verso) of the title page after the copyright symbol—©. Put all three items of the publication information in parentheses:

(New York: Grove Press, 1982)

Then place the page reference after that.

Periodicals form the basis of much of your documentation. In citing a periodical, include the three main areas of information—author, article title, publication information. Cite the author as you did for a book, and put the title of the article in quotation marks. Under publication information, give name of the periodical, the date of the issue, and then the page reference:

Author, "Title of the Article in the Journal," *Journal Title,* issue I, date, page reference.

Begin the citation of anonymous publications with the title.

---

[3]See how you automatically looked here; and now you have lost the flow of the sentence.

Other Sources.   Although books and articles will provide most of your information from secondary sources, you may have gathered facts and ideas from other places and need a way to identify their sources for your reader. Table 10–2 provides guidelines for citing common reference material, other than the usual printed documents.

Second Reference and Subsequent References.   In notes be brief and clear in citing a source for the second time. Usually the author's name, followed by a page, is sufficient:

[4]Author's Last Name, page.

Table 10–2   Citing Information from Uncommon Sources

Source	Information Order & Type
Manuscripts (material typed or handwritten by the author)	Location of the papers and other material; identifying number, if any.
Lectures	Speaker, lecture title, sponsoring organization, place, date.
Films	Title, Distributor, date (if applicable, writer, director, producer, performer).
Radio and TV Programs	Title of Program, Network, or station, and the city, date of the broadcast.
Recordings (Tape, Record, Videotape, etc.)	Title, artist, manufacturer, catalog no., date.
Information Services (data bases, etc.)	Author, title, name of the service, page, identification number.
Interviews—personal and telephone	Telephone or personal interview with: Person interviewed, date. (If appropriate, you may include other information such as place.)
Letters—Published	"Recipient," date, title of book, editor, place, publisher, date, page.
—In a file or archive	Writer to recipient, date, location.
—Personal	Letter from _____, date.

However, if you use several works by the same writer, give a short title here too:

[5]Author's Last Name, Short Title, page.

If you cite one work extensively, make a statement in the first reference that you will document the source in the text of your report. For example:

[6]Author's Name, Title (Publication Information), page.
Further references to this book appear in the text.

In the text, the next reference will follow the quoted material and look something like this:

(Author, page).

As with all citations, the key to success lies in being clear, brief, and consistent.

Alternate Citation Methods.   The endnote and footnote method is the familiar, commonly accepted, and most used way to cite published sources of information. However, scientific and technical publications employ different approaches to citation, such as the author-title, author-date, and number system. For example, you may find a book or article by Black cited in one of these ways:

(Black, 35–37)
(Black, *Techniques,* 35–37)
(Black, 1982, 35–37)
(22, 35–37)

In each case, the book or article has a list of the books and articles in a central bibliography or list of references. In the number system, the last method, each reference is assigned a number and the number is used in the text. Most readers and scholars prefer a system that uses the name of the author instead of a number because if the wrong number appears, the reference becomes useless. With the use of at least the writer's last name, errors that might occur allow the reader to find the article through the author citation.

Before you cite articles, look at the suggested methods of documentation that your discipline commonly uses. Some useful general guides for documentation in addition to the *MLA Handbook for Writers* are:

GPO Style Manual (Washington, D.C.: U.S. Gov. Printing Office), (annual).
American National Standards Bibliographic References, ANSI Z39.29–1977 American National Standards Institute (1430 Broadway, New York 10018, 1977).
The Chicago Manual of Style, 13th ed. (Chicago: Univ. of Chicago Press, 1982).
Style Manual 1978. American Institute of Physics (335 East 45th St., New York 10017).
Publication Manual, 2nd ed. American Psychological Association. n.d.

Most professional associations issue their own publication guidelines or refer you to standards that the group has adopted for papers and articles.

Bibliography or References

In a long report you may wish to include a list of all the works that you used in preparing your statements, even ones you did not actually quote or refer to directly. In that case you should call your list a bibliography—an alphabetical list of the works you used. If you included only works used directly in the text of your report, call it a List of Works Cited, or List of References. In any case your list differs from the endnotes and footnotes because it is organized alphabetically, and the form of entries differs slightly. A representation of the differences between notes and bibliography citations is shown here.

The appendix of this book includes a bibliography of some of the material related to the chapter subjects. Refer to that list of books and articles, as well as the

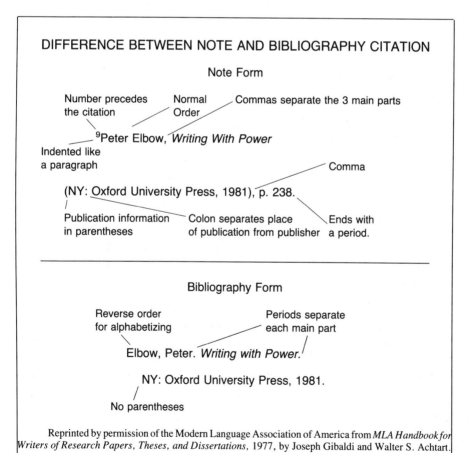

DIFFERENCE BETWEEN NOTE AND BIBLIOGRAPHY CITATION

Note Form

Number precedes the citation / Normal Order / Commas separate the 3 main parts

⁹Peter Elbow, *Writing With Power*

Indented like a paragraph

Comma

(NY: Oxford University Press, 1981), p. 238.

Publication information in parentheses / Colon separates place of publication from publisher / Ends with a period.

Bibliography Form

Reverse order for alphabetizing / Periods separate each main part

Elbow, Peter. *Writing with Power.*

NY: Oxford University Press, 1981.

No parentheses

Reprinted by permission of the Modern Language Association of America from *MLA Handbook for Writers of Research Papers, Theses, and Dissertations,* 1977, by Joseph Gibaldi and Walter S. Achtart.

lists in this chapter, not only as a source of information, but also as an example of the alphabetical listing of published information on a specific topic.

Other Types of Documentation

Include information and data in a report through an appendix, attachments, or exhibits. Many inexperienced writers use these devices to dump all the information they have gathered, providing several dozen pages of irrelevant material. Just as the biological appendix serves no crucial function in your body, the appendix of your report should not include any information vital to the understanding of the text of the report. Appropriate appendix material might be a sample questionnaire, computer print-out, data table, statistical summary, letter, or a report on a related topic.

As a practice of efficient writing, put important information in the report. Avoid appendix-like material if possible. Information does not speak alone, and data included for its own sake may not clearly show what you wanted. No one appreciates a routine report with a 50-page appendix and a dozen attachments. Rather than a demonstration of diligence, such bulk shows misunderstanding of the use and purpose of reports.

PREPARING THE FINAL COPY

Before preparing the final copy of the report, become familiar with similar reports submitted previously. Note their arrangement of information and their formats. But keep in mind, too, that many old-timers in companies interpret previous reports as a strict "company policy" on the method of preparation. Most often routine reports fall into a pattern, but that myth of a company policy can destroy the freshness that you want to inject even into a routine presentation. Look at the format of a report as a stimulus for the one you are now writing. If the information you must report would be more effective in a different place, put it in an effective position. Remember, though, that a report that covers several items or activities and is distributed widely in the organization, falls on the desk of managers who scan reports. Scanning means they automatically flip to the part of the report that concerns them. Some even come to expect a certain page of the report to contain the information they need. Be aware of this report-reading practice before you change the format. If necessary, make a note in the new format so scanners can find the information they are interested in.

When you give the report to a typist or secretary, make sure you go over the format of the document beforehand. Seek the secretary's advice on the best arrangement and format for the entire report. The report is still yours and a secretary should not bear the blame for a poor presentation.

PROOFREADING

No one expects you to write a rough draft with no errors, or even have a final draft with no mistakes. But professionals do expect to receive a document that has no errors in spelling, punctuation, and grammar. For that reason make a habit of proofreading material before you send it out. If you find some major error, or even minor ones, ask the typist to rework the page. That does two things for you: it establishes the level of quality you expect from the typist, and it lets your readers know that you take an interest in them by sending a clean, polished report.

You may never have to perform near the level of accuracy of professional proofreaders, some of whom can pick out an error on a page as you walk by with a paper in your hand. Nevertheless, here are practical approaches to proofing your own report.

Read Copy Aloud.   Find someone to read the draft to, or better yet, have them read it to you. Note points that the reader stumbles over as good candidates for revision. Reading aloud reveals stiff, pretentious, and often funny constructions, as well as loaded, biased, or slanted words and phrases.

Consult a Dictionary Often.   If a word sounds odd to you in the context of your report, look it up to see if your meaning fits the accepted definition. Most of the time the word and its spelling will be correct, but look anyway if you are in doubt.

Read Punctuation Marks as Signs.   Punctuation (see appendix) serves your reader as road signs do a driver. They tell when to stop, slow down, or when something is coming. In proofreading, consider the marks of punctuation as dramatic directions, rather than a list of rules to justify placement of a comma or a semicolon. Let the content of the sentence dictate whether or not punctuation makes the sentence clearer.

Read from Last Word to First, Backwards.   The best way to check any document for spelling is to look at the last word on the last page and ask if it is spelled correctly. If not, look it up. If so, go to the next to last word and so on until you get to the first word. Reading backwards makes sure that you don't understand what you read. That is the point. When looking for spelling errors, this technique assures that the content of the report does not distract you from misspellings.

Pay Attention to Verbs.   As we said earlier, the verb is the foundation of good writing. Careful attention to the form and use of strong verbs, even at this point in the writing, will often reveal some major grammatical errors.

In proofreading, make sure that you check:

- punctuation
- capitalization
- spelling, including abbreviations

Also check:

- word choice
- arrangement of ideas

- sentence construction
- transitions between sentences, paragraphs, and sections
- verb form, tense, consistency
- appropriate and informative title

EVALUATING A REPORT

Use the Report Evaluation that follows to identify strengths and weaknesses of your own report and in reports of others. This scorecard is a longer version of the one used previously and will take more time to use. Both are tools to help you write a better document by allowing you to identify parts that need work. Use these forms with caution and some understanding of the nature of writing. Let these questions remind you of points that need to be covered or revised.

REPORT EVALUATION

Rate a sample report, your own or another's, by answering the following questions. Use a plus (+) for yes, and a minus(–) for no. Put a "0" if the question does not apply.

Purpose

Have you or the writer:

1. Defined the problems or goals?	1._____
2. Compiled the necessary information?	2._____
3. Checked that information supports conclusions?	3._____
4. Mentioned previous or related studies?	4._____

Audience

Have you or the writer:

5. Determined important information about the reader?	5._____
6. Identified why the reader will read it?	6._____
7. Anticipated the reader's questions?	7._____
8. Defined the reader's attitude toward the subject?	8._____
9. Altered the presentation to fit the reader's needs?	9._____
10. Checked the presentation against any company guidelines?	10._____

Organization

11. Given the report a useful title?	11._____
12. Introduced the subject in the introduction?	12._____
13. Arranged information clearly?	13._____
14. Provided bridges between the parts of the report?	14._____
15. Included enough background information?	15._____

16. Eliminated irrelevant facts and ideas?   16._____
17. Presented illustrations, headings, and white space effectively?   17._____

Conclusions or Recommendations

Have you or the writer:

18. Supported conclusions with details in the report?   18._____
19. Stated conclusions clearly?   19._____
20. Arranged conclusions according to importance?   20._____

Style and Diction

Have you or the writer:

21. Exactly fit the language with the intended meaning?   21._____
22. Used a tone suitable for the purpose and the reader?   22._____
23. Pointed out significance of facts and details?   23._____
24. Chosen words for economy without sacrificing clarity?   24._____
25. Positioned illustrations, tables, graphs, and charts close to their discussion?   25._____
26. Made the paper readable and interesting, as well as accurate and clear?   26._____

Revision

27. Fulfilled the purpose?   27._____
28. Proofread for errors in grammar, punctuation, and spelling?   28._____
29. Eliminated wordiness?   29._____
30. Provided a variety of sentence lengths and patterns?   30._____
31. Checked headings as useful subject guides?   31._____
32. Determined if each word will be clear to the reader?   32._____
33. Made sure the bridges between parts of the report are adequate for smooth reading?   33._____

Before Signing Off

Have you or the writer:

34. Checked the introduction for a clear statement of the purpose, scope, and plan of the report?   34._____
35. Finished before the deadline?   35._____
36. Written something to be proud of?   36._____

SUBTOTAL   _____

SUBTOTAL   _____

TOTAL   _____

(+36 highest,
−36 lowest)

SUMMARY

- Long projects require a strategy that enables you to complete the job accurately and on time. Develop a disciplined approach to writing by clearing time and space for your work. Select a manageable approach to the material by focusing the information through detail, time, or space. Functional analysis helps focus the report, value analysis reveals the true purpose of the document, and a systems analysis will help you select an organizational approach. Use some representation of the whole work to help you see the overall organizational pattern.

- Use an outline to begin developing the ideas. Methods of development include: chronological, sequential, step-by-step, descriptive, geographical, analytical—cause and effect, comparison and contrast. Logic and reasoning—inductive and deductive development—prove effective in organizing longer papers.

- Longer papers and reports demand that you gather strong evidence. You can use research material found in libraries, company reports, observation, experiments, interviews, questionnaires, and computer data bases.

- Plenty of white space, subject headings, and clearly labeled diagrams, drawings, and pictures add to the impact of a report.

- Edit and revise the report by putting yourself in the reader's shoes. Also use the cut-and-paste method, the flagged constructions method, and Peter Elbow's advice to rework material effectively, quickly, and well.

- Provide clear, consistent references and notes in the text itself, in footnotes, or in a list at the end of the report. Place material not crucial to the report in an appendix.

- Prepare a final copy of the report that conforms to organizational standards and guidelines, but also meets the needs and expectations of accuracy and clarity. Proofread the final document for mechanical or grammatical errors. Also make sure that the language is accurate and professional, avoiding slanted, biased, or loaded words. Evaluate your final report to identify strengths and weaknesses in it to improve future efforts.

EXERCISES

1. Select a report that you have written, or one that you received. Evaluate its effectiveness by closely determining its approach to the audience, true purpose, organization and arrangement of information, and selection of facts. Write your evaluation and recommend improvements to its presentation. Also, be sure to detail the examples of strong writing that should be retained and incorporated into future reports.

2. Take a report that you have received and edit it with one or more of the techniques we have discussed. Compare the original with your revision.

3. Consider the kinds of reports you write. Decide which methods of organization and development best suit each situation. Write a report on your investigation to an appropriate person in your organization, arguing for what you consider a more effective way of writing reports.

4. Prepare an outline of a report you have to write in the near future. Take that outline and arrange it in at least two different ways, reflecting different approaches in developing ideas. Decide which outline promises the clearest presentation for the report that you must write. Explain your choice.

5. Comment on or revise the sample report, New East Telephone PBX Strategies, that appears in Appendix B.

6. Revise this section of the report below. Explain briefly what you did and why.

BANK CARD SECURITY DEPARTMENT

Responsible for prevention of losses due to fraudulent use of lost or stolen Bank Cards, and control and reduction of losses when such incidents occur.

Banking office personnel are an integral part of the loss prevention process. Through proper authorization of Bank Card transactions, securing and recording Cardholder identification, and careful scrutiny of suspicious customers or transactions, significant loss prevention and reductions may be achieved.

Fraud

Bank Card fraud occurs when a person, by an act of misrepresentation or guile, uses a credit card illegally to deprive the true owner of services or property.

Causes of Bank Card Frauds

1. Economic Climate—Tight money environments, periodic economic recessions increase likelihood of fraudulent use.
2. Cardholder Resistance—Cardholders are sometimes remiss in reporting loss or theft of cards.
3. Cardholder Collusion—Highly deceptive and often difficult and costly to uncover. Card may be used when usage is denied, or card may be handed out for use by another.
4. Public Apathy—white collar crime.
5. Criminal Justice System's low enforcement priority because Bank Card fraud is traditionally viewed as white collar crime.
6. Merchant Collusion.
7. Urban area's sensitivity to crime.
8. "Low Apprehension Risk" of Bank Card fraud.

Sources of Bank Card Fraud

Organized Crime—Provides outlets for stolen cards. These are obtained from addicts, prostitutes, gamblers, and pick pockets.

Petty Thieves—Obtain cards during commission of other crimes.

Opportunists—Persons who have access to Bank Cards such as Merchants, Bank Card industry personnel, postal workers, and all persons recognizing the Bank Card for its potential worth by misuse.

Fraud Applicants—Applications for cards, using false Cardholder information.

Control of Fraud

1. *Initial Controls*
   Cardholder Applications
   Merchant Applications
   Plastic Vendors
2. *Processing Controls*
   Data Input
   Card Production
   Card Mailing

---

3. *Usage Controls*
   Follow-up Letters
   First Use Notices
   Overlimit and Collection Reports
   Authorization
4. *Fraud Analysis*
   Card Loss Reports
   Authorization Bulletins
   Fraud Activity Reports for Merchants/Cardholders
   Enforcement through Investigations and Arrests.

---

7. You are in charge of a section in a large company. Your immediate supervisor calls you in early Friday morning and tells you that the work of your section will be observed the following Monday by three executives from an out-of-town institution. Thay have indicated that they want to "drop around for an informal discussion of similar goals and problems while they are in town for a convention."

Your boss then asks you for a short, written statement outlining how best to handle the visit. Your paper will be used at a three o'clock meeting called to discuss the visit. It should help to focus discussion and expedite preparations.

8. You are in charge of a large project involving many individuals and costing several million dollars. You have detected professional misconduct in one of those individuals under your supervision. In a report that you know will eventually find its way out of the institution, explain exactly what you have found and recommend specific action to be taken to eliminate the problem.

9. For the past six months you have been having trouble with one of your employees, Johnny Walker. When he began work he seemed well qualified for the job, and in fact you wrote a favorable evaluation of his performance after the first three months. But after that, something went wrong.

Mr. Walker began to miss work. He said it was a virus, but rumors were that he had a drinking problem. His absences became frequent: seven days his fourth month, six in the fifth, and nine in the last. On eight occasions Mr. Walker showed up for work intoxicated, and you sent him home. His work has been unsatisfactory, and riddled with mistakes.

During his last three months you called Mr. Walker into your office four times for long talks. On the fourth you told him he was in danger of dismissal. Today he again showed up for work drunk. Before you sent him home, you told him you were recommending his termination. Now you must follow through.

You must submit a report to the Director of Personnel, recommending that Mr. Johnny Walker be fired. You must clearly state your recommendations, summarize your supporting reasons, and provide justification for your actions.

SUGGESTED PROJECTS

1. Prepare a list of reference materials that might be useful to you in writing a report. For each entry, provide a brief annotation. That is, describe in a short paragraph the kinds of information, the method of presentation, and your judgment of the quality of each source. Arrange the citations with their descriptive paragraphs into groups that you find useful. Clearly label each of these groups. Write an introduction to your annotated list that would explain to someone else what the intent of the list is and how you think it could be used.

2. Select a company that you know nothing about. Find a general description of that company in several reference sources. Write an evaluation of the company based on the material that you have surveyed. Then write a comparison of the relative value of each reference source that you used. You may select a similar fact that all sources share as one specific point of comparison.

3. Further suggested projects follow the summary of Chapters 9 and 12.

Part III

Practical Communications

Pictures, graphs, spacing, speech, computers, résumés—each of these becomes increasingly important for professional communications. The artist's conception of a space station pictured here focuses an audience's attention, and reduces the amount of text necessary for understanding. Graphics enhance your reader's positive reception of the message.

The speed of contemporary business now requires more oral presentations of ideas. This section covers graphics and oral presentations. It explores the advantages and drawbacks of a computer as a tool for the writer. In addition, this part of the text discusses résumés, cover letters, and interviews—your first professional written and oral communications.

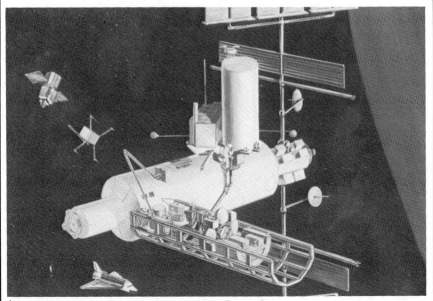

[1]Used with permission from Grumman Aerospace Corporation

11

Graphics and Oral Presentation

No one in a business setting expects you to create illustrations like an artist or to perform like a professional actor. But everyone expects a professional effort. This chapter introduces the types of visual aids that accompany a professional report or oral presentation, and describes how to approach, organize, and execute them.

As you gain experience and take on wider responsibility, you will write more reports for a broader audience. Charts, maps, diagrams, and photographs communicate complicated information to a nontechnical audience. Though you may have no artistic ability, you must be aware of the types of graphic presentations available for use in business reports, and what each type of illustration can accomplish effectively.

As your experience and responsibility expand, you begin to lead more meetings and group discussions. According to a survey, most Americans fear speaking before a group. Nevertheless, you will need to know how to plan and deliver an effective presentation, even though you may have had little experience as a public speaker.

GRAPHIC MATERIAL IN REPORTS

Desirable Qualities in Graphics

A well-planned and carefully built graphic helps the reader to see your point quickly and clearly. Appropriate graphic aids sharply focus the discussion. A useful graphic illustration:

- concentrates on a particular point
- rests on a central theme or idea
- eliminates nonessential lines and words
- labels elements clearly

- includes useful titles, captions, and reference numbers
- keeps to a simple, familiar format
- carries its own weight
- makes a self-contained statement with no dependence on the text.

Graphics should focus the reader's understanding of the material. For that reason, be straightforward in presenting information.

Don't mislead.

Illustrations can be drawn in such a way that they manipulate the reader's perception by altering the scale, making proportions of lines and spaces seem to show a larger contrast than exists. But such manipulation cannot fool everyone, and eventually it erodes your credibility.

Figure 11–1 is a slightly misleading illustration from an annual report. The bar graph is professional and artistically pleasing, and in an annual report, professional appearance is the primary consideration. But look closely at the figure and you will see that, in the vertical scale of dollars, each line represents increments of 70¢. Other illustrations in the same report also use the 70¢ scale. Such division is unusual, because dollars lend themselves to more familiar divisions, such as quarter segments. Here 25¢, 50¢, or 75¢ increments seem more logical.

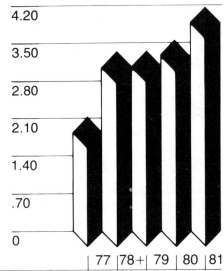

Primary Income per Share Applicable to Common Stock*

Dollars

4.20
3.50
2.80
2.10
1.40
.70
0

*Restated to reflect a 100% common stock distribution on June 8, 1979 and a 50% distribution on September 12, 1980.
+Includes extraordinary items of $1.20 per share.

77 |78+| 79 | 80 |81

FIGURE 11–1   A Slightly Misleading Graphic

The shape of the bars themselves reveals another slightly misleading element in this illustration. The point at the tip of each is not really the point that references the dollar figure. The position on the bar that makes the reference is the imaginary horizontal line drawn through the two widest points. Note that the "0" point on each bar runs through the widest portion, and part of the bar runs below the "0." In effect, a glance at the 1981 bar leaves the impression that a per share income is closer to $4.20, when the figure should be somewhat less than $4.00.

Although Fig. 11–1 shows a graphic pleasing to the eye and fits into the artistic theme of the whole report, it nevertheless presents a confusing message. The line between confusing and willful misrepresentation can be crossed without difficulty. However, most illustrations are clear, direct, and straightforward.

In writing a report, use graphic aids for clarity and brevity. Not that a picture is worth a thousand words, but a good picture is worth more words than it takes to explain the concept represented graphically. As a general working principle, a good graphic will displace at least its own weight in words. To use graphics effectively, you should be familiar with the types of illustrations that are available, and the situations that are appropriate for each.

Types of Graphic Aids

Table 11–1 provides an introductory list of the types of graphics used in reports, the applications of each, and a representation of the appearance.[1]

Highlighting techniques
An effective illustration in a report, or for projection, demonstrates these graphic elements:

- focus
- enhancement
- captioning

Focus refers to the way the piece of art draws the audience's attention to the central theme, and how it demonstrates that main idea graphically.

Enhancement refers to the appropriate use of such devices as line weight, typeface, shading, cross-hatching, color, arrows, silhouetting. Fig 11–2 a and b offers a comparison of two illustrations. The first, part a, lacks effective focus and enhancement. The second, part b, has much more impact because it is clear and direct.

Captioning refers to the title that you give an illustration. Fig 11–2 c and d demonstrate the difference between a weak title and one that works hard:

[1]Adapted from originals in *American National Standard, Illustrations for Publication and Projection* (ANSI Y15. 1M-1979), page 2. Used with permission from The American Society of Mechanical Engineers.

Table 11–1 Types and Uses of Graphic Aids

Type	Representation	Uses	Drawbacks
Area (Pie)		Shows relationships of parts to the whole; ratios; proportions	Difficult to label; small parts get lost in the mass of lines
Bar. Horizontal Vertical		Dramatic picture of contrasts and comparisons; scaled information	Unremarkable or weak differences come across with little force
Curves (Line)		Two or more lines show contrast; pictures trends; illustrates plotted points; future performance can be extrapolated	Use weighted and colored lines; more than three lines cause confusion
Tables	Exec. '60 '70 '80 BBA 25 78 192 MBA 3 41 107 PhD 0 3 10 Total 28 122 309	Understandable and useful presentation of numerical data; organizes statistics; columns and rows allow easy comparison.	Needs explanation; Requires the reader to study it closely. Poor for trends; Significant numbers need highlighting.
Computer Graphics Area, Bar, Tables. Curves		Several versions from the same data are possible in a brief time; reproduction; speed; accuracy; reduced cost of artwork	Cost of machines; takes some "getting used to."
Organizational Charts		Shows the chain of command; quick determination of responsibility and company authority	Can be too brief concerning duties; depends on job titles for understanding
Flow Chart		Gives a picture of an action or a process that is not visable; makes time visable	Can oversimplify; can lead to confusion if not accurately executed
Schedule	J F M A M J Phase I ⊢—⊣ Phase II ⊢——⊣ Phase III ⊢⊣ Reports ▫ ▫ ▫	Condenses time of various parts of a project; shows relationships of one part to another; helps reader to see "when" clearly	Can reveal poor planning to the reader; looks silly for very small projects
Drawings and Photographs		Saves a physical description; provides a picture or rendering of an object	Requires high quality reproduction; original art work and photography can be expensive; pictures for their own sake work against clarity

(a)

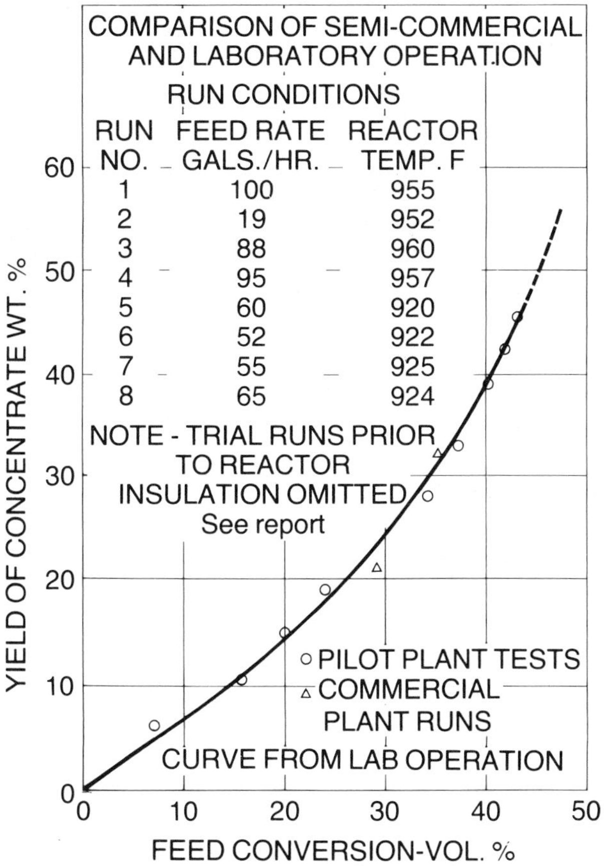
COMPARISON OF SEMI-COMMERCIAL AND LABORATORY OPERATION
RUN CONDITIONS
RUN NO. FEED RATE GALS./HR. REACTOR TEMP. F
1 100 955
2 19 952
3 88 960
4 95 957
5 60 920
6 52 922
7 55 925
8 65 924
NOTE - TRIAL RUNS PRIOR TO REACTOR INSULATION OMITTED
See report
○ PILOT PLANT TESTS
△ COMMERCIAL PLANT RUNS
CURVE FROM LAB OPERATION
YIELD OF CONCENTRATE WT. %
FEED CONVERSION-VOL. %
0
10
20
30
40
50
60

FIGURE 11–2 (a) An Example of a Poor Illustration.

Placement in Reports

Considering the planning and effort that a graphic requires to create and complete, take time to determine where you want it to appear in your report. Generally, illustrations can appear:

- in the text itself
- collected in an insert section
- collected in an appendix

Information that is peripheral to the discussion, or that is presented as support data for the technical part of the audience, can appear effectively in the appendix. Also, drawings that are too big for the standard paper of the report can be placed in a separate folder, or folded into the text.

PLANT AND LABORATORY COMPARISON—

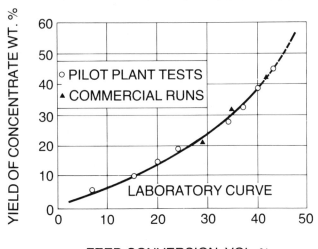

(b)

FIGURE 11–2   (b)   An Example of a Good Illustration.

The most useful position of a graphic is on the same page or on a page facing the text that refers directly to it. Often the placement of these illustrations in proximity with the text can become a logistical problem that you need to work out with the typist or with the printer. If the illustration is several pages away, that forces the reader to flip the pages back and forth as your comments refer to information represented in the graphic. This can frustrate the reader and become a barrier to an understanding of your report.

Recall the last time you read a report and had to search for an illustration or drawing, but could not find it easily. With that in mind, spend extra time making an illustration *aid* the reader. Place graphics as close to the text of the discussion as possible.

ILLUSTRATING YOUR OWN
REPORTS

Much of what we have discussed concerning graphics deals with illustrations prepared by professionals. However, you can create graphic aids for your own report

Instead of this weak caption:

(c)

Try one that works hard, like this:

(d)

FIGURE 11–2 (c) An Example of a Poor Caption (d) An Example of a Good Caption

without professional help. If you can count and draw a straight line, you can make simple charts or graphs to accompany the text.

In making illustrations, start early. Think of the drawings as *visual islands* that allow your reader to rest and contemplate material. Such visual representations can enhance the theme of the report and, in so doing, can provide continuity and coherence. Make your final drawings neat and in ink. Label them clearly. You can use a lettering template like LEROY or WRICO, or commercial letters such as Lettra-set and Instant-Lettering. For graphs you can also use commerical shading material like Zip-a-Tone or Contak. Other materials, such as figures of houses and other uniform drawings, are also available. Visit a commercial office or art supply store to inspect the material they have and acquaint yourself with others you might use in building a graphic representation.

Figures 11–3, 11-4, and 11–5 are examples of graphics created by report writers. Each example furnishes a clear visual representation, even though none of them was the work of professional graphic artists.

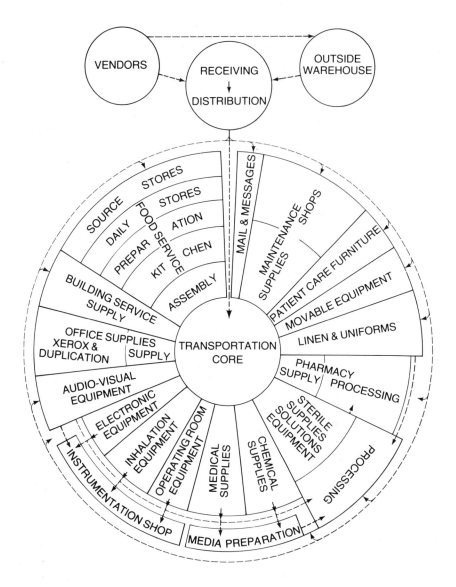

FIGURE 11-3   Flow Diagram for Materials Management in a Medical Center

ELECTRONIC PRINTING AND GRAPHICS

Charts, illustrations, and text generated through a computer and stored in its memory can be reproduced electronically, without using conventional methods of printing. Electronic printing merges the technology of the computer with that of the copy machine, and its applications suit the needs of business and industry as well. Internal

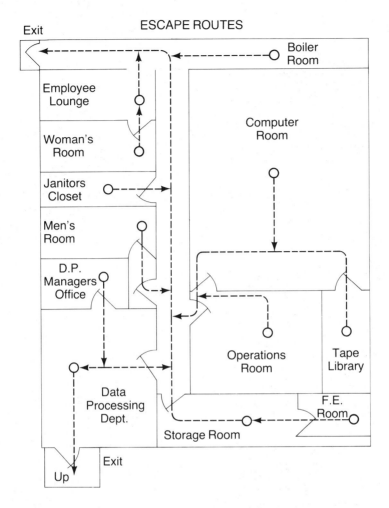

FIGURE 11–4   Emergency Exit Plan for a Computer Facility

business reports, price lists, and technical and training manuals that have a short utility, and usually require small print runs that can be expensive to produce through a commerical printer, lend themselves well to electronic printing.

Figures 11–6 to 11-10 are a few examples of the kinds of reports and graphic representations that can result from a computer and printer set-up.

COMPOSITION OF THE TEXT
AS GRAPHIC AID

Headings, subtitles, and generous margins make your writing easier to read. Using the layout of the text as a graphic aid anticipates the way most people first encounter your writing. They thumb through a report quickly to get an overview of it, noticing the headings, illustrations, and captions on charts and graphs. Plan your report with this in mind so that they capture your reader's attention and invite a closer reading of

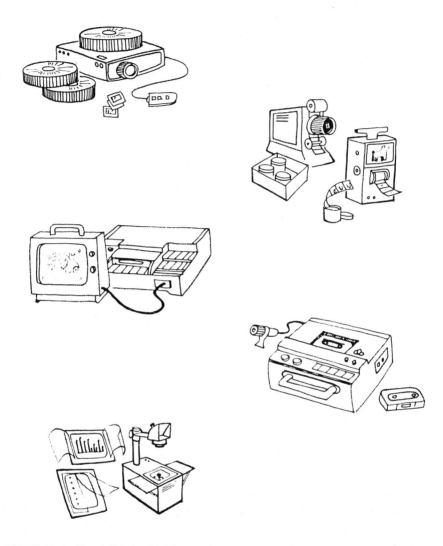

FIGURE 11–5   Visual Aids for Oral Presentations

the whole report. To hook the reader with headings and captions, write them carefully. Headings demonstrate the organization of your report and reflect your knowledge and understanding of the material.

Examples of Graphics

Figures 11–11 to 11–17 show several common types of graphic representations—charts, graphs, flow diagrams, schedules, drawings. Each of these examples, when used appropriately, make a report more dramatic, useful, and readable.

mn manufacturers

In thousands, except per share	Third quarter ended 9/30		Nine months ended 9/30	
Consolidated Income Report	**1981**	**1980**	**1981**	**1980**
Net Sales	$ **49,955**	$ **43,441**	$ **146,370**	$ **128,826**
Cost and Expenses:				
Costs of products sold	31,188	30,721	94,481	94,363
Selling expenses	1,023	940	3,168	1,962
General & administrative expenses	2,995	3,010	9,727	9,620
Interest & debt expense	816	712	2,740	2,360
Misc. income (expense)−net	(520)	(492)	(1,572)	(1,327)
Total Expense	$ **35,502**	$ **34,891**	$ **108,544**	$ **106,978**
Income before taxes	14,453	8,550	37,826	21,848
Income taxes	4,452	2,633	11,650	6,729
Net Income	$ **10,001**	$ **5,971**	$ **26,176**	$ **15,119**

Stock Report	**1981**	**1980**	**1981**	**1980**
Average number of common shares outstanding	2,482	2,397	2,508	2,367
Earnings per common share − Primary	$ **.83**	$ **.22**	$ **2.13**	$ **1.77**
Earnings per common share − Fully diluted	$ **.79**	$ **.21**	$ **2.03**	$ **1.71**

FIGURE 11–6   Example of Computer/Printer output: Quarterly Report enhanced with boldface type (Courtesy of Comvestrix)

pb productions

Product	M	T	W	T	F	WT	WF	MTD	MTDF	YTD	YTDF
600 S1	12	7	23	17	27	86	80	255	290	3,217	3,120
600 M2	31	10	26	31	34	135	140	408	420	5,183	5,460
600 L3	2	14	17	8	13	54	50	171	150	2,064	1,950
Total 600 Series	**45**	**31**	**66**	**56**	**74**	**275**	**270**	**834**	**810**	**10,464**	**10,530**
1200 S1	18	19	10	24	15	84	60	303	190	2,471	2,340
1200 M2	25	33	30	17	39	143	120	388	360	4,701	4,680
1200 L3	5	15	13	7	11	51	70	182	210	2,598	2,730
Total 1200 Series	**48**	**67**	**53**	**48**	**65**	**278**	**250**	**773**	**760**	**9,771**	**9,750**
1800 S1	12	13	10	8	10	53	50	159	150	1,939	1,950
1800 M2	4	0	7	8	15	34	25	81	75	991	975
1800 L3	3	6	0	2	3	14	10	81	30	403	390
Total 1800 Series	**19**	**19**	**17**	**18**	**28**	**101**	**85**	**321**	**255**	**3,333**	**3,315**

Weekly Sales by Region

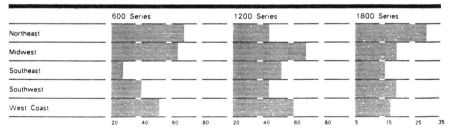

FIGURE 11-7   Example of Computer/Printer Formatted Sales Report, including Bar Graphs (Courtesy of Comvestrix)

ORAL PRESENTATIONS

Business and professional occupations involve talk. From the time you enter your office in the morning to the time you leave in the evening, you talk. Much of it is routine, but all of it reflects on you as a professional. Casual business conversations are extemporaneous. Formal ones require planning. Occasionally your position requires that you prepare a presentation for a meeting or a conference. Think of such

				DATE	INVOICE NO.	PAGE

TO: DUMMY DATA CORP.
1600 Tape Drive
White Plains, NY   12345
ADP Dept.
ATTN :   Mr. T. Ascii

DATE	INVOICE NO.	PAGE
03/21/81	11234 HS	1

PURCHASE ORDER #1234567890

TERMS: Net payable on receipt of invoice. To insure proper credit, return remittance advice with payment.

RUN DATE	JOB NO.	JOB DESCRIPTION	QUANTITY COPIES/IMPR.	SUB DESCRIPTION	UNIT PRICE	TOTAL
2/11	6123	XEROX 9700 SERVICES				
		DUM/ABCD tape #123				
			50/ 300	TOTAL 15000 IMPRESSIONS	.0420	630.00
			15000	VELO PAGES	.0030	45.00
		#124				
			50/ 100	TOTAL 5000 IMPRESSIONS	.0420	210.00
			5000	VELO PAGES	.0030	15.00
		#125				
			30/ 300	TOTAL 9000 IMPRESSIONS	.0420	378.00
			4500	3- HOLE PAGES	.0030	13.50
		#126				
			30/ 100	TOTAL 3000 IMPRESSIONS	.0420	126.00
			1500	3- HOLE PAGES	.0030	4.50
			1.0	SET-UPS	20.00	20.00
			1.50	CLERICAL HOURS	9.75	14.63
			30	VELO BINDS	3.00	90.00
			1	PICK-UPS	25.00	25.00
			1	DELIVERIES	25.00	25.00
				FORMS DESIGN		100.00
				JOB SUBTOTAL		1696.63
2/15	6215	XEROX 9700 SERVICES				
		DUM/INVENTORY tape #8896				
			30/ 10	TOTAL 300 IMPRESSIONS	.0420	12.60
			150	3- HOLE PAGES	.0030	.45
		#8897				
			30/ 50	TOTAL 1500 IMPRESSIONS	.0420	63.00
			750	3- HOLE PAGES	.0030	2.25
			1.0	SET-UPS	20.00	20.00
			2.00	CLERICAL HOURS	9.75	19.50
			1	PICK-UPS	25.00	25.00
			1	DELIVERIES	25.00	25.00
				JOB SUBTOTAL		167.80
				SUBTOTAL		1864.43
				SALES TAX		88.22

NET DUE: ➡   1952.65

FIGURE 11–8   Example of an Invoice Generated on Computer/Printer Equipment (Courtesy of Comvestrix)

COMPOSITE INVENTORY STOCK STATUS REPORT

Code	No.	Description										
JCC	417868	Glue - white, 3 oz	3.000	1.000	9.871		4.000	8.415	8.415			1,456
NZY	892303	Notebook - Composition	300	100	7.924		5.000	6.557	6.557			4,737
GSA	892340	Chalk - yellow, 12/box	4.000	1.000	11.753	4.000	5.000	10.831	10.831			922
LLN	672301	Ink, Stamp Pad - red	3.000	100	4.503	1.600	1.000	2.782	2.782			3,321
DTA	843713	Binders - 3 ring, blue	1.300	500	4.733		2.000	3.452	3.452			1,281
HMT	405607	Correction Fluid	1.200	100	3.450	500	2.000	3.513	3.413	1.000		437
MBY	515678	Lead - #9030-3H	5.000	1.000	12.222	4.000		6.430	6.430		3.000	12,792
FMA	309814	Calendar Refill	5.000	2.000	2.749	6.000	6.000	14.345	10.345	4.000		4,404
JBV	705706	Eraser - pink, #101	4.600	1.000	8.293		5.000	6.727	6.727		1.200	2,766
NSB	437758	Notebook - Steno	4.200	100	6.788	3.000	3.000	4.469	4.469			5,519
GMY	642102	Chalk - white, 12/box	4.000	1.000	13.455	1.000	3.000	12.456	12.456			1,993
KSA	409888	Ink, Stamp Pad - blk	4.150	300	7.063			5.833	3.833	2.000		1,230
DPR	690301	Binders - 3 ring, blk	1.300	500	3.567	1.200	2.000	3.567	3.567			0
GYA	725356	Clip Board	500	150	2.513			1.672	1.672			2,041
MAB	717444	Lead - #9030-2H	5.000	1.000	16.293	6.000		12.329	14.339	2.010		7,954
FGY	418328	Calendar Base	4.300	100	8.452		3.700	5.284	5.370	3.000		168
JBV	313804	Eraser, Chalk	2.500	500	6.751	2.496		4.370	4.370	1.000		3,877
NRA	832747	Marker - blue, grease	2.500	500	6.732			3.017	3.017			3,715
GMV	731890	Cement, Rubber - 4 oz	3.500	500	7.632		7.000	6.232	4.732	1.500		1,400
KNX	458374	Ink, Drawing - blk, 1 oz	5.200	500	9.323		1.000	9.316	9.316			7
DMT	074132	Rubber Band - #11, 4 lb	5.500	1.000	8.627	6.000		13.671	13.671			956
GSH	841039	Clips, Paper - Jum(100)	1.000	500	4.562		5.000	2.051	2.051			2,511
LMR	783439	Kimwipes - #900L	4.200	1.000	9.555		6.000	7.452	7.452			2,103
FGY	310293	Bookend - metal, 1 pr	4.600	200	5.624			5.024	5.024			0
HZQ	803417	Envelopes - Man., 9x12	3.000	1.000	7.439	8.000		11.952	6.452	5.000		3,487
MCZ	452362	Marker - red, grease	2.500	500	7.828			3.223	3.223			4,605
GBN	283684	Cards, File - 5x8(100)	3.000	1.000	8.403			6.415	6.415			1,988
KMR	673846	Glue - rubber, 4 oz	3.500	500	6.230		3.000	5.033	5.033			1,197
POC	415666	Opener, Letter	3.000	200	7.015			3.851	3.851			3,164
GSB	724833	Clips, Paper - #1(100)	5.000	1.000	18.415		10.000	15.722	15.722		3.010	4,703
LLP	648291	Kimwipes - 900s	5.300	1.000	12.410		5.000	8.732	8.732		3.400	3,678
FAE	783103	Blotter - white, 1 doz	2.500	300	6.239	900		3.039	3.039		374	4,474
HMX	308616	Envelopes - Man., 6x9	5.000	1.000	16.452		5.000	11.674	11.674			4,778
MCR	491383	Lead - #9030-5H	5.000	1.000	10.571	4.000	3.000	5.782	5.792	2.000		2,789
FMX	341298	Cards, File - 3x5(100)	3.000	1.000	8.403			12.303	14.303		2.000	3,973
JCC	417868	Glue - white, 3 oz	3.000	1.000	9.871	4.000		8.415	8.415		2.020	1,456

FIGURE 11-9 Example of Computer/Printer output without Enhancement (Courtesy of Comvestrix)

COMPOSITE INVENTORY STOCK STATUS REPORT

CAT NO	ITEM NUMBER	DESCRIPTION	REORDER POINT	REORDER QTY	OPENING STOCK	VENDOR RECEIPTS	STOCK ORDERS	COM STOCK	SALES ISSUES	BACK ORDERS	VENDOR RETURNS	STOCK TRANSFERS	CURRENT STOCK
JCC	417868	Glue - white, 3 oz	3.000	1.000	9.871	4.000	4.000	8.415	8.415				1.456
NZY	892303	Notebook - Composition	.300	.100	7.924			6.557	6.557				4.737
GSA	892340	Chalk - yellow, 12/box	4.000	1.000	11.753		5.000	10.831	10.831				.922
LLN	672301	Ink, Stamp Pad - red	3.000	.100	4.503	1.600		2.782	2.782				3.321
DTA	843713	Binders - 3 ring,blue	1.300	.500	4.733		1.000	3.452	3.452				1.281
HMT	405607	Correction Fluid	1.200	.100	3.450	.500	2.000	4.413	3.413	1.000			.437
MBY	515678	Lead - #9030-3H	5.000	1.000	2.222	4.000		6.430	6.430			3.000	1.792
FMA	309814	Calendar Refill	5.000	2.000	2.749	6.000		14.345	10.345	4.000			4.404
JBV	705706	Eraser - pink. #101	4.600	1.000	8.293		5.000	6.727	6.727			1.200	2.766
NSB	437758	Notebook - Steno	4.200	.100	6.788	3.000	3.000	4.469	4.469				5.519
GMY	642102	Chalk - white, 12/box	4.000	1.000	13.455	1.000	3.000	12.456	12.456				1.999
KSA	409888	Ink, Stamp Pad - blk	4.150	.300	7.063		3.000	5.833	3.833	2.000			1.230
DPR	690301	Binders - 3 ring. blk	1.300	.500	3.567	1.200	2.000	3.567	3.567				0
GYA	725356	Clip Board	.500	.150	2.513	6.000		1.672	1.672				2.041
MAB	717444	Lead - #9030-2H	5.000	1.000	16.293			14.339	12.329	2.010			7.954
FGY	418328	Calendar Base	4.300	.100	8.452	2.496	3.700	8.284	5.284	3.000			168
JBV	313804	Eraser, Chalk	2.500	.500	6.751			5.370	4.370	1.000			3.877
NRA	832747	Marker - blue, grease	2.500	.500	6.732			3.017	3.017				3.715
GMV	731890	Cement, Rubber - 4 oz	3.500	.500	7.632		7.000	6.232	4.732	1.500			1.400
KNX	458374	Ink, Drawing - blk, 1 oz	5.200	.500	9.323		1.000	9.316	9.316				7
DMT	074132	Rubber Band - #11, 4 lb	5.500	1.000	8.627	6.000		13.671	13.671				956
GSH	841039	Clips, Paper - Jum(100)	1.000	.500	4.562		5.000	2.051	2.051				2.511
LMR	783439	Kimwipes - #900L	4.200	1.000	9.555		6.000	7.452	7.452				2.103
FGY	310293	Bookend - metal, 1 pr.	4.600	.200	5.624			5.024	5.024				0
HZQ	803417	Envelopes - Man., 9x12	3.000	1.000	7.439	8.000		11.952	6.452	5.000			3.487
MCZ	452362	Marker - red, grease	2.500	.500	7.828			3.223	3.223				4.605
GBN	283684	Cards, File - 5x8(100)	3.000	1.000	8.403		3.000	6.415	6.415				1.988
KMR	673846	Glue - rubber, 4 oz	3.500	.500	6.230			5.033	5.033				1.197
POC	415666	Opener, Letter	3.000	.200	7.015			3.851	3.851				3.164
GSB	724833	Clips, Paper - #1(100)	5.000	1.000	18.415		10.000	15.722	15.722			3.010	4.703
LLP	648291	Kimwipes - #900s	5.300	1.000	12.410		5.000	8.732	8.732			3.400	3.678
FAE	783103	Blotter - white, 1 doz	2.500	.300	6.239	900	5.000	3.039	3.039			.374	4.474
HMX	308616	Envelopes - Man., 6x9	5.000	1.000	16.452			11.674	11.674				4.778
MCR	491383	Lead - #9030-5H	5.000	1.000	10.571	4.000	3.000	5.782	5.782		2.000		2.789
FMX	341298	Cards, File - 3x5(100)	3.000	1.000	8.403		4.000	14.303	12.303	2.000	2.020		3.973
JCC	417868	Glue - white, 3 oz	3.000	1.000	9.871			8.415	8.415				1.456

FIGURE 11–10   Example of Computer/Printer output with Enhancing Lines and Type (Courtesy of Comvestrix)

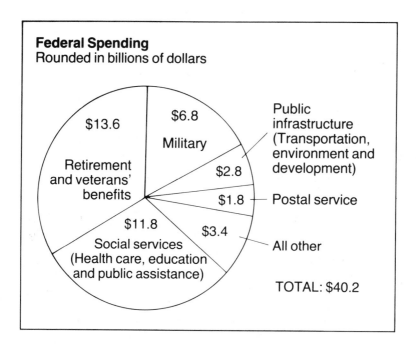

FIGURE 11–11   Example of Pie Chart: A Commonly Used Visual Aid in Oral
Presentations

gatherings as occasions of concentrated talk. Do your homework, and if you prepare,
you can gain confidence because the material is fresh in your mind, and it is news to
the audience. During the presentation speak slowly, clearly, and naturally.

Types of Oral Presentation

Routine office situations are the oral transactions that you think the least about
since most of them are extemporaneous. Some consideration of your approach to
daily business conversation, instructions to co-workers or subordinates, and casual
comments in passing will give you something to say so that you avoid awkward or
inappropriate comments. It is up to you to develop a style of conversation that fits
your personality and accommodates the demands of the workplace. In the office
conversation should demonstrate courtesy and professionalism. Remember that you
have to return to work, so speak to others with interest and understanding. Tact helps,
too.

The telephone   This is a major medium for much of the conversation you have
in the office. In several of its publications on an effective approach to telephone use,
telephone companies suggest that you:

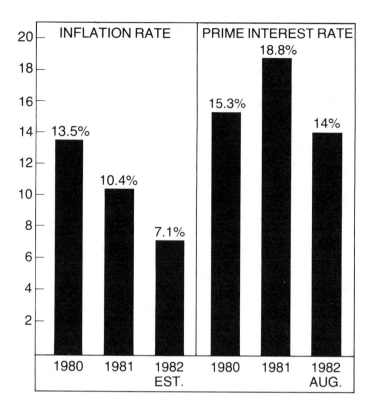

FIGURE 11–12   Example of Bar Chart: Another Commonly Used Visual Aid in
Formal Presentations

- be alert
- use simple language to be natural, and avoid slang or technical terms
- let your voice show emphasis and vitality, but not too loud, or fast
- pronounce words clearly and carefully to be expressive
- show interest, and use the other person's name occasionally to be pleasant
- use tact and good manners to be courteous

When calling, you present a more positive and active impression if you yourself call. This approach is also more personal and efficient since it guarantees that you are on the line, ready to talk. Clients and associates may be startled if you call directly, rather than through a secretary, but it leaves a favorable impression. Before you dial, though, gather all the papers and correspondence you will need so you can be ready to talk when the right person answers.

Identify yourself, even if the person knows your voice. Do them the favor of anticipating that they are busy and may not recognize you immediately. If they are not

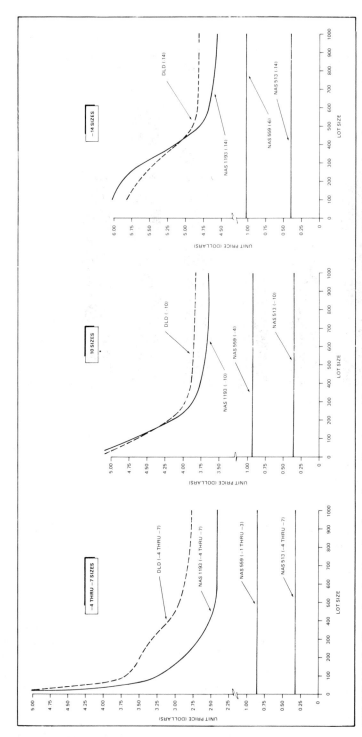

FIGURE 11–13 Example of a Line Graph, or Curve, Used as a Visual Aid in Presentations (Used with permission from Grumman Aerospace Corporation)

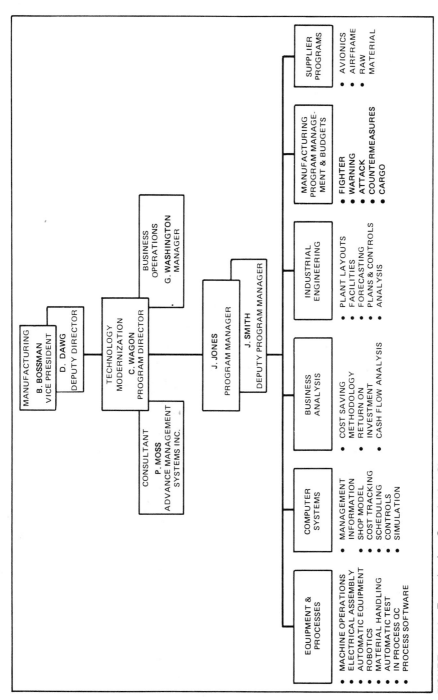

FIGURE 11-14 Example of an Organization Chart, (Used with permission from Grumman Aerospace Corporation)

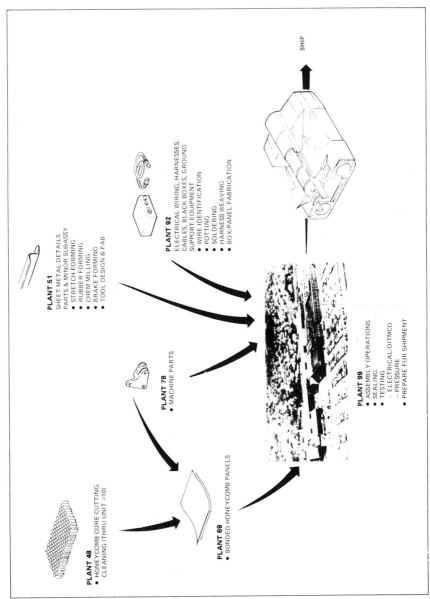

FIGURE 11-15   Example of a Flow Diagram Used to Describe the Interaction of parts in a Process (Used with permission from Grumman Aerospace Corporation)

PLANT 48
• HONEYCOMB CORE CUTTING, CLEANING (THRU UNIT =10)

PLANT 69
• BONDED HONEYCOMB PANELS

PLANT 78
• MACHINE PARTS

PLANT 51
SHEET METAL DETAILS
PARTS & MINOR SUBASSY
• STRETCH FORMING
• RUBBER FORMING
• CHEM MILLING
• BRAKE FORMING
• TOOL DESIGN & FAB

PLANT 92
ELECTRICAL WIRING, HARNESSES,
CABLES, BLACK BOXES, GROUND
SUPPORT EQUIPMENT
• WIRE IDENTIFICATION
• POTTING
• SOLDERING
• HARNESS WEAVING
• BOX/PANEL FABRICATION

PLANT 99
• ASSEMBLY OPERATIONS
• SEALING
• TESTING
  – ELECTRICAL/DITMCO
  – PRESSURE
• PREPARE FOR SHIPMENT

SHIP

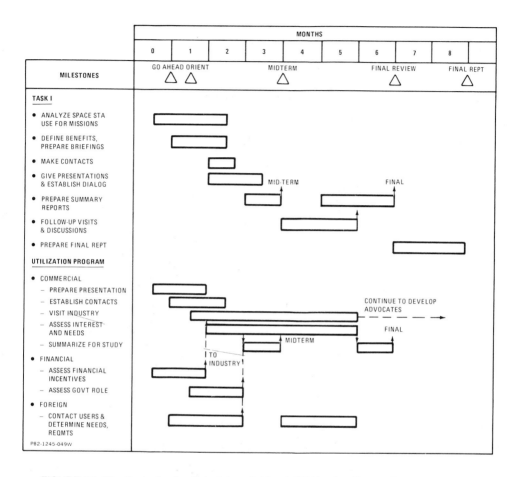

FIGURE 11–16 Example of a Schedule: A Visual Aid Used to Denote Time Elements of a Process (Used with permission from Grumman Aerospace Corporation)

in, leave a time for them to reach you and where to call. If you offer to call back, be sure to do so when you promised. Above all, on the telephone, or in person, be yourself.

Informal social situations. These allow you more time than do routine ones to consider your comments. You might think about what you would say in introducing a visitor to your office, addressing a weekly staff meeting, or talking casually during a break at a long conference. No one can provide you with lines appropriate for these occasions, but a flexible, courteous attitude towards others can provide positive impressions. If someone asks, for instance, about a recent football game, don't cut off their opening gesture by saying that you hate organized sports. Instead, ask them what they found interesting in it. Dick Cavett, in commenting on conversation improvement, advised that you begin with the word "you" to show your interest in the other

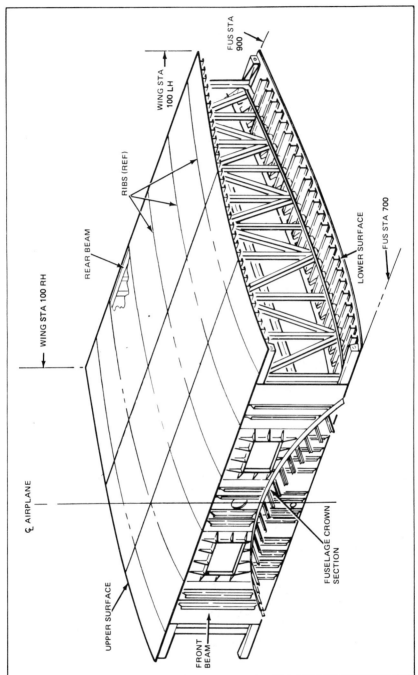

FIGURE 11–17 Example of a Line Drawing: A Visual Aid Used to Clarify Text Discussion (Used with permission from Grumman Aerospace Corporation)

person. Almost everyone likes to talk about their own experiences, and your interest in another's activities helps to start a conversation and to keep it going. In conversation, be courteous and natural. Phoniness rings like a cracked bell; everyone can hear the false notes.

Brief presentations.   These usually communicate formal messages, and for that reason demand more preparation. Several situations may require you to outline your comments in advance:

- a short welcoming speech at a meeting, conference or dinner
- formal introductions of people to a group
- comment at meetings or briefings

Because they are scheduled in advance, each of these occasions gives you some time to sketch and polish your remarks. To overcome stage fright, outline what you want to say on small note cards. Put one item on a card and then arrange the cards in an order that you consider understandable for the audience. Once you have done that, put the cards in your pocket when speaking, and refer to them only if necessary. Writing down your thoughts, arranging the cards, and having them in your pocket, if you need them, should provide you with confidence and allow your remarks to flow naturally.

Formal presentations.   These demand planning and require still more organization. Whether your presentation is a sales pitch to a client or a discussion at a professional conference, your audience expects and deserves a planned program. For instance, if you begin to talk and get to a crucial point that you had decided was best presented with a chart, and then discover that you had left off critical information that might appear slightly disorganized to the audience. Such omissions do occur, and the more experience you have with "platform skills," the better you will be able to fill in time or talk around a problem that results from incomplete preparation, or hitches in program coordination.

APPROACHES TO ORAL PRESENTATION

Whether the presentation you must give is brief or elaborate, a clear approach to it and a method for preparation should make the work easier. You can present information before a group in at least four ways:

- reading word for word from the text of a paper
- reciting lines that you have memorized
- improvising your remarks
- planning a natural speech

Reading directly from a paper.   This technique often disappoints a group. Many people have difficulty absorbing information that way, unless the paper was written with the intention of being a speech and not a paper. Though this approach to a presentation has undesirable qualities, often it cannot be avoided. For example, if you must make a statement to the press, you often provide reporters with a text of your remarks as a courtesy. In this case the text is usually read verbatum.

Reciting a memorized speech.   Unless you are an actor, trained in the delivery of lines to give the appearance of spontaneity, this approach make you appear wooden and lifeless. Memorizing the speech can actually exacerbate the anxiety associated with speaking before a group. In selecting this approach you always think of what happens if you forget your lines. You certainly do not want to start again when you stumble over your lines, like the laughable handyman in Shakespeare's *A Midsummer Night's Dream*. Doing that calls attention to your error. Only if you have had some acting experience, and were mildly successful with it, should you attempt to deliver a memorized speech.

Improvising, or speaking off the cuff.   If you have prepared a general outline of what you have to say this can be effective. Prepared improvisation may seem like a contradiction, but it is not really. The difference is only the amount of time for planning. Use the time you have to think about your remarks. The most famous improvised speech was Lincoln's Gettysberg address, the text written on an envelope as the President's train made its way to the battlefield cemetery. No one, of course, expects your comments to go down in the history books, but that does not mean that you cannot think about what you want to say during the time you have.

Planned extemporaneousness.   This is the quality that best characterizes effective oral presentation. In other words, you carefully select and arrange the content of your presentation, but give to it a naturalness and spontaneity that draws you and your audience together.

PREPARING A PRESENTATION

Any presentation, read, memorized, extemporaneously planned, or improvised, benefits greatly from preparation. Understanding the four stages of the process should help you create your presentation efficiently: plan, set-up, perform, close.

Planning

Think about the oral presentation as having three distinct sections or parts—the hook, the discussion, the close. These correspond roughly to the broad sections of any piece of writing—the beginning, middle, end.

The hook captures the attention of the audience. You might look at the beginnings of movies and television programs for examples of dramatic hooks. Although you won't need to entertain, like those programs do, you do need to capture the audience by stimulating their curiosity. A well-phrased question—Why has XYW Corp. consistently outperformed its competitors for the last four quarters?—can hook the audience by creating their desire for the answer. A startling example, figure, or statistic—such as, "Americans spend more on dog food than on aid to the disabled"—can stimulate the audience's curiosity by luring them to expect the answer in the presentation.

The discussion presents the real meat of the presentation. This section provides the facts, examples, ideas that prompted you to bring the people there in the first place. *The close* offers a conclusion so people finish feeling that something has been said or accomplished.

When planning, follow the steps you would in writing, that is:

- determine the purpose of the presentation and analyze the needs of the audience
- use a storyboard to outline the material
- write a script
- rehearse

Determine the purpose, analyze the audience.   The purpose of the oral presentation can usually be expressed in a few sentences. The time allotted affects your planning of an oral presentation. In front of a group, five to 10 minutes provide sufficient time to get a point across, but require you to concentrate and condense information. Ten to 15 minutes is really a difficult time since you can present only a limited amount of substantiation and examples to support the idea. Half an hour to an hour is really a full-scale production and requires a great deal of planning and attention.

When analyzing the audience for a presentation, consider, as you did for writing, who they are and what they are interested in. The size of the audience is an important factor in preparing a presentation. Some techniques work better with a group of 15 than with one of 500. What might work very well with a small group can be lost with a large one. In presenting information:

- net the audience with a story, either real or imagined
- go easy on jokes, unless you have a natural talent
- include interesting facts, examples, statistics, quotations
- use visual aids for a purpose, not for their own sake
- create a picture in their minds

Use a storyboard to outline the material.   Approach this task by first selecting a sequence that fits with your message, your goals, and the audience involved. Sequences can, like the principles of arranging written information, follow one or a combination of these: priority, topic, time, comparision and contrast, cause and effect, narrative, inductive or deductive reasoning.

The storyboard, which we discussed in relation to proposals (Chapter 9), works well in organizing and preparing the parts and sections of a presentation. The storyboard allows you to combine the text and graphic material you have in an orderly way. It allows you to coordinate the visual aids that you will use with text and comments.

Business and government agencies more frequently require presentations in place of short reports. The storyboard technique helps you organize the text of your presentation with the appropriate visual aids. Such preparation allows you to perform with planned extemporaneousness.

For example, if you were to make a simple five- to 10 minute-oral presentation of the Checklist for Meeting Preparation, you could write a script for it that would incorporate 14 viewgraphs. A modified storyboard/planning sheet allows you to show both the illustrations and the text of the explanation that accompanies each viewgraph (Fig. 11–18).

Write a script of your comments.   Sometimes, especially with more difficult and complicated information, you might write out what you think you should say to see how the unarticulated thoughts "sound." This practice helps you to focus the information, and to select words, phrases, and sentences that condense information for the reader.

In doing this, keep in mind that you write down the material, not to memorize your lines, but to discover ways to express thoughts efficiently.

CHECKLIST FOR MEETING PREPARATION

_____ Have you reminded the people you wish to attend of the time and place of the meeting?

_____ Have you checked the room for proper ventilation, light, enough seating and tables?

_____ Is the room clean and set up appropriately?

_____ Is the flipchart or chalk board clean? Enough chalk, markers, and erasers?

_____ If you plan to use film or slides, is the projector ready, tested for sound and focus? Have you arranged the slides in the magazine? Provided a spare bulb, or spare projector?

_____ Have you arranged handout material in sequence?

_____ Have you left instructions for secretaries to hold messages and calls until the break, or the end of the meeting?

_____ Have you prepared charts and other material in advance that you plan to use in the session?

_____ If you use a tape or video player, have you tested it and made sure the cassette is set at the beginning of the material that you will use?

_____ Have you arranged for a coffee break?

_____ Are pencils, pads and other materials on hand for use?

Graphic Type _____

Text
Function _____

Graphic
Number

_____

Key Point:

Graphic Type _____

Text
Function _____

Graphic
Number

_____

Key Point:

Storyboard/Presentation Planning Sheet   # _I_ of _7_

Graphic Type _Vu- Graph_

Text   _Title +_
Function _Intro. focus_

Graphic
Number
①

Checklist:

Preparing
for
Meetings

Key Point:

• Provides the
  lecture title

• Functions as
  visual focus for
  Introduction

Graphic Type _Vu-Graph_

Text
Function _outline_

Graphic
Number
②

Preparing for meetings
• Participants
• Facilities
• Presentation Aids
• Amenities
• The Check List

Key Point:
The lecture will
cover four
items necessary
for successful
meetings

Graphic Type _Vu-Graph_

Text Function _Focus on Point #1_

Graphic Number ③

Preparing for meetings
• Participants
    (Highlight with color)

Key Point:
Meetings involve people, so make sure they attend.

Graphic Type _Vu-Graph_

Text Function _Enhance Point #1_

Graphic Number ④

• Participants
— call them
— remind of time
— give clear location

Key Point:
Busy people often need a gentle reminder.

Storyboard/Presentation Planning Sheet   # __3__ of __7__

Graphic Type _Vu-Graph_

Text Function _Focus on Point #2_

Graphic Number ⑤

Preparing for meetings
• Participants
• Facilities (Highlight with color)

Key Point:
Meetings require appropriate and adequate space and equipment

---

Graphic Type _Vu-Graph_

Text Function _Enhance Point #2_

Graphic Number ⑥

• Facilities
— room cleaned
— proper ventilation
— enough seating and tables appropriately arranged
— clean flip chart or board

Key Point:
Plan, because people remember overheated rooms, crowding, and dirt.

Graphic Type _Vu- Graph_

Text Function _Focus on Point #3_

Graphic Number
⑦

Preparing for meetings
• Participants
• Facilities
• Presentation Aids
(Highlight with color)

Key Point:
Aids should speed The presentation of ideas, not detract from Them.

Graphic Type _Vu- Graph_

Text Function _Enhance Point #3_

Graphic Number
⑧

• Presentation aids
– charts prepared
– chalkboard / flipchart
– tested projectors
  • spare bulbs
  • slides checked
– tape player set at volume and position
– handouts in sequence

Key Point:
Aids demonstrate your mastery of The material, your planning, and your respect for The participants time.

Graphic Type _Vu- Graph_

Text Function _Focus on Point #4_

Graphic Number
⑨

Preparing for meetings
• Participants
• Facilities
• Presentation aids
• Amenities
(Highlight with color)

Key Point:
People remember more Than just The content of discussion

Graphic Type _Vu – Graph_

Text Function _Enhance Point #4_

Graphic Number
⑩

• Amenities
– hold calls and interruptions
– coffee break
– Pencils, pads, hand out materials

Key Point:
Planned details create an atmosphere of professionalism

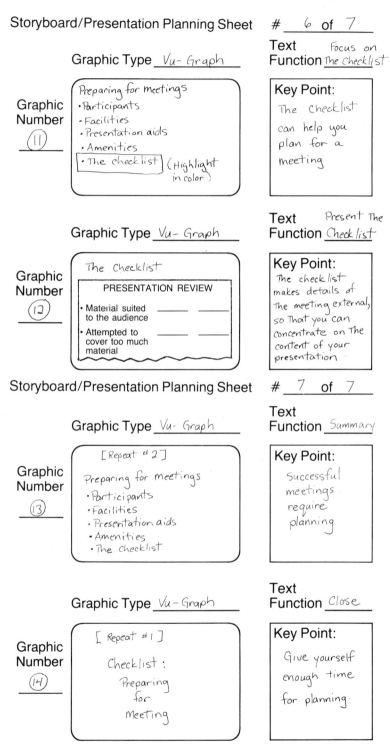

Graphic Type _Vu- Graph_

Text
Function _Focus on The Checklist_

Graphic
Number
⑪

Preparing for meetings
• Participants
• Facilities
• Presentation aids
• Amenities
• The checklist (Highlight in color)

Key Point:
The checklist can help you plan for a meeting

Graphic Type _Vu- Graph_

Text
Function _Present The Checklist_

Graphic
Number
⑫

The Checklist

PRESENTATION REVIEW

• Material suited
  to the audience   _____ _____

• Attempted to
  cover too much
  material   _____ _____

Key Point:
The checklist makes details of The meeting external, so That you can concentrate on The content of your presentation

Storyboard/Presentation Planning Sheet   #   7 of 7

Graphic Type _Vu- Graph_

Text
Function _Summary_

Graphic
Number
⑬

[Repeat #2]

Preparing for meetings
• Participants
• Facilities
• Presentation aids
• Amenities
• The checklist

Key Point:
Successful meetings require planning

Graphic Type _Vu- Graph_

Text
Function _Close_

Graphic
Number
⑭

[Repeat #1]

Checklist :
    Preparing
       for
    Meeting

Key Point:
Give yourself enough time for planning.

FIGURE 11–18   Example of Storyboard Sequence Used to Develop the Structure of a Lecture or Sales Presentation

Writing a script for the presentation also will reveal the material you have and what you need. In addition it will uncover language that might be too technical and obscure. You get better results if you keep language simple, but not simplistic. Use difficult words sparingly in speaking, and limit the length of sentences. Remember that the audience listens to your voice, so speak naturally and distinctly.

Rehearse   Make a practice of going over the material in advance. You can rehearse by yourself before a mirror, use a tape recorder, or a video recorder to enable you to look at your performance critically. If you have colleagues, give the presentation to them before the actual performance. Be sure that they will give you an honest appraisal and offer you useful suggestions for improvement. Many professional organizations, in their guidelines to members who must make presentations at conventions and conferences, urge them to rehearse before a critical audience so that the convention presentation will go smoothly and within the time limits. If you can do so, a critical rehearsal is a beneficial practice.

Set-Up

Some of the items in setting up the presentation might send you back to the initial steps. Oral presentations are the products of dynamic efforts that require you to rethink, rearrange and reevaluate material. Welcome the opportunity to make adjustments, changes, and major renovations. In setting up, consider:

- the time of day the talk is scheduled
- the shape and size of the room
- the number of people in the audience
- the facilities (screens, projectors) available
- refreshments, if necessary
- material to be prepared as handouts
- charts, graphs, slides, pictures, that you will use

Arrange, if you can, to rehearse in the room where the meeting will be held. If that is not possible, at least arrange to look at it beforehand. The effect of this makes you comfortable with the physical setting so that does not contribute to any jitters you might have before the actual performance.

Perform

Professional actors work very hard on their stage appearance, and no one expects your oral presentation to fall into that category. But an audience is entitled to a planned, professional performance. Above all, be natural. Orchestrated movements make you appear as if you were a puppet being manipulated by strings. Instead, make contact with your audience. Use your eyes to single out people in the group and talk to them. Look them in the eye. If that makes you uncomfortable, try looking at an imaginary spot on someone's forehead. This way they will get the impression you are looking into their eyes. Eye contact is essential to building a strong bond with the people you address. In turn, their response to you is much more positive.

When speaking, modulate your speech, pronounce words clearly, and express yourself enthusiastically. Genuine enthusiasm is contagious and helps to involve the listeners in your material. The Greek root of enthusiasm means "to be inspired by a god." If you are able to communicate your passion for your subject, then you can pass that inspiration on to others.

Also, your body movements communicate feelings and unspoken messages to the audience. Make sure your gestures are appropriate, and natural. I once listened to a lecturer who had the annoying habit of making a broad, sweeping arc with his left hand at the end of each sentence. I cannot recall the content of the lectures, only the annoying gesture. If you want people to remember the content, make appropriate and natural gestures.

Be sure, too, that your appearance is professional. Dress properly for an occasion, depending on the audience and your own good judgment. The language you use should be professional, no cliches. Avoid jokes unless you are a natural wit. If you must tell a joke, tell it. Don't waste time with a "funny thing happened on the way to the office" build-up.

Use language to indicate to your audience that you will move on to another part of the material. You can do this with pat words like "Next," and phrases like "On the other hand." You can also solicit questions about material you have just covered as a way to signal that you are done with that part and wish to get on to the next. You can signal a transition by referring to a visual aid.

In using visual aids, introduce them, read them to the audience, making sure the material on them corresponds exactly with what you say, and allow the audience enough time to absorb the figure, slide, or chart.

Close

After the presentation, make sure you close the session yourself. In relationships, this is called termination. No matter how you look at it, you must end the gathering. So

- leave the group with something, a parting thought—a brief summary, conclusion, quotation, memorable statement
- tie up the loose ends
- terminate with professionalism, not flashiness.

You can lead the way out.

Fielding questions   Probably the most difficult part of any presentation is fielding questions. You might try to make control of the way questions come by establishing some ground rules. You can answer questions in three different ways:

- as you go
- at the end of the presentation
- at breaks between sections of material.

Of these, the first might prove the least satisfying, since you not only run the risk of constant interruption but also of answering a question that you will address later in the talk.

If you do not want questions during the lecture part of your presentation, say so either in the beginning with something like, "I'll be happy to answer questions at the end of the presentation," or "I'll invite your questions several times during the talk." Another less obvious way to handle this is to wait for the first question, and then say that you might answer it during the course of the presentation and, if not, encourage the person to ask once more at the end. Remember the person who asked, and later ask: "Did I answer your question?" Usually they give you a nod, or ask the question again.

Often you might prepare yourself to handle the member of the audience who has a "hidden agenda." That person focuses on an opening in your material as a platform for his or her own lecture. The best way to handle this situation might be a "limp response." Don't argue or engage in a battle, just say that the point is well taken, and then get on with your material. Once you discuss a hidden agenda, you have given that person control over the rest of your presentation. Maintain the control you have, but do it with tact.

If you cannot answer a question, say so. But suggest to them that you will seek an answer, or that you know where it may be found. No stigma haunts a lack of knowledge; only a lack of energy to find the answer.

Also in answering, be courteous. If the question is dumb, be understanding, and don't turn the situation into a joke. If you ridicule any member of the audience, all of them may turn against you. Make no jokes, slurs, or snide remarks. Once you do, you give up the right to courtesy, and any vultures in the group can snip at you at will.

Another way to field questions that are serious, but of limited interest to the whole group, is to ask the person to join you after the presentation. As with the presentation as a whole, answer questions honestly, directly and clearly.

The politician's tactic of "not answering" a question, but launching into their own ideas that may be indirectly related to the general topic, might work well on the stump, but it is a sure way to anger and annoy the people present at your meeting. It suggests that you may have something to hide and are avoiding the issue for that reason.

EVALUATION CHECKLIST

After each presentation, make a practice of a critical review of what you did. Find strengths to repeat next time, and weaknesses to avoid or change.

Try answering the Presentation Review questions as a way to evaluate your own performance. The checklist Rating an Oral Presentation provides an instrument you can use to examine another's presentation performance.

PRESENTATION REVIEW

	Yes	No
• Material suited to the audience	____	____
• Attempted to cover too much material	____	____
• Depended too much on notes	____	____
• Felt prepared	____	____
• Spoke clearly	____	____
• Read extensively from notes	____	____
• Stumbled over words	____	____
• Discussed visual aids adequately	____	____
• Used all of the graphics prepared	____	____
• Covered the visual display after its use	____	____
• Audience understood the graphics quickly	____	____
• Hooked the audience	____	____
• Made eye contact	____	____
• Held their attention	____	____
• Finished on time	____	____
• Fielded questions professionally	____	____

RATING AN ORAL PRESENTATION

For each category, decide if items are average (0), above average (+), or below average (−)

	−	0	+
1. Organization: first things first, purpose stated clearly, audience addressed appropriately	____	____	____
2. Language: direct, simple, clearly delivered, quickly understandable	____	____	____
3. General Impression of the Information: convincing; enough data and examples	____	____	____
4. Visual Aids: each adds to audience understanding, appropriate for facilities	____	____	____
5. Mechanics of Presentation: voice, eye contact, gestures, body language	____	____	____
6. General Impression of the Presentation	____	____	____
TOTAL	____	____	____

VISUAL AIDS FOR PRESENTATIONS

Your presentation gains power and impact with carefully designed and clearly labeled visual aids. The number of ways to present ideas visually may at first put you off. Just a list of them suggests that no one person could, or should, use them all in every presentation:

- handouts
- flipcharts
- chalkboard (whiteboard)
- viewgraph (overhead projector)
- slides
- film
- videotape
- audio tape and records

Each of these media can serve well in a presentation, but you must decide what you want to demonstrate and emphasize with them. Select one that suits your purpose. Table 11–2 will help you decide what is appropriate for your material. It lists several

Table 11–2 Visuals for Presentations

Type	Applications	Drawbacks
Handouts: • copies of the text • outline of presentation • copy of charts, slides, data, statistics, tables	• show planning • provide illustration • allow audience to focus on presentation and not on taking notes • gives audience a reference to the material in the presentation	• can distract the audience if you distribute the printed matter too early
Charts, Graphs, Tables	• allow the condensed presentation of data, statistics, trends, comparisons, contrasts	• might require some professional help • useful in small groups • if inappropriate, can distract the audience and confuse them
Flipchart	• allows you to prepare drawings, illustrations and sketches in advance • provides a portable writing surface in rooms without adequate facilities • useful for brainstorming sessions	• not effective for medium or large groups • distracting if an example is not covered after use

Table 11-2 Continued

Type	Applications	Drawbacks
Chalkboard (whiteboard)	• excellent for open-ended discussion, demonstrations that require audience participation, and brainstorming	• takes time away from your talk as your write information down
Vu-graph (overhead projector)	• allows a quick way to reproduce for projection drawings and even pages of text	• ineffective with large groups • highly detailed drawings come off poorly • draws attention away from you • dimmed lights can make an audience less attentive
Slides	• excellent reproduction of detailed drawings, artwork, and photographs • suited to larger rooms • useful in specialized subjects that change rapidly	• more time to prepare • slightly more expensive • dimmed lights make an audience drowsy
Film	• useful for repeated presentations • dramatic • entertaining • informative • appropriate for general treatment of material that does not change too often	• expensive to make • requires equipment and a technician • can waste time if it is not appropriate • lights need to be dimmed
Videotape	• more portable and easier to work with for a non-professional • dramatic, informative, entertaining • appropriate for general and specialized topics that might change rapidly • captures attention	• too similar to TV • need a large screen for large groups • can appear amateurish
Audio tape recordings	• provides the speech of others; the pacing and delivery • brings in the people without their presence • stimulates the imagination like radio	• provides no visual focus without accompanying slides or charts • can tire an audience

aids and the applications and drawbacks of each. The choice of visual aids is subjective. Of course, choose a medium that you are comfortable with, but also one that is appropriate. Some basic points to remember when you use a visual aid:

- continue to face the audience, and talk to them, not the image on the screen or chart
- display the aid when you discuss it, not before
- when you are through with it, cover it up, or turn up the lights
- turn up the lights as soon as you finish with slides, films, videotapes, and viewgraphs
- read the material on the visual verbatim in an effort not to confuse the audience
- pass texts and other handouts around as needed, or at the end

Often a picture can misrepresent information. And graphics and visuals can distort data. Figure 11–19 is an example of a visual representation that a letter in *The New York Times* described as "cleverly manipulated to distort the truth." The letter comments further:

> The chart entitled "Your Taxes" makes the President's tax plan look better in comparison with the Democrats' plan than it really is. There is no scale on the vertical axis of the chart and, although the red space between the lines depicting what President Reagan called "the tax money that will remain in your pockets" if the plan is passed covers a very large area, it is impossible to determine how much money it actually represents. . . .
>
> (Letter, Adam Saravay, *The New York Times,* 8/6/81, p. A22. By permission of Adam Saravay)

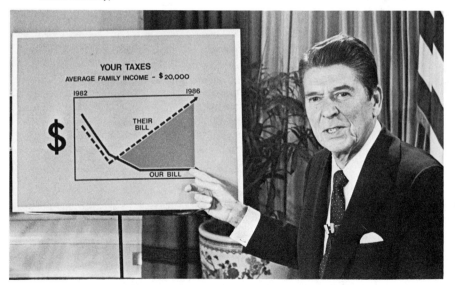

FIGURE 11–19   One "Letter to the Editor" in *The New York Times* described this chart as "cleverly manipulated to distort the truth."

© 1982 *The New York Times.* Photo by Geroge Tames, NYT Pictures.
Photo used with permission. Letter portion used by permission of Adam Saravay)

Of course, this kind of subtle misrepresentation can remain undetected. Even understated deception, once uncovered, can turn your audience against you. It undermines you as a credible source of information. Your audience will begin to suspect other information in your presentation. Honest, straightforward visual aids contribute to a strong presentation and reflect well on you and your organization.

SUMMARY

Graphics

- Good graphic illustrations concentrate on a particular point, develop a central idea, use simple language and format, use clear labels and titles, and carry their own weight.
- An effective graphic demonstrates focus, enhancement, and captioning.
- Each type of representation—pie, bar, line, flow chart, organizational chart, schedule, drawings and photographs—has strong applications and drawbacks. Determine what you want to write and select an appropriate graphic.
- Place a graphic in the text itself as close as possible to the words that refer to it. An appendix or insert section is also appropriate for graphics.
- Whether you make your own or have others create them for you, start early.
- Computers generate many routine graphics and reproduce them through electronic printing devices.

Oral Presentations

- In routine conversations in the office and on the telephone, be alert, natural, expressive, distinct, pleasant, and courteous.
- For short presentations, plan what you will say, and maintain a courteous and natural approach.
- You can present information before a group by reading from a text, memorizing and reciting a text, improvising, or planning a natural speech. Planned extemporaneousness achieves positive responses in oral presentations.
- A presentation should hook the audience, discuss the topic, and close. In planning, determine the purpose, analyze the needs of the audience, outline the material, write a draft or script, and rehearse. A storyboard offers a dynamic planning tool for presentations.
- Set-up for the meeting. Consider the lay-out of the room, the number of people you expect, equipment you need, charts, and other visual aids.
- In performing, be natural, make eye contact, modulate your voice, make natural gestures and body movements, use simple and professional language, and field questions courteously and honestly. Select appropriate visual aids for your audience and for the information you wish to communicate.
- At the end, tie up loose ends, leave the audience with a summary or conclusion, and terminate the session professionally. After the presentation, evaluate your own performance.

EXERCISES

1. Select a report that presents a great many numbers and figures in the text. See if you could present the same information graphically with charts, curves, tables, or drawings.

2. From material you have received in the mail, from newspapers, or on the job, select five graphics that you consider well done, and five that seem more a visual hinderance than an aid. Write a brief evaluation comment on each and submit the examples and comments as a report. Be sure to include your own conclusions about the quality of the visual aids.

3. Find three examples of computer generated graphics and three conventional ones of the same type. Compare their similarities and differences. Find out the production costs of each, and submit a report of your findings in which you suggest whether or not your company should invest in computer graphics.

4. Take statistical data that you have gathered, or that you use from a source such as the Bureau of Labor Statistics, and build an appropriate graphic. Write a brief explanation of what you did, why you selected the format you did, and whether the finished representation fulfilled the goals you had in the beginning.

5. Attend a sales or marketing presentation in your city. Note the way the material is arranged, performed, and supported with visual aids and other material. Write an evaluation of what you have seen, determining qualities that should be incorporated as desirable, and identifying points that you should avoid.

6. Prepare remarks that you would make in introducing a respected professional you have invited to address a meeting.

7. You have been asked to brief a meeting of several groups that are involved with different aspects of a common project. Make your remarks an overview of the project that launches a discussion of the next phase of the effort.

8. Using a long report that you have written, prepare an oral presentation. Decide what visual aids, handouts, and other material you might need to get the information across clearly.

9. A group of visitors from another company has come to your headquarters. You have been asked to welcome them to your area of the company and to say a few words at the lunch break in the cafeteria. Plan these remarks, and draft them in sufficient detail so that someone else could, if necessary, make the speech.

10. Observe closely someone you believe to be a superior presenter. Isolate the qualities and behavior patterns you feel make that person an excellent performer. Determine if those acts can be learned, and if so, how you would go about acquiring them.

12

Writing in the Automated Office

A full page ad in *The New York Times* carried this caption: "Every day the American businessman travels to the 19th Century." The picture was of a typist before an antique typewriter and a clerk filing at an oak cabinet, both dressed as if they had escaped the pages of a Charles Dickens' novel. The graphically presented point, of course, implies that few, if any, improvements have occurred in the way we manage the essential product of an office—paper. The ad introduced new equipment to automate the office.

OFFICE AUTOMATION AND WRITING

The emergence of *OA,* a quarterly supplement to *Computerworld* devoted to office automation, demonstrates the growing attention directed toward office technology. According to *OA* editors, office automation as a concept did not exist five years before its first issue in 1982. The supplement concentrates on ways of integrating into the office, equipment that will help produce more efficient information handling, a function traditional and essential to the office.

Forms of information include:

- verbal: personal conversations, meetings, telephone
- written: typed or by hand
- printed: set by computer or professional typesetters
- structured: processed by machines

Office information must be handled. That is, it must be:

- created
- distributed
- stored

As far as writing is concerned, word- and text-editing equipment can ease the burden of retyping, correcting, and distributing typed copies of letters, memos, and short reports. They are perfectly suited to common office needs that require a flawless document be sent quickly to a limited audience of 25 to 100 people. Producing such a small quantity at a commercial printer would be prohibitively expensive, and much too slow. The information in such a document usually has a short effective life, so rapid duplication and distribution are necessary. The capability to print letter-quality copies on a text processor meets both requirements for speed and accuracy. In addition, the original document can be stored and updated easily with most equipment.

As our economy becomes more and more service and technology oriented, veering away from heavy industrial production, the office "industry" will increasingly be the target of efforts to enhance traditional office roles: to coordinate and communicate the management of money, time, effort, ideas, and materials. Offices will be dominated by computer terminals placed at work stations. In the same way that the telephone and the typewriter became indispensible office tools, the computer has come out of the back room and now is in full view.

The introduction of computers into a general office environment has not been without problems. People who used them extensively at first complained of headaches, backaches, and eyestrain. Much of this physical discomfort was the result of poor design. Newest equipment designs place primary emphasis on the use by people. Such machines are described as "user-friendly." This innovative design emphasis reflects the application of ergonomics, the study of human interaction with machines. In addition, equipment that is easy to operate, and physically comfortable as well, has helped to eliminate some psychological barriers office workers have to automation. Computer terminals, of course, will not solve all office problems; often they create a few. Nevertheless, the electronic data and text processing technology has profoundly influenced the way we routinely produce letters and memos, keep accounting records and files, and manage large volumes of information.

Applied to writing, the computer speeds the creation of a clean, error-free final copy, and it gives documents a professional appearance. For business letters, especially those that must be flawless, this power saves endless hours of retyping, and thus saves money. Electronic text processors increase expectations concerning the speed, accuracy, and quality of producing a written document. Though the computer technology may threaten some office workers, its application to the world of work cannot be willed away. Like it or not, the computer is tied to the office.

Users of Automated Writing Equipment

Simply, automated writing has two types of users—casual and professional. By far the larger of the two groups, the casual users write on the system infrequently. They may never become totally familiar with the system, or proficient in its use. A system designed with the casual user in mind is easy to operate, and requires a minimum of training. It provides the operator with built-in directions, or prompts, that guide, answer frequent questions, or solve common problems with the machine.

The second group, professional text or word processing operators, are people who use the equipment full-time. They, of course, are expected to show much higher proficiency and productivity than the casual user. A professional operator should be able to select appropriate typefaces, formats, and spacing. In short, the professional should be able, if called upon, to function the way a production layout editor does in publishing. The equipment at which most professionals work is usually connected to a letter-quality printer, and often is also connected to photocopying or photocomposing machines. This demands that the operator understand all aspects of the finished product, and the capabilities of the individual machine.

Data Entry

The efficient application of text processing equipment depends upon data entry: that is, the building of a file of information to work with. In the case of writing, the lines of text to be printed make up the data file. To enter text into a machine, the most common method is the keyboard to "key-in" the information. Another method still under development employs an optical character reader (OCR) to scan a typed or printed page, "digitize" each letter, and enter that information into the computer. This would benefit writing by eliminating the need to type information a second time into the memory from a corrected draft.

A writer could, of course, generate the first draft directly on the machine, but that is time consuming and costly for the casual user. In most cases, the first draft differs significantly from the final draft, making substantial changes necessary. Major draft revisions can take just as long to perform as retyping the entire document. Some companies use draft typing before going to word processing, especially on the write-ups of new projects or unique messages.

The many functions of electronic text processing, however, aid writing.

Getting Acquainted with the Machine

It takes from eight to 40 hours to get used to the way the machine works. Even though the machines are designed to be simple, no one learned to drive a car, or even cook eggs, without some instruction. Early promotional material for electronic office equipment suggested that anyone could sit down at a terminal, turn it on, and the prompting signals would enable efficient use of the machine almost immediately.

Like a car, the computer requires some adjustment. Of course, if you have worked with one machine, you can quickly adjust to another, in the same way that you can drive a Ford if you know how to drive a Pontiac.

However, in acquiring practical experience with the machine's text processing power, you may initially find some frustrations adjusting to the video display terminal (VDT). For instance, a typewriter has fewer keys. The terminal's keyboard has special control and function keys which allow you to jump from page to page, or move around on a page to make changes, additions, and corrections. Some older terminals do not separate these control and function keys from letters and numbers, which can lead to confusion. Such equipment was used primarily for data processing. Text processing was later added to it. However, most machines currently provide a user-friendly keyboard that physically separates letter and number keys from control and function ones.

No matter what machine you use, the method of writing changes slightly. For example, if you have worked with paper and typewriter, the display of characters on the TV-like screen, takes some readjustment. If you are used to paper, you might reach forward to move lines up or down, or to make corrections. That response soon disappears, along with the fear that the words you wrote have vanished forever when you transmitted several lines to the storage disks.

You may also have difficulty writing at a terminal located in the open spaces common in contemporary business offices. It may take time to adjust to writing in a room with others around. That is simply the nature of the location of the terminals. A terminal, a stand-alone or one that is part of a large network, could be placed in the privacy of an office almost as easily as a telephone.

Unlike a typewriter, the data entry and editing functions depend on the cursor, a highlighted moveable box, and the arrowed keys that move it up and down or left and right. The first time I tried to move the cursor to add a word at the end of a line, I depressed the space bar as I would on a typewriter. I erased the entire line since the space bar represents a character, in the same way a "Q" or a "7" does. Mistakenly, I used the wrong symbol to direct the machine. The arrowed keys move the cursor, not the space bar. Since the computer is a literal machine it cannot do more than it is instructed. Some aggravations accompany any introduction to a new method or machine, so expect some as you become accustomed to writing on a computer. Like breaking in a pair of quality leather shoes, the initial period may give you some painful blisters. Once they are gone, however, the shoes fit comfortably.

New Tools of an Old Trade

A carpenter has quality tools designed to perform specific tasks and to last a lifetime. For the weekend handyman, for example, one hammer is serviceable for most jobs. But the professional has several: one for fine cabinet work, one for brass nails, a plastic mallot for positioning parts with finished surfaces, a heavy one for driving spikes. Writers, too, have many tools, and a text processor represents one that allows them to perform more rapidly and professionally. The tool itself cannot make a better writer, but it can enhance the human skills and talents. Think of the machine as a tool with definite and appropriate functions, applications, advantages, and drawbacks.

BASIC FUNCTIONS OF AUTOMATED EQUIPMENT

Word and text processing machines are offered by numerous companies and have various editorial, storage, and security features. Some of the functions, however, are basic; others advanced.

Editorial

All systems designed to process text can perform the functions of the typewriter, and most have powers that go far beyond it. They allow editorial changes, storage of text, and security. Any text processor offers editorial power through commands, such as: insert, delete, replace, move, merge, sort. Depending on the machine and the language that it uses, basic word processing should allow you to:

- insert a line, or several lines, in a text
- find and replace a word or phrase each time it appears in the text
- format a page: center titles and lines; set paragraphs and spacing; select type faces
- provide automatic paging with headers or footers on each page (running heads and feet in books)
- print or copy: command a printer or copy machine
- arrange information from the memory according to alphabetic or numeric sequence
- generate form letters and boiler-plate documents

Security

Most text processors provide some degree of security for the material filed in the memory. In the creation of a file, or a set of records that you store electronically, you determine whether access to that particular file is unlimited or highly restricted. In other words, the information on the memory disk or tape remains secure. If you allow general access to your file, on some systems you can specify not only the type, such as reading but not copying, but also the general category of people allowed access. No matter how elaborate, though, the security of computer files is ephemeral.

Memory

Although the computer has no human style memory, the word is used to describe the capacity of electronic devices to store data. For writing, that means lines or pages of text. The storage capacity is expressed with a number followed by "K"; for example, 16K or 32K. The greater the number, the more storage is available. Writing, especially longer documents, requires quite a bit of memory, since each letter of a word occupies a portion of the memory disk or tape. Depending on the equipment you use, a disk for a stand-alone word processor can store the equivalent of 10 to 30 typed pages. Since exposure to heat, dust, moisture, and magnets destroys the information on the disks and tapes, they require special handling.

ADVANCED FUNCTIONS OF TEXT PROCESSING EQUIPMENT

Common devices available are either stand-alone systems or terminals that are part of a computer network with a central processing unit. A stand-alone system consists of a video display terminal, a screen and keyboard, that is connected to a printer. Depending on the design, this equipment may also be connected to a computer network.

Most large offices have a computer that processes data. Such systems accommodate text processing terminals and printers. The two form a network of data and word processing and provide a powerful combination for office uses. With the availability of microprocessors, much of the power of large computers is possible with relatively small systems.

Advanced functions depend on the programs that drive the machine. Advanced text processing functions include:

- math and data processing
- dictionary capabilities
- records processing
- flexibility to accommodate new programs
- communications

Math

A typical math program helps to show:

- percentage and statistical calculations
- negative numbers
- verification of totals from the memory
- relationships between columns of figures, then the calculation of further information

A terminal linked to a data processor can merge calculations and text. That capability is important for business and professional communications that require the evaluation, discussion, and analysis of numbers and statistical data. Machines with advanced functions enable the writer to use the memory to gather statistical information from records, called data bases. A data base can be one generated by an individual or company, or leased or purchased from an agency or organization.

Dictionary

Another option for text processing is a dictionary that will check spelling automatically. Individual users can create a short list of technical terms, or personal names to check for spelling. A program based on a standard dictionary is also available to check for errors. Such an option speeds the creation of the final draft by comparing the words with those in the dictionary. A caution with this, though; the dictionary program will not be able to indicate the proper use of the words in context. So if you used *affect,* the dictionary program would not warn you that you should have selected *effect* instead.

Records Processing

A basic program selects and sorts items from an established file of data. Sorting can be by letter (alphabetic), by number (numeric), or by a combination of the two (alphanumeric). A sophisticated program enables you to produce a report based on a previously determined layout, providing numbers where appropriate. The program also maintains statistics, such as the number of records read and written, and provides running totals on columns in multipage documents.

Advanced programs can also:

- provide an index of documents on file
- generate a table of contents from that index
- create a key-word index

Such programs can function as electronic directors, allowing you to simultaneously search the file of data for key words, particular authors, and specific time limits.

Communications

A terminal that is part of a computer network can communicate. One operator can send messages electronically. Instead of writing a memo, the machine can eliminate the typing of it, and send the message directly to another location that could display the words on a screen, or to a printer that would type an original copy in another part of the building, or halfway around the world.

The communication power of electronic text processors has major advantages:

- reduces the need for travel and the associated costs
- provides information to several people in numerous locations almost instantaneously

Electronic connections, interfaces, or software packages allow you to expand the communications ability of many machines. For example:

- electronic mail—the ability to send messages person-to-person within the network
- telex connection—extends message capability to any location that has a telex receiver
- photocomposition connection—ability to set type electronically

292

Office equipment manufacturers see future office tasks as a combination of data and word processing, printing, phototypesetting, and telecommunications. That vision requires a computer network that provides the power to communicate electronically with members of your organization in widespread locations. Some analysts see the office as we know it dissolving in favor of terminals placed in the home of individual employees and managers, connected by telephone lines to the main computer. When we consider the potential saving in time and money by eliminating the need for central offices and the need to commute to them, the idea becomes attractive and sensible. Such an arrangement might allow families to raise children and work simultaneously, since much work could be done at home. Whether the terminal is in the home or in a glass office tower, the skills of writing remain the same. A business writer must invent, arrange, and present ideas that achieve results.

ADVANTAGES FOR THE WRITER

Since each machine approaches the task of word processing in a slightly different way, I'll concentrate on the general functions they share. Three broad functions—editorial, memory, mechanics—offer advantages over typing.

Editorial

Word processors and computers with word processing programs can insert or delete characters, words, lines, paragraphs, and sections of text quickly and easily, eliminating the need for repetitive typing and messy corrections. With such power comes the demand for more skillful and knowledgeable writers. Also, it assumes that writers know what to keep and what to throw out. In the past, if you were lucky, an editor or a gifted secretary performed copy editing for you. But the word processor places more power, and thus more responsibility, in the hands of each individual writer by eliminating some people who previously participated in the production of a document.

Often a line of text communicates more clearly if it is moved to a more appropriate place. To move a block of text, the writer or editor had to cut and paste, literally cutting a page apart and pasting the rearranged sections together. This provided a draft for retyping or if it was carefully done, a "camera ready" document, one clean enough to be reproduced on a photocopier or by photo-offset printing. The capacity to cut and paste electronically allows the writer to generate a draft, altering it several times before final printing. Several drafts can precede the physical printing of words on the page.

Nevertheless, the effectiveness of the tool depends on the writer's ability. If the writer can efficiently manipulate the machine, it enhances his power. Otherwise, one of the lesser mentioned effects of automation occurs. While electronic manipulation of text and data dramatically improves the productivity of about a quarter of the people who use it, the rest show little benefit and some actually become less productive. To take advantage of the computer's power, cultivate your ability to write.

Text processing changes the editorial process slightly. In the past, editors and writers made corrections, provided additions, and deleted material on paper, then had the draft retyped. The machine has eliminated the need for retyping, since you can make changes in the text and see them on the video display before printing a copy on paper. But if you are used to working with paper, you can instruct the machine to print a copy. Then you can make editorial corrections and transfer them to the machine.

The editorial capacity allows you to edit a text with greater speed and mechanical accuracy than retyping the draft over and over. The ability to rearrange material in a document is especially welcome in longer projects that require revision as new information is added.

Use of text processing leaves the initial steps in writing unchanged. The machine cannot invent a point of view, nor generate an idea independently. But it certainly makes secondary steps easier, faster, and more accurate by eliminating the need to painstakingly retype successive drafts and revisions. Additions can be incorporated without worrying about the correctness of the rest of the document.

Merge, Memory, and Form Letters

Depending on the type of machine that you work with, memory can range from a few lines to several hundred pages. The ability to store text electronically has an economical application to boiler-plate or repetitive writing. Users of stock phrases in form letters and routine business reports, as well as in contracts, leases, and agreements, benefit greatly from the computer's capacity to store text, and to modify that stored text for particular situations.

The production of multiple copies of information, for form letters, also represents a productive and time-saving application of electronic writing technology. Most computers used for data processing allow you to merge the stored text of your letter with a list of names and addresses. In this way you free people from typing personal correspondence, much of which, in a business setting, is merely formulated to fit a particular situation. By merging a prewritten letter with a prewritten list of names and addresses, you limit the possibility of typing errors to only those that appear in the original lists and texts.

Most of us recognize these computer-generated letters. Our name appears in the middle of a paragraph as if the whole letter were typed individually. The effect transforms an impersonal document into one that looks personal. This way form letters have a slight psychological advantage with the reader. This moderate plus may soon be lost as more companies use computers for form letters and other widely distributed documents.

Every computer and electronic word processor of commercial size can generate form letters through merging lists and text. Individual machines, however, require that you as a writer learn how to command the equipment to perform that task.

Electronic File Cabinet, Spreading Out, Seeing

Of greater interest for a writer than the merge application to form letters appears in the storage power of the electronic memory. It can serve as a small file cabinet, allowing you to exchange information from past files with current projects.

Such files, however, may be less efficient than conventional writing in one way. A computer terminal cannot display several pages of information simultaneously the way you can if you lay pages or note cards out on a large table, a cork board, or the floor. Most writers insist that the ability to look physically at large amounts of factual or technical information provides an opportunity for discovery, which is necessary when writing even the most ordinary piece. Spreading out most or all of the text allows a writer to create intellectual connections between facts and to see literally a pattern that had not previously been evident.

That technique will be slowed, if not obliterated, by the physical limitations of the display screen. Without looking at the material as a whole, we may present it in a rote, raw, unordered manner. If that is the case, the information that we have might be better left unwritten. Writing is, after all, an organized presentation which should give the reader a clear picture of an idea or fact. Disorganized professional writing converts ideas into mere data.

Effective business and professional writing orders information, bringing light and sense.

Mechanical Aids

Text processing equipment allows you to select different typefaces without retyping the entire document, or without setting the material in type. Such versatility allows you to give your work the appearance of printing. If you are lucky enough to work with terminals connected to photo-typesetters, you can select several type styles and generate sample pages more economically and quickly than with printing.

Most word processing equipment can make such routine format changes as:

- spacing
- margins
- justified right margin
- paging
- paragraphing
- location of columns of figures

Each of these features is accomplished with a simple command. Altering the format electronically saves time and eliminates expensive and unproductive retyping. The ability to locate and change a word throughout a document, called global replace, is also efficient and requires no retyping.

If your terminal is connected to data processing equipment, you can find information and present it in your writing. A graphics program enables you to generate tables, lists, graphs, and curves. Graphics make technical and statistical information easily digestible and more clearly understood. (See Chapter 11.) In the near future the convergence of photocopying, word processing, and data processing will allow writers with little artistic ability to create graphic displays to accompany the text.

Since most math equations must be set by hand and use numerous special symbols, they are expensive to print. The American Mathematical Society has developed a program called TEX for electronic typesetting of mathematical equations. TEX is just one example of how computer technology applied to the mechanical production of documents provides greater speed, accuracy, and economy.

WRITERS' REACTIONS

Many people working with computers see benefits and disadvantages. In an article on writing with computer equipment in *Writer's Digest,* a monthly magazine devoted to the concerns of writers, several writers responded with impressions. Three of the writers agreed that word processing increased "wordage." But some found quality instead. A writer of historical romances said, "My style seems to be greatly improved because I can make word changes with ease," adding that first drafts appeared more polished than those composed on the typewriter.

Another writer observed that electronic text editing and typing makes it much easier to be sloppy with the choice and use of words. According to him, "You may become a faster writer, but you might also be a little bit less cautious in your use of words."

DRAWBACKS OF
ELECTRONIC EQUIPMENT

Finished Look of an Unfinished
Draft

The machine can make you faster. But will it make you better? As far as the etiquette of writing—spelling, punctuation, capitalization—is concerned, the text editing power of a word processor helps you to write more accurately. In several university writing programs, instructors found that the use of the computer improved student spelling. Not that the machine did it; but to write with it, the students had to type their information. That necessity increased their awareness of words, which helped them to spell more accurately. Also, compared with handwriting, they could see words clearly on the screen. However, making the mechanics of writing more accurate is a double-edged sword.

A danger lies in the look of the final product. The word processor can make ordinary typed work look very professional, with margins that are justified on the right and left. It will print a flawless page with no strikeovers, or visible corrections. What you write can look so finished that it has the appearance of material professionally set in type.

Danger may be too strong a term for the power of the computer to process words and present them in a readable format. But in the office of the not too distant future, the ability to blur the distinction between the writer, typist, editor, and printer will be further erased by text processing terminals which are connected to phototypesetting equipment. For instance, one graphics system can use the information generated by computer graphics to drive a photocopier. (See Chapter 11, Figs. 11–6 to 11–10.) In effect, that furnishes the ability to join the functions of writer and graphic artist. Power to perform such complicated tasks seems awesome, and it is.

In many ways the advances in technology applied to writing have placed us back several hundred years. Given access to the machines that are available now, any individual who will write in the future could write, edit, and print a whole book. To do that well assumes that a writer must possess wide-ranging talents. The power of computers applied to writing may make a few people extremely productive.

Computers also have the potential to generate a great deal of fat, useless paper.

Barriers to Automated Writing

Efforts to automate the office with electronic mail, meetings, memos, and reports sent and received on video terminals, create as well as solve problems. The change forces managers and executives who are talkers and not typers to sit in front of a keyboard and type a memo onto it. Most of them do not want to "look like a jerk," as *The Wall Street Journal* put it in a June 24, 1980, article. Most consider the computer rigid, and they feel they could work on the mail, reports, rescheduling, or memo writing in a far less restricted way than the machine permits. Some executives simply turn their display terminals to the wall and pull the plug, choosing not to type on the machine the memos and messages they usually dictate to a secretary.

Sending written messages on the computer requires a change in work habits because writing requires more thought and organization than speaking. When you speak you appear to make sense, but writing often brings back mercilessly the weakness of an idea. An uneasy feeling that a written message sounds dumb, and therefore makes a writer look foolish or even incompetent, provides a major barrier to the use of the computer. Also, it is a technology that is based on a written culture, rather than on an oral one. Increased dependence on radio and TV for information suggests that the ordinary person is less of a reader and, therefore, less of a writer. These people are casual users of text processing and are now expected to look professional on paper or the VDT. Often their fear of writing sent them into business and technical occupations in the first place. Many took science and math courses because they did not like the idea of writing long papers or essays. Once on the job, they found more memo, letter, and report writing than they desired or anticipated.

Clearly, anxiety associated with an automated office results partially from fear of being replaced by the machine, in addition to the fear of writing. After all, the communications features of some computers allow messages to be sent all over the world, bypassing jammed telephone lines and shortcutting the unpredictable mails.

With this in mind, the act of writing becomes even more important, since many messages and letters can be written directly on the computer and transmitted to other terminals.

Whether with a computer or not, writing demands more than an ability to spell correctly and write grammatically. It is an outward reflection of thought, and must shape ideas for the reader. When people speak, they digress. They respond to questions, facial expressions, body movements, and sounds that are not words. However, writing eliminates the opportunity to interact with our audience as we would in conversation. Writing pressures us to be organized and to present information in a meaningful way so the reader will understand our message.

But writing cannot, and should not, be expected to replace the social functions of meetings, informal discussions, and other situations that allow people to react freely with one another.

TECHNICAL PROBLEMS BEYOND YOUR CONTROL: THREE PERSONAL EXPERIENCES

I expected that my background as a writer, as well as my familiarity with the typewriter and my positive attitude toward the machine as a powerful tool, would make writing with it easy. When I first composed material strictly on a computer terminal, I experienced three small occurrences to watch out for in using the computer. These are examples of problems that lie beyond your control. Knowing that they exist allows you to prepare for the frustration you might initially experience when a file is lost, the computer goes down, or electric power fails.

Lost Files

Material I worked on was lost for a while. For one day, before I was assigned a personal access code, I entered material into the machine on a general code. About four pages of manuscript information was stored there. I placed the other information in another file under my own code. I did not know that the general code files were purged weekly to make room for current information. Any information left dormant longer than two weeks was eliminated, like my four pages. Luckily, a file tape existed, and after some checking, was reloaded and the information recovered. Of course, I was only hampered a little, since that material had to be reworked anyway, but it could have been a completed portion. It could have been lost entirely.

Downtime

During a composing session, the computer went down. That means that nothing goes in and nothing comes out. The machine stops. It is similar to taking a trip in your car and as you ride along the engine acts up, then stops. You go nowhere until the motor is repaired. When such breakdowns happen with electric and manual typewriters, they are out for several days at the repair shop. However, so many people need the computer to be able to work at all, that the machine is fixed within minutes or hours to make the time lost as small as possible.

Power Failure

On an April Saturday the weather was terrible, so I decided to continue writing at the terminal. While adding some information to a data file, the rain began to pour down. Thunder boomed; lightning flashed. The lights went off. The image on the video terminal went dark green, with only a bright, small dot in the middle. An alarm buzzer at each terminal squealed for only a few seconds, like a pig being slaughtered. Then darkness and silence. The system manager said casually in the dark silence, "I think the computer is down."

All the information on the screen was, of course, lost forever. I managed to remember the substance of the material and rewrote it later. While that loss represents a setback, writers have always fought against forces that destroy the words set down on the page. Floods, thunder storms, wind, snow, fire, the Post Office, children, cats, dogs, rats, editors, and typists have all destroyed the work of writers. Now we can add the computer to that list.

My three examples are personal experiences, but common ones. As computers gradually replace typewriters, the list of instances and anecdotes will grow. We will see the machines more clearly as powerful tools of the trade, not a panacea for office and professional productivity.

As writers use computers more and more frequently, some of the myths that surround them will begin to disappear. Any new technology creates a myth because of its relative rarity, and the limited number of people familiar with it. We are all in a sense like Dorothy before the throne of the Wizard of Oz, shaking with fear of his power and mystery until the dog Toto pulls back the curtain to reveal a life-sized man at the controls of the presence of the Great Oz. The more we work with and the more we know about the machines that are an integral part of our workplace, the less mysterious and god-like our perceptions of them become.

THE ILLUSION OF PAPERLESS WRITING

An early draft of this chapter was titled "Writing in the Paperless Office," parroting a phrase that referred to future electronic office communications. After interviews with people involved, "paperless" did not discribe the situation adequately. Computers created a paper-filled office. The power to generate a clean draft copy almost at will tempts many people to print a draft copy after each session.

Simply because the ability to print a draft is present, many writers find they create several copies. Previously, they would not consider the wasteful and expensive practice of using a typist to make a clean draft of working papers. When an organization transferred its editorial and typing to word and text processing it experienced a paper explosion. They began to see six drafts, where before they had three. And the stylistic revisions in those drafts increased, creating less productivity instead of more. A consultant traced the problem to one of attitude. Preoccupied with the transition to the computer text processing, managers ignored the psychological and professional needs of the editors and typists, as well as the attitudes of the professionals who generated the information.

Briefly, the writers became sloppy. The editors found themselves working with much more disorganized copy than they had previously received. They were forced to revise and correct random ideas. Mistakenly, the professionals writing the draft material felt the computer made the editor's job much simpler. After all, the company purchased the machines for that reason. Such lazy writing forced editors to spend much more time on the drafts than they had before, cancelling the advantage the computer gave in removing the tedium of constant typing. Productivity declined dramatically. They would have scrapped the whole idea if the machines were not so expensive. Instead, they discovered at formal meetings held to address the problem that the writers expected the machine to work miracles, and that the editors and typists were now inundated with more unconsidered and vague first drafts. The problem was solved when the writers and editors realized the liimitations of computers for the act of writing. The machines, they discovered, do not alter essentially the demand for clear writing.

Automation belongs in the office, and we would be foolish to ignore its proper use there. The notion of an automated office, not a "paperless" one, is more realistic. Office automation fits the new technolgogy accurately and will allow us to think effectively about what computers do for, and do to, the act of writing.

WRITING WITH A COMPUTER

The computer may change some of the previous ways that people wrote messages to one another. But the essence of writing remains the same. The machine's power allows us to be more productive, or sloppier.

During an address at Harvard, T. S. Eliot commented that it might be interesting to see a doctoral dissertation on the effects of the typewriter on the creation of a novel, poem, or story. No one took up the challenge, not because the idea is vacuous, but because it would be a fruitless effort.

Psychological adjustment and physical differences notwithstanding, writing on a computer remains writing.

SUMMARY

- Word and text editing equipment can ease the burden of retyping, correcting, and distributing individually typed copies of letters, memos, and reports. The computer will soon become as commonplace in the office as the telephone and the typewriter.
- Becoming acquainted with a particular machine requires from eight to 40 hours. Text processors offer a writing tool with definite and appropriate functions, applications, advantages, and drawbacks.
- Basic functions of the machines offered by numerous companies have editorial, storage, and security features. Advanced functions depend on the available programs that drive a particular model. Typical features, however, include: math functions, spelling correction, records processing, communication.
- Three broad functions—editorial, memory, mechanics—offer the writer distinct advantages over typing.
- A clear disadvantage lies in the finished look of an unfinished draft.
- Automating the office often creates some anxiety among office workers. In addition, machines are subject to power failure. Downtime and mishandled information also create problems.
- The computer may change ways people previously wrote to one another, but the essence of writing remains the same.

SUGGESTED PROJECTS

1. For a term project, investigate three models of computer-aided writing equipment available for office use. Write a memo to your company in which you evaluate these machines against the needs of your organization and the cost of their installation.

2. If you have never written with a word or text processor, write a memo, letter, or short report of three pages with one. Submit the result with another memo describing your new writing experience.

3. If you have used a text processor in the past, write an evaluation in which you suggest added features that you, as a writer, would welcome.

4. Select several form letters you have received. Compare those generated by computer with those that were not for tone, impact, style, accuracy, and interest.

5. Write an evaluation report of the computer's impact on writing in the workplace. Survey representative business associations, professional groups, technological organizations, and employee unions. Use their observations and responses in reporting your findings. If appropriate, make recommendations for actions that would improve the use of text processing.

6. Compare the cost of writing routine letters by computer and by conventional methods. For your investigation find three organizations now using computers, and three that do not write that way. Prepare a simple series of questions that you will ask all six organizations. For example, "What is your estimate of the cost of a routine letter, includng overhead items?" Determine which applications suit the computer, and which do not, based on your survey of the six companies.

13

The Résumé,
Letter of Application,
and Job Interview

Your first professional writing assignment is the letter and accompanying résumé that you send to an employer. To prepare a résumé, think of it as a sales tool that offers your experience and qualifications to a company you want to work for. When you think of a résumé that way, many of the formats and résumé guides do not set you apart from other applicants. They are formats to make it easy for personnel staff, administrative assistants, and secretaries to screen you out.

Since competition for good positions with reputable firms is keen, an employer might get 50 to 100 résumés for a single job. To manage that number, they whittle down the candidates to the top five or 10.

If at all possible, get the job first, then write a résumé later for the personnel department to use. That can happen through contacts you have made. You may have had an internship with an organization during college and they liked the way you worked with their staff. Suggest to the manager who supervised your training that you will soon graduate and begin to seek full-time employment.

Another way to get a foot in the door is for you to interview the company. In other words, contact someone who performs the job you desire at a company that you are thinking of working for. Call up and ask if you could talk about the demands of the position since you would like to know more about it to help plan your career. Some may tell you to get lost, but others will welcome your interest. When you go, really talk about the job, and not your lack of one. Then a week or so later, you can write a letter thanking that person for the information you obtained. Then you might ask if they anticipate any expansion or training positions in the future. In this way you learn whether or not the job is what you want, and at the same time you have made a personal contact in the company. Personal contact is important since further letters from you cease to be unfamiliar. You have given the prospective employer an idea of your personality and something that helps yours stand apart from the stack of letters asking for work. That brings us to the writing of the cover letter and the résumé.

COVER LETTERS THAT SET
YOU APART

Generally a single-page, well-constructed letter can replace writing a résumé and a cover letter. Most of who you are and what you offer a prospective employer in experience, education, and professional training can fit into an economically expressed letter. The specific content of that letter, of course, depends on your background and activities related to the job you seek.

Nevertheless, have the attitude that you *offer* an employer your talents and skills, rather than you have expectations and goals they can fulfill. In other words, when selling your qualifications you must go back to those questions listed previously for analyzing your audience. In this case, an employer wants you to do something for the company, and is usually not too concerned with your goal of being the company president some day. So you offer your ability to perform as a professional. That attitude signals that you know at least the first thing about the job you seek, as well as some information about how the workplace operates.

Some mechanical points also characterize applications. Any job-related letter, whether a résumé accompanies it or not, should touch these bases:

- address a specific person by name and title
- describe the job you seek and mention how you learned of it
- detail the skills and talents you offer
- present work experience relevant to the job
- suggest the next step by mentioning an interview

Address a Specific Person by
Name and Title

When you do this you write a "bullet" letter, one addressed to the person in charge of the opening. For example, write to a Mr. Harold F. Coley, Director of Billing Services, rather than some vague title like The Director of Accounting. Many business people feel that if you cannot perform the research to learn their name, then you will be less efficient on the job than someone who can. The way you ask for a job in your letter says a great deal about your professionalism. Often finding the name of the person you should address in your letter requires nothing more than a simple phone call to the company. When the switchboard or PBX operator gives you the information, be sure to ask for the spelling. Misspelling a person's name or the name of the company places a psychological handicap on your candidacy, suggesting that you are careless, lazy, or worse—incompetent. You would have to be superior indeed to overcome those poor impressions.

Describe the Job You Seek;
Mention Your Source

Somewhere in the beginning paragraph, or very early in the next paragraph, tell the person where you learned that an opening was available. That statement reflects another aspect of your professionalism. The local newspapers are not the only sources for jobs. You will learn of job openings from:

- professional journals and special newspapers; for example, *Computerworld,* a weekly devoted to the computer professions, runs pages of ads for all types of individuals.
- the job placement and career counselling center at your school.
- state employment office; most states have an agency that provides a free service.
- employment agencies; private agencies are in the business of finding you a job, so they will charge you for their service, either a flat fee, or a percentage of the income of the job they secure for you. Be careful with agencies. Read all their promises carefully before you sign any agreement. Make sure you understand exactly what you are getting for your money. Be sure you ask whether you or the employer pays the agency fee. Deal with a reputable, established agency. Ask them for references, and if they refuse, look elsewhere. Also call the Better Business Bureau and ask about the firm's reputation. You could also contact the local association of personnel professionals for an evaluation of the agency's track record.
- the local library. The library carries newspapers from many cities, as well as professional journals and magazines. It may also have an employment opportunity center. In New York City, for example, the larger branches of the public library have a section devoted to business directories, bulletin boards with job information, as well as books on preparing résumés, taking aptitude tests, succeeding at interviews. Such a facility, though, is generally geared to people with a high school education or a minimum amount of college.
- co-workers and friends; one of the best ways to find out about a job is to ask people you know—members of the family, friends, people you work with or meet during the day. Cultivate these contacts.

Survey the Skills and Talents You Offer

In order to present yourself professionally and effectively, you must first take stock of yourself. It sounds very easy for me to say that, but the most difficult part of résumé and job letter preparation happens to be your objective confrontation with what you can and cannot do.

Take a clear, long look at your background and experience, and write what the New York State Labor Department *Guide to the Preparation of a Résumé* calls "an asset list." Include in that list your:

- employment history
- educational experience
- personal strengths and weaknesses
- leads to jobs, contacts, friends

Employment history.   Make a list of all the jobs you have ever held, including full-time, part-time, vacation, freelance. For each one provide:

- talents and skills you gained or improved there
- job title and list of duties
- time spent on the job
- reason for being hired, and reason for leaving
- what you liked and disliked about the job
- what part of the job you did best
- what references can come from that job
- what personal attributes helped you perform that job successfully.

In preparing the list you have to identify areas in which you do well, and the ones where your performance might be improved. Your personal inventory of your job history might be uncomfortable, especially if you have had little or no experience, or if your experience has little or no direct connection with the job you apply for. In that case, be very creative in the way you think about what you have done and express the skills you offer the employer that substitute for a "track record" of experience. No one expects you to have 10 years' experience at age 21, but employers do react well to a professional attitude.

Personal attributes—strengths and weaknesses.   Since most states legally restrict an employer from asking you questions about your national origin, color, race, creed, sex, marital status (married, single, divorced, cohabiting, separated), age (except to determine if you are over 18 or under 65), or physical handicaps, you do not have to answer such questions. However, you may volunteer any of that information. If you do not want an employer to know personal information and they ask you for it, be prepared to ask tactfully why they need to know—without accusing the prospective employer of breaking the law. The employer then must explain to you why they asked an illegal question.

All that aside, the survey you make of your background is an essential step in your honestly identifying your selling points. For example, your past experience might suggest that you have imagination, initiative, leadership, willingness to follow orders, compatability. You might identify personal attributes like reticence, aggressiveness, moodiness, as well as your ability to cooperate, adapt, and show up for work dependably. You might also identify your ability to communicate clearly both orally and in writing, an important quality you can incorporate in the sales presentation of your skills and talents.

In compiling your list of personal assets and liabilities, the employer's needs indicate which ones you should emphasize. For instance, a student who patented a guitar pick applied for jobs concerning research for TV programs. His experience as an aid in the library of a major network helped him look professional on the résumé, but the mention of the guitar pick might reveal other interests. He used the research he performed to develop the guitar pick, together with the paperwork for his patent on it, to show his range and imagination in research. An accomplishment such as a patent set him above others and allowed him to interview for better positions within his company, and to prepare himself for the job search after graduation.

Suggest the Next Step

To be forceful and effective in finding a position your cover letter must open a path to the next step in the hiring process—an interview. Be tactful, but don't be shy. Instead of closing:

> I appreciate whatever attention you may give to this matter and look forward to corresponding with you.

or,   Thank you for your consideration, and I anticipate your call.

or,   I look forward to discussing the position with you.

Why not take control of the opportunity by providing the employer with something specific:

> I'll take the liberty of calling next week to arrange for an appointment.

or,   I am free to come to your office for an interview before 10:00 a.m. Monday to Thursday mornings. I will call Thursday afternoon to find out if those times are appropriate for us to meet.

You may feel a bit awkward or self-conscious by offering to call and set up an appointment. Actually you save the person you have written the time and cost of calling you or writing you a letter.

Also, provide your phone number. That invites the person to call. An editor once called me about a minor point in publishing an article, and when I asked why the urgent call, he said, "Well, you have that phone number on the letter and that just invites someone to have the issue resolved right then by telephone." The same can work for you, too.

You can also include a postcard addressed to yourself that promotes a response. In *Writing that Works,* Roman and Raphaelson suggest something like this in responding to a box number:

> _____ Please call my secretary for an appointment.
> Her name is: _____
> Phone: _____
> _____ Sorry, you're not quite right for this job.

Generally a box number gives a company anonymity. It provides them with the ability to advertise for a job without implying to the competition that they need certain types of professionals.

More often than not, however, and considering the high cost of placing a display ad in a major newspaper, the employment notice actually serves a dual purpose. It lets people know about the job, and about the activities of the company. In effect, the employment ad promotes the company as well. With that in mind, a box number might suggest something else. Personnel agencies often place large ads in the newspapers to gather résumés in an effort to show clients that they are putting forth an effort for them. Expect less of a response when you submit your credentials to a box than to a company that puts its name on the ad.

WRITING A RÉSUMÉ THAT GETS READ, NOT DISCARDED

Large organizations that receive 2,000 or more résumés a week use the personnel staff to screen applicants. A screener goes through about 100 résumés in an hour. In other words, the staff member spends 30 seconds on a single résumé. The goal is to find the five or so applicants appropriate for the job. In most cases, the personnel department does not have the authority to hire, beyond the entry level. But you have no chance for an interview if you have been screened out. Let's look at some positive ways to get past the screener.

Getting Your Résumé Out of the Pile

Since from your point of view the résumé is a sales tool, a way of winning an interview, emphasize the experience and background you have which is related to the advertisement. Employers see résumés as useful tools, but they use them to help screeners go through all applicants, and to help an interviewer ask questions about your background and experience. With screeners in mind, career counsellors and personnel agencies suggest that when you write a résumé you:

- use clear, jargon-free language to present your skills, talents, and relevant experience
- respond exactly to each item called for in the ad
- emphasize the qualifications you offer
- show that you are professional

Incorporating information about yourself into a résumé demonstrates how difficult it is to write about yourself. The kind of writing needed to secure a job may be the most difficult of all.

For that reason, poor résumés are easy for screeners to spot. Here are some items to avoid in constructing yours:

- Avoid showing your ignorance of the business world. If a job asks for two years' experience as a programmer for retail sales, and you have that, don't blow it by talking about the fraternity social functions you organized in college.
- Don't put a photograph of yourself on the résumé. A picture of you might put the screener off and it is illegal in some states to request one. Pictures are only necessary if you are a model, actor, or "personality" whose looks are important for the job. For the rest of us, performance is more important than looks.
- Avoid irrelevant jargon or technical terms.
- Avoid misspelled words.
- Avoid unreadable or fancy typing.
- Avoid smudges or dirty copies.
- Avoid having your résumé professionally set in type on expensive, colored paper. This indicates to the screener that you are a "professional" job hunter, a dilettante who would rather look for work than work. Such a résumé suggests that you needlessly spent a great deal of money on packaging.

WRITING A RÉSUMÉ

A résumé represents facts about your work experience, education and other information relevant to the job you seek. If it has not yet occurred to you, you should have several résumés, not just one. Don't make an all-purpose résumé, and send that out to the president of every company you know. That's a sure waste of your money for paper and postage, and a sure waste of the employer's time.

No perfect résumé exists—even your own changes as your education and experience become more varied. Experts disagree drastically on how to construct a winning résumé. Some insist that it is no sales tool at all, merely a formal way to inform the employer of your professional existence. Others insist, just as vehemently, that the résumé is indeed a sales tool, and suggest some radical ways to prepare the document. As we have mentioned, it is both, as well as a way to set the tone for the interview. Generally, touch the following bases in your résumé:

- Present the essential information first—name, address, and telephone.
- Clearly describe the talents and skills you *offer* the company. Most often résumés call this a "job objective," or if they want to suggest professionalism, "career objective." Instead, think of this as an offer to the employer, rather than a goal that you have for the employer to meet. Call this element "offering" or "career qualifications."
- Make a clear list of past experience, beginning with the most recent. Include everything that you have ever done in the initial compilation. If you need to cut out information, eliminate peripheral jobs such as the two weeks you mowed lawns for the neighbors to buy a bicycle. But do include, for example, a summer internship at a chemical plant, if you seek similar work.
- Education. Just as with the jobs, list your education, beginning with the most recent. If you are enrolled in a program, but have not yet received a degree, mention the date and type of degree you expect; for example, B.B.A. (expected 19XX). Mention of high school education usually signals that you are not aware of the professional rules of the game. However, if you have gone to a special school, like the Performing Arts or Aeronautical High Schools in New York, say so.

- Publications, awards, honors, patents, professional organization memberships. If you have ever published anything, list it. The same goes for the other items in this list.

Essential Categories

Name
Address
Telephone Number
Career Offering, or Career Qualifications
Previous work history, including all dates.
Special skills:

- computer programming and program languages like FORTRAN or BASIC
- foreign languages
- blueprint reading
- computer aided design/manufacture
- text processing
- laboratory skills

Be sure the skills you mention are relevant and that you really have them. If you have a license that verifies the skill, mention that too.

Education.   Mention the degree, school, and date. Do not list courses by name since this only wastes space; and, anyway, what does a course title like "Accounting I" tell an employer? If, as a new graduate, you must detail your college curriculum, concentrate on the skills and knowledge gained, rather than the course titles or even the grade point average. But if you earned a Phi Beta Kappa key, mention it.

Activities.   Include activities only relevant to the position. If you apply for an entry level opening as a electronics engineer, and you have built your own digital computer in your garage, then by all means mention that. If you played professional baseball for a few years, mention that. You may have already been contacted to work for a company, not because you can do well with the accounting department, but on afternoons, in the summer, you will be able to help the company baseball team better its league standing.

References.   Either give the names of the people who have agreed to write, or have already written, a letter for you or don't mention references at all. Employers assume they can have letters of reference if they ask.

Unnecessary Items

Race, religion, nationality, marital status

As we stated, laws forbid an employer from asking you these questions. It's up to you to provide that information or not.

Height, weight, health	For most jobs these items are irrelevant. Put them on your résumé only if you are applying to be a jockey or for modeling. Employers expect that you will look and dress appropriately. Questions about health are only pertinent when you have the job and begin to fill out forms for medical and other benefit programs.
Photograph	As we mentioned, no picture is called for.
Salary	If the ad called for "salary requirement," put in something like: competitive or negotiable. If you put in a dollar value, that provides the screener with a way to screen out your résumé. Discussion of salary should come at the interview.

Hobbies
High School
Volunteer work (unless pertinent)

FORMATS FOR RÉSUMÉS

Personal Format

Since the résumé has the three applications that we discussed—your sales tool, the employer's screening tool, the interviewer's springboard—present information about yourself so that all three audiences are satisfied. The best format for you depends on the position sought and on your qualifications. Prepare several résumés with different formats in advance to help you respond effectively to the positions you seek. Formats include: Historical or Chronological, Analytical, Functional, Expanded, Creative.

Historical or chronological   Personnel professionals call this format the "obituary" since it reads like one. Most of you have seen this format. It is recommended if you can show a steady progress toward a goal. Jobs are presented first, from the most recent to the first job you had. Education is also presented in reverse with the highest degree first. This type of presentation makes it easy for the screener. If they require five years' experience, and you have only two, your résumé is out of the running.

Analytical   This one takes particular job skills and presents them rather than the job title. That is good for you, for instance, if you really did much more than your title implies. If you worked for a small firm as an accountant, your real experience might be comparable in kind with that of a financial analyst. Relevant data of employment usually come at the end of the résumé.

Functional   Presents your qualifications, using both job titles as well as the actual functions of the positions. For new college graduates whose job history is diverse and limited, this approach to the presentation of your credentials is effective.

Creative   Most people should avoid this format. But if you want a position as an industrial designer, then make your résumé a design project. The form and the content should demonstrate your creative and professional talents.

Professional Format

If you participate in a conference or are assigned to a large project, you will be asked to prepare a professional résumé. Rather than seeking a job, this form presents your background and experience as part of a company project or program, either to a prospective client or a government agency. In the past, companies had your old résumé on file, pulled it out and included it, usually with no changes, in the material on the project.

However, more mindful organizations, sensitive to the needs of their clients and customers, ask that you prepare a résumé tailored to the project. The approach is similar to preparing a job-seeking résumé. Since you will be part of a large project you might be asked to submit a résumé with space and style considerations. For example, here are excerpts from a memo circulated to people on a project, showing them how to submit a résumé. The three formats reflect three levels of involvement on the project. Each person was given a copy of their résumé that was on file, and the memo indicated which of the three formats the individual was to follow.

1. Should be 1 to 1¼ pages (typed double spaced). The first paragraph should describe your job on the proposed program. The second paragraph should describe your applicable background and experience in reverse chronological order. The last paragraph should present education, honors, and awards in reverse chronological order.
2. Should be ½-page typed double spaced. The first sentence should describe your job on the proposed program. The middle portion should describe background experience (in reverse chronological order). The closing sentences should present education, honors, etc. in chronological order.
3. Should be brief, telegraphic phrases not exceeding three lines, typed double spaced. (Your program title will indicate proposed program responsibility.) Highlight relevant past assignments.

Other professional situations might call for a brief outline in paragraph form. For example, if you plan to run for an office in an organization you might submit something like this:

Debra Kostantos is currently the supervisor of publications at Monolithic Corporate Research Laboratories in Long Island City, New York. She was previously a Senior Editor for the division where she handled software user documentation, training manuals, brochures, and audio-visual presentations. Before joining Monolithic, Debra was a senior high school English teacher as well as the editor of a weekly county paper, *The Johnson County Clarion.*

She holds an undergraduate degree in psychology and a master's in English Communications, both from State University. Most recently, Debra has served as Regional Manager for the New York Chapter of the society. She has also served as a judge in past society publication competitions and has been active in the Executive Council.

Professional résumés cover the same bases by presenting relevant background, experience, education, honors, and publications. Essentially they see you as part of a large project team or as a candidate for professional office. In both cases, remember that the résumé presents *relevant* information, not all information.

Why No Ideal Format

I have hesitated from presenting sample résumé formats in this chapter because the examples I choose might limit you in writing your own effective résumé and presenting yourself in your own terms. Several résumés are included here. (See Figs. 13–1 to 13–5) However, you should design one of your own that clearly and powerfully presents the skills and talents you offer.

THE INTERVIEW

Preparing for the Interview

Once you receive a positive response to your résumé and have been invited to come to the company to talk about your possible employment, you will have some time to let the anxiety of the situation build, or to expend that nervous energy productively. Instead of worrying, gather your thoughts and ideas, and concentrate on a positive attitude toward the job and the company. Do your best to convey professionalism, interest, and ability.

Not only is the interview your sales presentation of your talent to the company, but it also gives the interviewer the chance to see that you offer much more than your résumé and letter. Just as important, the interview gives you a chance to evaluate the company and the job you seek. You now have the opportunity to decide if the job meets your needs, and if the employer measures up to the realistic standards you expect. To prepare:

- Gather as much information as you can about the company and the job for which you are interviewing. You can spend an afternoon at the library looking at an annual report of the company, as well as brief descriptions in *Dun and Bradstreet, Fortune* magazine's listing of firms, and local newspaper stories on the company, if any.
- Review your offering, the talents, skills, education, and experience that you will bring to the job you seek.
- Prepare a list of questions you might be asked. Along with the questions, briefly outline the way you would respond to them.

```
                    LISA SUSAN SCHMIT
                 157 ST. STEPHENS STREET
               BOSTON, MASSACHUSETTS 02188
                     (617)-555-8165
```

CAREER QUALIFICATIONS
>   Technical Writer trained in preparation of technical
>   definitions and descriptions, manuals, catalogues,
>   part lists and instructional materials. Experienced
>   in evaluating and editing computer documentation
>   containing syntax formats.

WORK EXPERIENCE
>   **Northeastern University**   Boston, Massachusetts
>   **Reading and Writing Specialist, English Language Ctr.**
>   March-July 1982, January-March 1983
>   Created individual lesson plans for each student
>   assigned to the Reading and Writing Laboratory.
>   Designed materials for use in the Laboratory. Ran
>   the Laboratory for approximately one hundred students
>   for twenty hours each week, maintained records of
>   students work, and prepared written and oral reports
>   on student progress and the operation of the
>   Laboratory.
>
>   **Tutor of Foreign Students**   September 1980-Present
>   Integrated foreign students into a large urban school
>   and community while being a positive role model
>   educationally and socially.
>
>   **William M. Mercer, Incorporated** Boston, Massachusetts
>   September-December 1981
>   Data Processing and general office duties. Initiated
>   and implemented a CRT search system for office
>   personnel.
>
>   **American Girl Magazine**   New York, New York
>   1976-1979
>   Student Editorial Advisor and Model.

SPECIAL SKILLS
>   BASIC programming. PASCAL, Editing and Graphics
>   courses to be completed June 1983.

EDUCATION
>   **Northeastern University**   Boston, Massachusetts
>   Bachelor of Arts   June 1983
>   Concentration: English with minors in Technical
>   Communications and Economics.
>   Activities: Selected to serve on the Residence
>   Judicial Board, an impartial group of faculty, staff
>   and students who adjudicate discipline problems;
>   Northeastern News; Northeastern Yearbook staff.
>   **Xavier University**   Cagayan de Oro, Philippines
>   Foreign Exchange Student   1979-1980
>   Rotary International Foundation Scholarship

FIGURE 13–1   Sample One-Page General Résumé

```
Mary Muncie                                    JUNIOR COPYWRITER
139 Fifth Street
Jersey City, NJ  07304
Tel.: Area Code 201: 834-9009


          The following experience was acquired during the summers while attending
          school.  However, I feel that it has enough value to qualify me for a
          position as Junior Copywriter.

          6/75 - 9/75    Intern with Arden Advertising Agency, 630 Fifth Avenue.
                         Worked as general assistant in copy department under direct
                         supervision of copy chief, Mr. John Smith.  Principal duties
                         were typing and checking copy, filing, relief receptionist
                         and switchboard operator.
                         Was permitted to attend several copy conferences, and finally
                         wrote original copy for two small food accounts.  Both accepted:
                         one appeared in Woman's Day, and the other used as a direct-
                         mail piece.

          6/74 - 9/74    Market research interviewer for Markets Inc., Jersey City, NJ.
                         Surveys conducted under variety of conditions, and among varied
                         consumer groups; housewives, summer college students, retail
                         store and motion picture patrons.  Subjects equally varied;
                         readership studies, consumer products, audience reaction.
                         Last two weeks, helped compile data from one study to form
                         basis for national research project on the consumer spending
                         potential for electrical appliances.

          ADDITIONAL
          EXPERIENCE:    (Non-paid)
                         Advertising manager for college newspaper; increased regular
                         advertisers from seven to twenty, over a period of three years.
                         This not only helped to eliminate the deficit in our budget,
                         but enabled us to start a nest egg toward the purchase of
                         much needed equipment for the newspaper office.

          EDUCATION:     Delford University - Summa Cum Laude
                         B.A. -                Major: English
                                               Minor: Psychology

          EXTRA-
          CURRICULAR:    Reporter and Advertising Manager on College paper.
                         Member of debating team.

          PERSONAL
          DATA:          Am free to travel.  Will relocate.
```

FIGURE 13–2   Example of Specific Job Résumé

- Be prepared to list factual information about your jobs, education, and references. Before the interview ask your references if you may use their names.
- Find out the salary range of the type of job you seek from a professional association, or from newspapers and trade journals. Get a firm idea of where you fit in the range.

Some mechanical considerations:

- Show up at the right place *before time*. Allow yourself five to 10 minutes to relax and gather your thoughts before the interview. If you cannot help being late, you must call the interviewer, explain that you are delayed, and ask if the appointment can be set back

```
                              DONALD E. TERRELL
                              272 Hillcrest Road
                          Needham, Massachusetts  02129
                          Telephone:  (617)999-2592

Accounting Experience
Credit Manager:         NAD (USA) Inc.                 Spring, 1982
                        Norwood, Massachusetts
                        Appointed acting Credit Manager.  Responsible for analysis
                        of applications for wholesale credit, approval or rejection
                        of new credit accounts, supervision of collections of
                        accounts receivable, and approving or holding of orders
                        to be shipped under terms of credit agreement.  Position
                        reported to Vice-President of Finance.

Staff Accountant:       NAD (USA) Inc.               January, 1981 - Present
                        Norwood, Massachusetts
                        Worked on special projects at the request of Vice-
                        President and Accounting Manager, designed and developed
                        cross referencing accounting control system for return-
                        ing merchandise, designed system to deal effectively
                        and efficiently with both international and domestic
                        freight claims, reconciled bank statement to company
                        books, dealt at various times with posting of accounts,
                        accounts payable and  receiveable.  (Part time and Co-Op)

Bookkeeper:             Phil's Texaco and Junction     December, 1975 - October, 1978
                        Mobil Gas Stations
                        Needham, Massachusetts
                        Compiled day sheets, did customer billing, payroll,
                        reconciled bank statement to books, maintained credit
                        card account, and performed accounts payable function.
                        (Part and full time)

Payroll Clerk:          Malden Trust Company          April 1 - June 30, 1980
                        Malden, Massachusetts
                        Set up and coded payrolls for computer, proved out
                        payroll account, maintained savings certificates, and
                        proved account to bank's General Ledger. (Co-Op)

Education
Name of School:         Northeastern University
                        Boston, Massachusetts
Degree:                 Bachelor of Science in Business Administration, 1983
Major:                  Intensive training in accounting which included develop-
                        ment and analysis of financial accounting systems,
                        budget preparation and analysis, cost variance analysis,
                        design and development of computer software.

Name of School:         Massachusetts Bay Community College
                        Wellesley, Massachusetts
Degree:                 Associates in Business Administration, 1980
Major:                  Accounting
```

FIGURE 13–3   Sample of Résumé for Management Level Position

for a few minutes, or made for the next day. You must *call*, otherwise the oversight immediately eliminates you from the running.

- Dress appropriately. Appear business-like, neither too casual nor too formal.
- If you do not know, ask the receptionist the name of the interviewer and the correct pronunciation of it.
- Be pleasant, friendly, and professional.

```
32 Arizona Terrace
West Lynn, Massachusetts 54321
October 11, 19XX

Mr. Vincent Bredde
Wang Laboratories, Inc.
Department G-9
One Industrial Avenue
Lowell, Massachusetts 01851

Dear Mr. Bredde:

I am writing in response to your recent Boston Globe advertisement
indicating a need for a specialist in international marketing support.
I was intrigued by your advertisement because I feel that my background
is particularly suited to this position for several reasons.

For the past three years, as a member of a small management consulting
firm, I have been involved in marketing research and financial planning
projects for client companies.  The project-oriented nature of consulting
work has provided me with experience in a broad range of domestic and
international business issues.  I have also done some programming of both
microcomputers and small mainframes.  My background also includes exten-
sive travel throughout western Europe, including one year of residence in
Paris.  I am fluent in both spoken and written French, and possess
conversational ability in German and Spanish as well.  These experiences
have given me a familiarity with the international marketplace and
European cultures which would contribute significant support to Wang's
presence abroad.

The enclosed resume will give you further information concerning my
qualifications for this position.  As you can see, I will be completing
work toward my M.B.A. in June of 1983.  As part of this program, I have
done course work in international marketing and in marketing research,
and am thus familiar with many of the current issues faced by firms such
as Wang in the international marketplace.

I would very much like to meet with you and discuss in detail how I feel
I could contribute to Wang's marketing efforts in Europe.  I will plan to
telephone you on Thursday morning to arrange an interview at a mutually
convenient time.  Or if you prefer, you may reach me at home at 555-5226.
I look forward to talking with you in the next few days.

Sincerely,


Karen Jones
```

FIGURE 13–4a   Example of Cover Letter with Résumé for International Marketing
Position

Meeting Face to Face

Beginning.   When you walk in, greet the interviewer by name and shake hands firmly, but not aggressively. Look into the interviewer's eyes. Let the interviewer begin and control the conversation, but assert key points that you have identified in preparing for the interview.

During the interview, don't smoke. Speak clearly and avoid running on and on.

Allow the first few minutes to assess the interviewer. As we did in the chapter on audience analysis, try to understand the person asking you the questions. One

```
KAREN JONES ● 32 ARIZONA TERRACE ● WEST LYNN, MASSACHUSETTS (617) 555-5226
```

PROFESSIONAL OBJECTIVE

A position with advancement potential in domestic or international marketing or marketing research

EXPERIENCE

Senior Consultant, 1979 to present
Equus Associates, Inc., Lexington, Kentucky and Boston, Massachusetts

- Directed market and patron surveys at six major sports and pari-mutuel facilities. Responsible for questionnaire development, on-site survey management, development of data processing software, analysis of survey data, report preparation, and presentation of results to clients.

- Directly involved in marketing a software package developed by Equus Associates as a planning tool for sports and leisure-time concerns.

- Conducted comparative assessments of pari-mutuel regulatory policies on a state-by-state basis. Direct responsibilities included data collection, interpretation, and report preparation in such areas as the appointment and compensation of officials and sales tax policies affecting the horse racing and breeding industries.

- Participated in studies in three states documenting the total contribution of racing to the state and local economies. Responsible for compilation of raw data, calculation of the impacts of the various cash flows, and report preparation.

- Participated in several studies assessing the likely impacts on an industry of proposed new legislation.

Instructor of French and Spanish, 1974 to 1979
Woburn Senior High School, Woburn, Massachusetts

EDUCATION

Master of Business Administration candidate, June 1983
Northeastern University, Boston, Massachusetts
Concentration in marketing   (part-time status)

Bachelor of Arts in French (Magna cum Laude), 1974
Tufts University, Medford, Massachusetts
Includes one year of study at the University of Paris, France

SKILLS AND ACHIEVEMENTS

Fluency in French; knowledge of Spanish and German
Extensive foreign travel
Programming experience in BASIC, INTERACT, SPSS

INTERESTS

Tennis, running, travel, choral music

FIGURE 13–4b   Résumé that Accompanies Fig. 13–4a.

specialist in human resources asserts that each interviewer falls into one of these categories:

1. Sensors. Sensors are present-oriented and respond to things they can touch and feel. They are doers.
2. Feelers. Feelers rely on emotion, on gut feeling. They like human contact and enjoy people.

TRAINING & DEVELOPMENT SPECIALIST

A high technology central Massachusetts heavy manufacturer with 2000 employees is seeking a person with at least a Bachelor's degree in an appropriate discipline and two or more years personnel experience to design, implement and coordinate training at all levels of organization.

Position reports to Manager of Human Resources.

Excellent salary and benefits. Send your resume or detailed letter in confidence for immediate consideration.

W 87, Globe Office
Boston Globe, Boston, MA 02107

Equal Opportunity Employer M/F/H

In response to your advertisement in the Sunday Globe dated October 10, 19XX, I wish to apply for the position of Training and Development Specialist with your company.

I have five years experience planning, developing, and implementing management, sales, and technical training programs in both corporate and academic environments. Specific programs presented include Basic Supervision, Business Management, Time Management, Sales Management, Selling Skills, and New Product Marketing.

I managed the Training and Development Department for Radi-Com Inc. a major manufacturer of automatic test equipment, and was responsible for the development and delivery of supervisory, sales, and technical training programs to over 2,000 management, engineering, sales, and support personnel. Other duties included the design of a performance appraisal system, response to industry surveys, design of the company workforce planning system, and contributions to the company newsletter.

I will be happy to discuss further details of my experience with you in a personal interview.

I enclose a copy of my resume for your consideration.

Yours sincerely,

William Burrows

FIGURE 13–5   Sample of "Blind" Ad and Letter of Response to a Newspaper Box Address

3. Thinkers. Thinkers are logical, systematic, orderly, and structured. They are data-oriented.
4. Intuitionists. This type looks to the future, and is concerned with planning and setting goals.

The interviewer's desk, office, style of dress, and opening question should give you a good idea of the kind of person you are talking to so you can tailor your response to the psychological needs of the interviewer.

The sensor will respond well to you if you are concise, factual, and candid, stressing your ability to solve problems. He will likely answer the phone during an interview.

If the phone does ring during your interview, and it becomes obvious to you that the interviewer can't soon hang up, assert your presence, but tactfully. Say, for example, "It seems that you are terribly busy, and I can come back at another time." As you say this, prepare to leave by gathering your coat and papers. If the interviewer is really side-tracked by the phone call, this gives you both a way out. On the other hand, if the interviewer is testing your professional behavior, then you have passed the test by not sitting around unproductively. Chances are if the call was a test, the interviewer will ask that all future calls be held during your session. Either way, your asking for another appropriate time forces the interviewer's hand.

Feelers will stress interpersonal relations, and thinkers will expect facts. Be as specific as you can with them. Intuitionists look to the future, so stress future planning, not the past.

Of course, this categorizing of people might not help you since most people have some of the characteristics of each of these types. Interviewers are, after all, like the people you meet at school, at the office, and on the street. With that in mind, apply what you know about people and your experience in dealing with others in understanding the person who interviews you.

Listen.   The smartest approach to the beginning of the inteview is to listen. The interviewer will likely give you some clues to their style and to their expectations and attitudes toward you. Concentrate and be attentive. You will pick up many clues during the interview by listening.

Emphasize yourself.   At an appropriate point, begin to sell yourself by bringing up your strong qualities and how you see them fitting into the company. Also during this part of the interview you respond to questions. But also be prepared to ask the interviewer pertinent and reasonable questions about the job. Stick to questions about the company and the job, and stay away from questions of benefits, vacations, and salary until you have almost been offered the position. It is a wise approach for you to let the interviewer introduce a discussion of salary and benefits.

Here are some questions that interviewers like to ask. Some are ice-breakers, designed to help you relax in what is certainly a stressful situation.

- "Well, tell me a little bit about yourself."
  A favorite question that not only allows you to talk about something you know, but allows the interviewer to find out some personal information they cannot ask. But more importantly it allows you to present your style. Respond, but not with your life story, and make the statements relate to the job you seek.
- "What made you want to work for our organization?"
  A valid question that you should answer directly and positively by stressing the good things you know or have learned about them.

- "Why are you leaving your present job?"

With this also be positive about your reasons. You show your professional tact when answering this question, so be honest, but evenhanded. Never show hostility or anger at your present employer. Leaving for a better job is the American way, and most employers understand how you can reach a plateau with a company and want to move up.

- "How did you get involved in this field?"

Again show that you are professional and that you like what you do. Be honest on this one, too.

- "What do you do in your spare time?"

This question usually determines the degree of devotion to your work. You may be a workaholic with no spare time, or you may be a member of the local hiking club. Respond honestly. Avoid the cliché, "I work hard and I play hard."

- "Where do you see yourself in five years?"

This is another favorite question at interviews, and your answer may suggest to the interviewer how you have planned your career. Answer realistically by mentioning when you plan to finish degrees, and if you will add to your training and education. In a way this is really a foolish question that should be eliminated from the list. Mindful interviewers don't ask it because of its foolishness and pretention.

- "What are your strong points?"

This should be easy because you have prepared for it. The question invites you to make your "sales presentation" to the interviewer. Be positive and direct your statements to the job. Avoid boasting.

- "What are your weak points?"

Watch out for this one, since the interviewer ultimately makes use of weaknesses to eliminate you from the competition. Address the question, but positively. For instance, if you feel that you could improve your communication talents, mention that and say that you are now enrolled in a writing course or speech course to beef up your skill. Or you could stress weaknesses that have nothing to do with the abilities required by the job. For instance, mention that you cannot resist the cheesecake at a bakery in town, and you see that as a weakness.

- "What is your present salary?"

This signals that the interviewer wants to either count you in or screen you out based on pay. You do not have to answer this, but some response helps diffuse a confrontation and might force the interviewer to mention the dollar figure that the position you seek has been budgeted for. Say "competitive" or "appropriate for the position and experience." You could then turn it around by asking what has been budgeted for the position.

- "What do you like about your current job?"

Be positive.

- "Can you see yourself doing this for the rest of your life?"

Turn this around by mentioning that you expected the position would eventually lead to one with more responsibility and challenge.

- "Do you plan to continue your education?"

Let them know that you see continued education as an investment in your growth, both as an employee and as a person.

- "What do others think of you?"

This question is best handled by mentioning that you get along well with others at work. Mention that you have the respect of both bosses and subordinates. The question invites you to discuss your social relations on the job, a very important quality for the interviewer and employer. So be positive, but never boastful or unrealistically upbeat for fear of protesting too much and implying the opposite.

- "What would you do if you were asked to go to an office overseas for a week to solve a problem for that division of the company, and your son was graduating from school?" This question is meant to test your loyalty to the company. It is up to you to decide how to respond, but give some slack on this question, otherwise you will appear as if you are loyal to a fault, like the staff of the Nixon White House.
- "What can you tell me about your schooling?"

Be positive about the institution you attended and the courses and professors you had there, since your response is really a reflection on you. After all, you are a product of that school, and if you denigrate its facilities, programs, and faculty, you indirectly belittle yourself.

The Exit

Sooner or later the interviewer signals that the session is coming to an end. This may come directly or indirectly. He may stand and begin to walk to the door, hand extended to shake yours. Before you leave try to end on a positive note and:

- be sure that you have summarized briefly your important offerings
- express a positive desire to work for the company
- thank the interviewer for an interesting and instructive meeting

Before you leave be sure you have asked, "What is the next step?" or "May I call you next week so that we can continue the discussion?" You may also be asked if the interviewer can "show you around," and introduce you to other people you will work with or you might be given an assignment pertaining to the job. Each of these is a good sign that they are interested enough to introduce you to the people you will be working with.

On the way home.   Take time to stop some place and go over what happened. Write down your recollections of the experience. Be sure to record clearly the names of all the people you met and make a clear list of the key points you discussed during the interview. Add any other observations you had that might be helpful not only for the next step, but the next interview situation if you do not get the job.

Curbstone Analysis.   The New York State Department of Labor's *Guide to Preparing a Résumé* suggests that you use a technique of salesmen after a sales interview, the "curbstone analysis." That is, review what was said and the reactions, and what might have been said or unsaid. Ask yourself these questions after the interview:

1. How did the interview go?
     - What points did I make that seemed to interest the employer?
     - Did I present my qualifications well? Did I overlook any that are pertinent to the job?
     - Did I pass up clues to the best way to "sell" myself?

- Did I learn all I needed to know about the job? Or did I forget or hesitate to ask about factors that are important to me?
- Did I talk too much? Too little?
- Did I interview the employer rather than permit the employer to interview me?
- Was I too tense?
- Was I too aggressive? Not aggressive enough?

2. How can I improve my next interview?

Follow-up.   After five days you can write a short note thanking the interviewer for the pleasant afternoon. Also stress:

- the main points you discussed.
- mention tactfully, and with appropriate compliments, the people you met.
- provide any added information that has occurred to you. After a week, call to find out what happened.

REMEMBER, IT TAKES ONLY ONE YES—SO DON'T GIVE UP.

SUMMARY

- The letter of application and résumé not only provides a statement of the skills and talents that you *offer* an employer, but a sales tool as well.
- Successful employment letters demonstrate your performance as a professional, as well as your knowledge of the position you seek. They also address a specific person, describe the position, detail the skills you offer, present relevant experience, and suggest the next step—the interview.
- Effective résumés win an interview by clearly presenting your qualifications through the appropriate written statement of your skills, education, and experience.
- Although no ideal format exists, the right résumé satisfies three main functions— your marketing tool, the employer's screening tool, the interviewer's springboard. Possible formats include: chronological, analytical, functional, expanded, creative. The format for the professional résumé presents a potential client with your credentials as part of a project.
- To prepare for an interview, gather information about the company and the job, review what you offer, and prepare a list of questions that you might be asked.
- During an interview, listen, emphasize your strengths, answer questions honestly and positively, and be sure you ask about the next step. Afterward, perform a curbstone analysis of your interview. Then follow up with a letter, and call if you have received no response after a week or 10 days.

EXERCISE 13–1, RÉSUMÉS

1. Look for notices of openings in a newspaper, or a professional journal in the field you are planning to start your career. Write a letter responding to an ad. Include a résumé tailored to the copy of the advertisement. Send out the letter and résumé and see what happens.

2. Write only a letter that you would send. Do not include a résumé with it, but incorporate the relevant information in the letter.

3. Prepare several résumés for yourself. You might construct one of each of the following formats:
   a. historical or chronological
   b. functional
   c. analytical
   d. creative

4. Write a professional résumé that might be included with a report on a company project.

EXERCISE 13–2, INTERVIEWS

1. Write out your responses to the questions that appear in this chapter.

2. Pair up with a friend or someone in class or at the office. Give them your résumé and ask them to interview you for a job. Then reverse the roles. Comment on the responses you both made. You might record or videotape the meeting for later critical evaluation.

3. In *What Color Is Your Parachute?*, Richard N. Bolles suggests that you interview a person now working in the job you seek. The purpose is to give you some important information, as well as the possibiity of getting your foot in the door of a company. Conduct one such interview with someone who works at a firm you are interested in. Based on the discussion, write a short analysis and evaluation of the company, and the skills and talents you offer them. Decide if you are their idea of a useful employee.

4. Write a follow-up letter for an interview that you considered positive.

EXERCISE 13–3

Write your impression of the person who submitted this cover letter. Discuss the type of response you think this letter received.

IN THE NAME OF THE GIVER OF LIFE, THE
COMPASSIONATE, THE MERCIFUL

Early in my life, I decided that I wanted to make a lasting contribution to the health and welfare of mankind. Hence, throughout the acquisition of my education experiences, I have met and studied with many scholars of the day. This contact with its many cultural influences as well as my own intellectual versatility has (to a certain extent) prevented me from specializing in any established or particular discipline.

Nevertheless, due to the MERCY AND BLESSINGS OF THE LORD AND CREATOR OF THE UNIVERSE, I have been gifted with a rare insight, which has enabled me to penetrate the essentials of my accumulated knowledge.

As a result of my diverse background in applied sociobiological research, I believe that I have demonstrated my ability to participate with program directors and administrators in

the design, implementation, feedback, and institutional
implication of personnel services.

John W. Doah

EXERCISE 13-4

Comment on the following letter of application.

Dear Sir:
   Throughout my life I have always considered myself to be
somewhat of a free-spirited herald. Recognizing that great
experiences can be forms of ego-enhancement because they
fill our drives for self-justification and ego-
verification—continually choosing and discarding, I am
free.
   If this be true, then all substantive change whether it be
of institutions or of personal beliefs derives from changing
one's metaphors for what one is doing. So, by dropping all
"conceptual blinders" and critically rethinking my own
basic presuppositions, I have come to realize that the present will always (ideally) contribute to the building of the
future. Thus, I belong irreducibly to my time and it is for my
time that I should live. Consequently, any full view of reality must consider the important reality that exists in your
head and mine, for our total reality consists solely of both
"manifest" and "unmanifest" lines.
   Hence, throughout the years I have kept abreast of the
latest "socially approved wisdoms" found most acceptable in
classrooms. Although these "socially approved wisdoms" are
but merely administrative labels, I have constantly incorporated these historical, political, psychological and
socioeconomic implications in my decisions, thus providing
for new sources of insights, discoveries, and concep-
tualizations. Simply, because life is reality-process interrelationship oriented rather than verbal-definition
oriented.
   For example, early in my collegial days I entertained the
thought that I might comparatively be satisfied with a career
solely as a primary care physician in a rural area or a metro-
politan hospital. However, now that I am somewhat more knowledgeable about the metaphysical relationships involved in
"altered states of consciousness," my experiences have led
me through the "roots" of botanical medicine on an ethnophar-
macologic search for psychoactive drugs. Moreover, my appe-
tite has now inherently shifted towards the field of ethno-
psychiatry or cross-cultural psychiatry in health care.

In "The Struggle Against Irrationalism," the fact was emphasized that "human behavior, normal and abnormal, is the most complex field facing mankind." Hence, due to the complexities of modern day life, a growing need has developed for interdisciplinary training which relates sociocultural systems and human variation to physical and mental health problems. Accordingly, this path I have chosen culminates with a Ph.D., and for this purpose a Master's degree is usually required. On the other hand, if the individual is able to show an equivalent level of training in a behavioral science, he or she is considered as a candidate. As argumentum ad hominem, my qualifications are chutzpah, zest and temerity! (See the enclosed curriculum vitae.) And so, regardless of any clinical or practical experience, applicants must demonstrate strong commitment to intellectual goals and the mastery of academic skills.

To sum up my work experiences, instrumentally a certain amount of this erudition was dervied from a hospital. Definitively, this general hospital, as all medical centers, is a prototype of the multi–purpose organization; it is a hotel, and a school, a laboratory and a stage for treatment. To a considerable extent, a self–contained social universe. Next, it is my belief that:

1) Learning takes place best not when it is conceived as a preparation for life, but when it occurs in the context of real daily life.

2) Each "learner" ultimately must organize his or her learning in his or her own way.

3) "Problems" and personal interests are more realistic structures than are "subjects" for organizing learning experience.

4) Students are capable of directly and authentically participating in the intellectual and social life of their community.

5) That the community badly needs them to do this. This transforms the relevant problems of the community into the student's "curriculum."

Yet, en route to fulfilling my career objectives, financial problems have constantly beset me. Therefore, because of my socioeconomic status (ipso facto), it will take me years to accumulate enough credits to matriculate to professional school. Consequently, my health occupation courses have been the primum mobile of my biomedical education thus far. This preparation of short–term goals has allowed me to obtain necessary job skills, which in turn have enabled me to work full or part–time (when I can find employment), while continuing my education. Each of these disciplined intermediate goals has given satisfaction; thus, at the same time preparing me for the next intermediate goal. Sic itur ad asta!

14

Writing Confidently
and Clearly
on the Job

Rather than summarize the contents of the entire book, some suggestions for maintaining good writing habits should help you hold on to the skills you have gained, and contribute to your autonomy as a writer. Writing is an act of discipline, as well as a skill and an ability. On the job, write often so you continue to build your writing strength and confidence.

SUGGESTIONS FOR
MAINTAINING GOOD
WRITING HABITS

As an athlete maintains tone, so must a writer. One of the best ways to do this is to keep a journal. Write in it regularly, every day if possible, but not less than twice a week.

To begin, you might try writing for five minutes without regard to anything other than writing. You should not allow your pencil to stop until the five minutes are up. Start with descriptions of familiar things. As you continue, you can lengthen the time and write in more complex forms such as argument or exposition. However, the important thing is not what you write in your journal, but the act of writing something regularly.

Be aware of good writing when you read it, and try to incorporate its good qualities into your own. Notice poor writing in order to identify it in your own work and eliminate it through revision.

Leave notes for members of your family, friends, co-workers. Write letters to newpapers, TV, and radio stations to let off steam, but do not mail them.

Rewrite your memos, letters, and reports in order to make them clearer and more readable.

Express ideas in your own words.

Edit the writing of those who write for you. Discuss the quality of writing with them.

When you feel stuck, or when you need to meet a pressing deadline, make use of the brainstorming techniques—the 10-minute focused writing, forced associations, questions to discover fresh thoughts, an absurd revision of material, a dialogue with the intended audience, deadlining, value analysis, function analysis, and storyboarding.

SUMMARY

- Keep a journal.
- Notice the writing of others.
- Write often.
- Review your own writing critically.
- Rethink events, and express them freshly.
- Use brainstorming techniques.

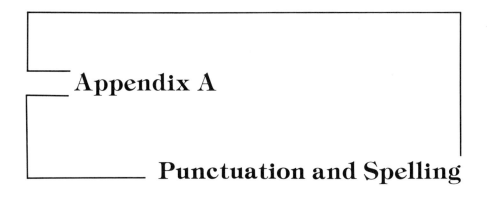

Appendix A

Punctuation and Spelling

PUNCTUATION

Rather than give you a set of rules, it is more fruitful first to discover a way to think about the function of those dots and other marks. If we consider them as signs, much like the ones on our streets and highways, their purpose clearly emerges. Without either road signs or punctuation, ambiguity and chaos grow. Take the following sentence:

> If you dont know that you dont know you think you know if you dont know that you know you think you dont know

Reading that is like driving through an unfamiliar city with no road signs and no map. We need a guide.

Punctuation provides your reader with directions. As the *MLA Handbook* explains: "The primary purpose of punctuation is to ensure the clarity and readability of your writing."

Periods end sentences. Think of them as a full stop.

Commas separate items in a series (morning, noon, and night) and coordinate adjectives (a professional, intelligent evaluation). They come before coordinating conjunctions that join independent clauses. They enclose parenthetical expressions and follow long introductory phrases or clauses. They are used in names (John Smith, Jr.), dates (July 10, 1982), and addresses (Dallas, Texas).

Colons indicate that a series of examples, an explanation, or a quotation follows.

Semicolons separate items in a series when some of the items themselves use commas. They also separate clauses when no coordinating conjunction is used.

Quotation Marks enclose the words of others, words referred to as words, and words purposely misused or used in a special sense.

Avoid the overuse of: italics, parentheses, and dashes.

SPELLING

Business, technical, and professional people expect accurate spelling. Although it is part of the etiquette, and not the substance of writing, spelling is the one item that almost everyone can easily judge and on which everyone seems expert when it comes to the writing of others. Since correct spelling is expected, rather than perceiving of it as a chore, think of it as a courtesy. It is a way of saying, "I care about how I present myself, my ideas, and the organization I represent to you."

Correct spelling requires work, and some short cuts often compound the effort. However, here are several methods to help you improve your spelling.

1. Pay attention to the way words are spelled as you read. Try to remember that picture of the word when writing.
2. Write a correct list of the words you actually misspell. Say the words aloud, emphasizing the troublesome part.
3. Distinguish between words that sound alike, but have dissimilar meanings and uses. Although I have included a list of sound-alikes, consult a dictionary as the final authority on usage.
4. Pay special attention to spelling when adding a suffix or forming the plural.
   a. Most words ending in a silent -e drop the -e before adding a suffix which starts with a vowel.
      move, moving
      bake, baking
      write, writing
      Exceptions: *the -e retained for meaning in words like *hoe, hoeing* and *singe, singeing*
      *keep the e if the suffix begins with a consonant
      *words ending in -ie usually drop the -e and change the -i to y before adding -ing
   b. Usually we add an s to make a noun plural.
      Exceptions: *Words ending in *f* or *fe*, add only the s as in *chiefs, gulfs*, and *safes*. Others, drop the *f* or *fe* and add *ves* as in *shelf, shelves*.
      *Words ending in s, sh, ch, or x usually add *es:*
      excess   excesses
      dish   dishes
      pitch   pitches
      box   boxes
      *Words ending in y add s if a vowel precedes it, as in *buy, buys*.
      However, if a consonant does, the y changes to i before adding *es, duty, duties*
5. Keep a college-level, hard-cover dictionary on your desk or nearby. Good dictionaries provide more complete spelling rules, and an indication of words with two acceptable spellings, for example *judgment, judgement*.

Spelling Exercise

Many of the commonly misspelled words appear in the letter below. Identify all the words spelled incorrectly by circling them.

Dear Mr. Hackensaken:

I would like to accomodate your request to complete the preceeding questionaire regarding Mr. Knbhobs work, but I feel it would be a hinderence to his career were I to do so. It would be a privelege to give you better assistance and full-fil a committment I feel toward your applicant by writing personally. I would not exagerate were I to tell you that his conscientous maintainance and managment of his unit let us sieze the collosal governor's trophy last year.

Incidentally, it was at his insistance that I was able to pursuade the hygeine department to innoculate the entire state after the existance of hog fever was discovered. This operation required unparalelled assistance from the auxillary units, and it would be permissable to say that similar useage of these units has often faltered because hte government interferred with the patients' psycological evaluations. He was able to sheperd his unit to success by sueing the agency; this was much preferrable to publicly challenging their wierd policies.

In all sincerety, I don't think he will disapoint you; he has been a bouyant, uplifting force here and I'm sure these successes will be transferrable to your organization.

Yours truly,

P. J. Norgan

Here is the same letter with the words correctly spelled. How many of the 35 words did you catch? This exercise is part of a computer manufacturer's demonstration of the power of the word processor to correct spelling rapidly.

SUGGESTION: To check for accurate spelling, read a page of text from the bottom up. That is, begin with the last word, see if it is correct, go to the next, and the next. . . . Reading backwards makes sure that you look only for spelling, and not for errors in content, grammar, or usage.

Dear Mr. Hackensaken:

I would like to <u>accommodate</u> your request to complete the <u>preceding</u> <u>questionnaire</u> regarding Mr. Knbhobs work, but I feel it would be a <u>hindrance</u> to his career were I to do so. It would be a <u>privilege</u> to give you better <u>assistance</u> and <u>ful-</u>

fill a commitment I feel toward your applicant by writing personally. I would not exaggerate were I to tell you that his conscientious maintenance and management of his unit let us seize the colossal governor's trophy last year.

Incidentally, it was at his insistence that I was able to persuade the hygiene department to inoculate the entire state after the existence of hog fever was discovered. This operation required unparalleled assistance from the auxiliary units, and it would be permissible to say that similar usage of these units has often faltered because the government interfered with the patients' psychological evaluations. He was able to shepherd his unit to success by suing the agency; this was much preferable to publicly challenging their weird policies.

In all sincerity, I don't think he will disappoint you; he has been a buoyant, uplifting force here and I'm sure these successes will be transferable to your organization.

Yours truly,

P. J. Norgan

WORDS OFTEN MISSPELLED

abbreviate	analyze	believe
absence	apparatus	benefit
accept	apparent	benefitted
access	appearance	biscuit
accessible	appointment	born
accident	appreciate	borne
accidentally	architecture	bureau
accommodate	arctic	bureaucracy
accumulate	argument	business
achieve	ascent	
acknowledge	assassin	calendar
acquaintance	assent	campaign
acquitted	assistance	candidate
address	athlete, athletics	can't
affect	author	catastrophe
aggravate	average	cellar
all right	awful	cemetery
alley, alleys		characteristic
ally, allies	balance	chauffeur
already	balloon	chimney
altar	bare	client
alter	baring	cloths
altogether	barrel	clothes
always	battalion	colonel
amateur	bear	column
among	bearing	commence

committee
comparative
comparison
compel, compelled
competitive
complement
compliment
conceive
condemn
condescend
confidential
connoisseur
conscience
conscious
conspicuous
continuous
convenient
cooperate
corollary
corps
corpse
correlate
council
counsel
counterfeit
crisis
criticism
criticize

deceit
deceive
decent
decision
defendant
descent
desert
desirable
dessert
diary
disappoint
disastrous
disease
dissipate

economics
efficient
eligible
eliminate
embarrass
employee
engineer
environment

equal
equally
equip, equipped
equivalent
especially
exaggerated
exceed
exhibit
exhilarate
existence
experience
extraordinary

familiar
fascinate
February
finally
financier
foreign
forfeit
formally
formerly
forty
friendly
fulfill
fundamental

gauge
genius
government
grammar
grievous
guarantee

harass
height
heroes
hygiene
hyposcrisy

illiterate
illusion
immediately
incident
incidentally
indispensable
individual
inevitable
ingenious
ingenuous
innocent
intelligent

intercede
interfere
irresistible

laboratory
legitimate
leisure
library
license
lightning
livelihood
lose

maintain
maintenance
maneuver
mathematics
medicine
miniature
minute
miscellaneous
muscle
mutilate
mysterious

naive
necessary
niece
noticeable

occasion
occasionally
occur
occurrence
official
omission
opinion
opponent
opportunity
optimist
origin
outrageous

pageant
pamphlet
parallel
parenthesis
peaceable
perform
permanent
perseverance
personnel

physical
picnic
picnickers
politician
possession
potatoes
precede
precedence
precious
prejudice
president
principal
principle
privilege
proceed
procedure
professor
prominent
psychology

quantity
questionnaire
quiet
quite

realize
receipt
receive
recommend
refer
referred
relieve

repetition
representative
resistance
restaurant
rhythm

sacrifice
scarcely
schedule
science
secede
secretary
seize
separate
sergeant
siege
similar
simultaneous
sincerely
skillful
society
sophomore
stationary
stationery
statistics
strategy
studying
succeed
supersede
suppress
symmetrical

synonym

temperament
temperature
temporary
tendency
therefore
thorough
tragedy
transferred
translate
truly
typical
tyranny

unanimous
unconscious
undoubtedly
unnecessary
until
usual
usually

valuable
vegetable
vengeance

weather
weird
whether
writing

WORDS THAT SOUND ALIKE

accede   to agree to
exceed   to go beyond; to surpass

*ac*cept   to receive willingly
*ex*cept   (verb) to omit; aside from

*ac*cess   admittance
*ex*cess   surplus

ad*a*pt   to adjust
ad*e*pt   proficient
ad*o*pt   to choose

*ad*dition   something added
*e*dition   one version of a printed work

adheren*ce*	attachment
adheren*ts*	followers
a*d*verse	opposing
a*v*erse	disinclined
advi*c*e	(n) information; recommendation
advi*s*e	(v) to recommend; to give counsel to
*a*ffect	to influence; to pretend
*e*ffect	(verb) to bring about; (noun) result
al*lowed*	permitted
al*oud*	audibly
al*l*ude	to refer to casually, or indirectly
*e*lude	to avoid, escape, or baffle
al*l*usion	an indirect reference
i*l*lusion	an unreal vision
alt*ar*	a place of religious worship
alt*er*	to modify
appra*i*se	to set a value on
appri*s*e	to inform
attendan*ce*	presence
attendan*ts*	escorts; followers; companions
ba*i*l	security; the handle of a pail; to dip water
ba*le*	a bundle
b*are*	naked
b*ear*	(n) an animal; to carry; to endure
bas*e*	(n) foundation; (adj.) mean
bas*s*	lower notes in music; a fish
bas*es*	plural of base and basis
bas*is*	foundation
besid*e*	next to
besid*es*	in addition to
b*illed*	charged
b*uild*	to construct
bo*a*rder	one who pays for his meals and often his lodging as well
b*o*rder	edge

calendar	a record of time
calender	a machine used in finishing paper, a cloth
colander	a strainer
canvas	(n) a coarse cloth
canvass	(v) to solicit; a survey
capital	a capital city; money
capitol	the building in which a government legislates
Capitol	the building in which the Congress of the United States meets
cite	to quote: call before a court
site	a location
sight	a view; vision
coma	an unconscious state
comma	a mark of punctuation
command	(n) an order; (v) to order
commend	to praise; to entrust
commence	(v) to begin
comments	(n) remarks
complement	that which completes
compliment	praise; recommendation
cooperation	the art of working together
corporation	a form of business or organization
correspondence	letters
correspondents	those who write letters
corespondents	parties in law suits
council	an assembly
counsel	(n)advice, adviser; (v) to advise
consul	a government official
decent	proper; right
descent	going down
decree	a law
degree	a grade; a step
deduce	to infer
deduct	to subtract
defer	to put off
differ	to disagree
deference	respect; regard for another's wishes
difference	dissimilarity; controversy

de*p*osition	a formal written statement
di*s*position	temper; disposal
devi*c*e	an instrument
devi*s*e	to bring about
dis*a*pprove	to withhold approval
dis*p*rove	to prove the falsity
dis*b*urse	to pay out
dis*p*erse	to separate; to scatter
di*ss*ent	a disagreement; to disagree
de*sc*ent	a downward step, decline
*e*licit	to draw forth
*il*licit	unlawful
*eli*gible	fitted, qualified
*ille*gible	unreadable
*e*merge	to rise out of
*im*merge	to plunge into
*e*minent	famous; prominent
*im*minent	impending, threatening to occur at once
imm*a*nent	operating within; actually present in
emanate	to originate; to issue from a source
envelo*p*	(v) to cover; to wrap
envelo*pe*	(n) a wrapper for a letter
equable	uniform, even
equi*t*able	fair, just
*erasa*ble	capable of being erased
*irasci*ble	quick tempered
everyone	all persons, collectively
every one	each thing or person without exception
exp*a*nd	to increase in size
exp*e*nd	to spend
exp*a*nsive	capable of being extended
exp*e*nsive	costly
ext*a*nt	existing
ext*e*nt	measure
f*a*rther	refers to actual measurable distance
f*u*rther	refers to time or degree (as, further thought)

formally	in a formal manner
formerly	previously
forth	away; forward
fourth	next after third
hear	to perceive by ear
here	in this place
heard	past tense of hear
herd	a group of animals
hypercritical	overcritical
hypocritical	pretending virtue
incidence	range of occurrence
incidents	accidental happenings
ingenious	clever
ingenuous	naive
intense	acute; strong
intents	aims
interstate	between states
intrastate	within one state
its	possessive form
it's	contraction of it is
lead	(n) heavy metal
led	guided (past tense of *to lead*)
lean	(adj.) thin; (v) to bend
lien	a legal claim
leased	rented
least	smallest
lessor	one who grants a lease
lesser	smaller; inferior
loan	that which one lends or borrows
lone	solitary
lessen	(v) to make smaller
lesson	(n) an exercise assigned for study
mail	correspondence
male	masculine
ordinance	a local law
ordnance	arms; munitions

p*ara*meter	a quantity with an assigned value; a constant
p*eri*meter	the outer boundary
person*al*	private
person*nel*	a staff of workers
pr*ecede*	to go before
pr*oceed*	to go forward
p*er*quisite	privilege
p*rere*quisite	a preliminary requirement
p*er*secute	to oppress
pr*o*secute	to sue
princip*al*	capital sum; head of a school; (adj.) chief
princip*le*	a truth; a law
pr*o*spective	anticipated
p*er*spective	a view in correct proportion
prophe*c*y	a prediction
prophe*s*y	to predict
real*ity*	actuality
real*ty*	real estate
re*c*ent	late
re*s*ent	(v) to be indignant
sometime	(adv.) at an unspecified time;
some time	*Some* is an adjective modifying the noun time. (If little could be placed between some and time, the words should not be run together.)
sp*a*cious	having ample room
sp*e*cious	outwardly correct but inwardly false
station*a*ry	not movable
station*e*ry	writing paper, etc.
th*a*n	used in comparison; e.g., greater than
th*e*n	refers to time; e.g., now and then
the*ir*	ownership (as, their hats)
the*y're*	contraction of they are
the*re*	in that place
v*e*racity	truthfulness
v*o*racity	greediness
w*ai*ve	to put away, forego
w*a*ve	to gesture with a hand

w*ai*ver	the relinquishment of a right
w*a*ver	to hesitate
who*se*	possessive form of *who*
who'*s*	contraction of *who is*
you*r*	possessive form of *you*
you'*re*	contraction of *you are*

Appendix B

Sample Report

New East Telephone PBX Strategy

December 7, 19XX

Submitted by

Karen E. Smith

By Karen E. Smith with her permission

BACKGROUND

The communications industry, a regulated utility for seventy-five years, is now competitive. This seventy-five billion dollar business is expected to double over the next five years. Dozens of companies are struggling to win bigger pieces of the U.S. market by applying new technologies.[1] The office of the future will have a joint telephone/computer system. Therefore, the successful communications vendor must integrate the voice capabilities of the telephone with the storage and retrieval power of the computer.[2]

The anti-trust settlement that American Telephone and Telegraph (AT&T) worked out with the Justice Department allows them to compete in these new fields and go after the high revenue computer business. AT&T will give up its costly, low profit local phone company operations in exhange for this freedom. They want to compete without restraints and are willing to shed $87 billion, about two-thirds, of their assets. (See Exhibit 1.)[3]

The entire Bell System will be restructured by January 1984. I interviewed Joe Rose of XYZ Planning Institute, the consulting firm handling the AT&T reorganization. They discussed the following important points, referring to an Organizational Chart. (See Exhibit 2.)[4]

1. American Telephone International (ATI), Advanced Mobile Phones (AMPS), and Long Lines (ATTIX) will remain wholly owned subsidiaries of the parent company, AT&T.
2. American Bell Incorporated (ABI) is a recently formed business subsidiary of AT&T. It will provide a shared data communications service and maintain the entire line of business PBX products in the country.[5]
3. The twenty-two Bell operating companies (BOCs) will be independent of AT&T and divided into seven U.S. regions.
4. The Directory Division of AT&T, commonly known as the Yellow Pages, will be part of the BOCs as a form of financial assistance since this is a lucrative segment.
5. Bell Research Laboratories will not be available to ABI or any of the operating companies.

BREAKING UP THE ASSETS OF A.T.+T.

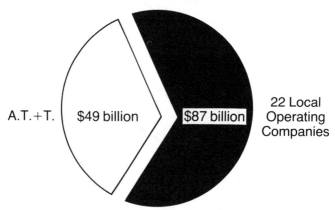

A.T.+T. $49 billion $87 billion   22 Local Operating Companies

EXHIBIT B-1   Pie Chart Showing the Distribution of Assets of AT&T

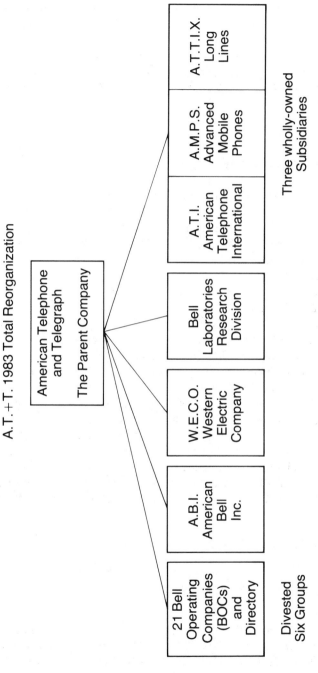

A.T.+T. 1983 Total Reorganization

American Telephone
and Telegraph

The Parent Company

| 21 Bell Operating Companies (BOCs) and Directory | A.B.I. American Bell Inc. | W.E.C.O. Western Electric Company | Bell Laboratories Research Division | A.T.I. American Telephone International | A.M.P.S. Advanced Mobile Phones | A.T.T.I.X. Long Lines |

Divested
Six Groups

Three wholly-owned
Subsidiaries

EXHIBIT B-2   Organizational Chart Showing the Proposed Reorganized Structure of AT&T

This is the most important anti-trust development since the Supreme Court ordered the dissolution of Standard Oil Trust in 1911.[6] To become a high-technology, innovative competitor in the market place, AT&T trades the 100 year role as the main supplier of basic telephone services to the nation's homes and offices. But what is the fate of the twenty-two (including Directory) BOCs that AT&T has agreed to divest by January 1984? According to some industry observers, AT&T is "casting [them] adrift on a leaking raft."[7]

THE NEW (BELL) OPERATING COMPANIES

The BOC will be significantly affected by the divestiture in many ways. Entrance to the electronic information processing businesses is prohibited.[8] Another restriction is that AT&T will take over all equipment already in use on customers' premises and the BOC may not supply equipment in the future.[9] The BOCs are permitted to supply equipment for new customers but are excluded from the telecommunications manufacturing market.[10]

Once the exclusive local communications franchise, the BOCs are now under attack by cable television, emerging radio services, and other firms. No part of the BOC revenue is safe from competition.[11] With no cross-subsidies from other phone services, such as toll and equipment revenue, previous installation undercharging will cease and local phone rates will increase.[12]

Local phone companies expect rate increases of 100% on the average.[13] For example, New East Telephone (NET) filed requests for rate boosts in Massachusetts[14] and in Maine.[15] Both requests include business customers with special services such as Private Branch Exchange (PBX) switching systems.

Aggressive business strategy is the salvation for the freed BOCs. They must overhaul operations and be the best provider of local access services. Of the nation's more than 142 million phones installed, the local phone companies have 80%.[16] They possess built-in expertise in management and technical training. However, aggressive sales techniques and new marketing strategies are necessary for BOC survival.

NEW EAST TELEPHONE—A NEW BOC

Background: The Northeast PBX Market

A PBX or Private Branch Exchange is an electronic telephone switching system employed by large business customers. The PBX provides a sophisticated communication system including such features as data transmission, energy control, and property management. The cost can start at $100,000 and be pushed easily into the millions depending upon the number of people it will serve and the features desired.

In addition, there is a substantial monthly charge which is a source of revenue for the vendor.[17]

If we focus on the market in the five-state region, we see a total of 6912 PBXs in service as of May 1982. Of this total, 4312 or 62.4% were installed by New East Telephone Company (NET). The remaining 2600 or 37.6% were installed by any of the 60 competitors. (See Exhibit 3.) The market is expanding rapidly and has grown by 50% in the last six years.[18] Growth is expected to accelerate tremendously due to the communications revolution.

Since the early 1970's, PBX service is no longer a legal monopoly of AT&T. The entrance of many new companies forced firms to become price competitive and to push for leading edge technology.[19] NET's dilemma is now compounded by the divestiture agreement of AT&T. They must rapidly implement marketing strategies to maintain status as a leader in an expanding PBX market.

CORPORATE PURPOSE AND OBJECTIVES

As a monopoly, NET's purpose or mission was perfectly clear and simple. The purpose has now lost its appropriateness due to the new conditions in the environment. NET is no longer *the* telephone company but rather another vendor with a leader's share of the market. NET must develop a new purpose based on the following five key elements:

A. *History*—NET has a reputation for being service oriented and highly responsive to consumer needs. Although they are now to be market driven, NET has to capitalize on this solid reputation and include it in their new statement of purpose.

B. *Current Preferences*—Stockholders and upper management want to maintain market leadership by strengthening their PBX business in whatever way necessary.

C. *Environmental Considerations*—The communications industry is expanding rapidly, creating large potential for PBX vendors. Competition is forcing the industry into faster technological growth. Deregulation will untie NET's hands in the use of marketing tools.

D. *Organizational Resources*—NET is the market leader and will be released from AT&T in a secure financial position. It is the only communications company (as part of AT&T) to design, manufacture, deliver, install, and service its own PBXs, providing "end to end" responsibility.

E. *Distinctive Competences*—AT&T invented the PBX and dominates in its ongoing development. NET, once backed by the Bell Labs and supported by AT&T, has a reputation for providing the best communications systems in the world. This is their strongest point. Even after the separation, NET will still carry the AT&T reputation.[20]

Generally, the company's new mission is to change from a previously engineering dominated complacent monopoly to a market driven, aggressively competitive market leader. The objective is to offer the best product with the best service at a competitive price. With the purpose and objectives established, the company

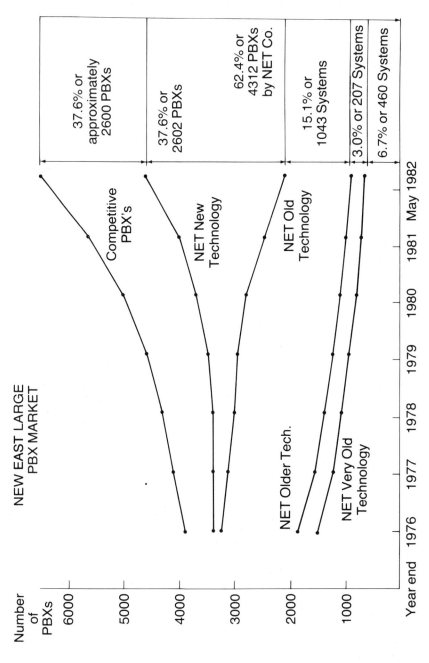

EXHIBIT B-3 A Line Graph Illustrating Historical Changes in Regional PBX Systems

must do two things: 1) Act on possible growth opportunities and, 2) devise a set of competitive marketing strategies appropriate for a market leader. I obtained information from PBX Marketing at New East Telephone Company. They answered 73 specific marketing audit questions[21] which I've summarized into the following:[22]

I. Growth Opportunities

A. Market Penetration
The company has set up a huge, aggressive, and well trained sales force. Sales people specialize in a particular industry and receive part of their salary on a commission basis. Accent is placed on stimulating current usage, and attracting competitor's customers and non-PBX users.

B. Product Development
The company has developed a most advanced PBX. It is available in fifteen different feature packages to suit all needs, adaptable to any type of business customer of any size. Designed as technological building blocks seven years ago, the PBX system architecture permits customers to add advanced functions as they need them, without purchasing new equipment.[23]

C. Horizontal Integration
The local telephone companies have been signing contracts with other PBX vendors which allow the use of the competitive products as part of the telephone companies' product line. Interconnection is now legal. This gives a much broader selling base for the competitior and eliminates some competitive threat to the phone company. For example, part of the NET PBX product line is manufactured by GTD Corporation.

D. Concentric Diversification
NET has added new products that are technologically related to the PBX. An example is an advanced paging system that can be an extension of a PBX and can also provide "talk back" capability to the paged party.

II. Market Leader Strategy in the PBX Market

A total of sixty competitive PBX firms are constantly challenging NET's strengths or attempting to exploit its weaknesses. As a dominant firm, NET's objective breaks down into the following three categories:

A. Make the total market grow larger.
A dominant firm gains the most from any increase in market size. NET has tried to find new users by advertising that their PBX can save a non-user money and time. New PBX capabilities were developed to appeal to those who need more than a simple piece of conversation equipment. An example is the Energy Feature Package PBX which shuts off connected devices on a scheduled basis. The Property Management Interface device enables management to check on the status of rooms in a building, such as a hotel, to detemine if occupied, clean, in need of cleaning, or whatever condition applies.

Increased usage is promoted by advertising that phones can substitute for business travel. The term "telemarketing" was coined as a way for PBX users to save expenses by increasing the use of their telephones and the PBX features.

B. Protect current market share.

NET concentrates efforts in the following three areas:

1) Innovative Strategy—NET publicizes new product ideas and customer services. Phone Center Stores as a means of distribution is a well known innovation. NET emphasizes cost cutting discoveries such as energy saving features.

   The great majority of business customers today are individual entrepreneurs, partnerships, and medium sized firms. They have a need for information management unlike the billion dollar multi-national corporations.[24] Decisions about communication solutions are based on different economic factors. A new PBX, the Prelude, sells for 25% less than the Dimension 100, yet offers a number of additional features not previously available on the smaller scale PBX. The enhancements are sophisticated and user friendly.[25]

   When asked why the Dimension PBX is the market leader, the Business Marketing Vice President, replied, "Other competitors may be able to match aspects of our PBX but no one has integrated so many capabilities as we have."[26]

2) Fortification Strategy—As a regulated utility, NET undercutting competitors' rates was prevented. Utility commissions insisted that NET rates be compensatory. Within these restrictions NET revamped pricing structures implementing variable term payment plans (VTPP) which give customers rate protection of long-term obligations.[27]

   The company has broadened the product line to satisfy very small users (less than 100 phones) with the Horizon and very large users (over 2000 phones) with the Dimension Custom.

   NET should aim to please the small customer as well as the large. Successful small businesses grow and remember quality products and dependable service. Attention to the little guy will help change NET's image as the uncaring monopoly. Adjust sales incentives to increase motivation to sell the smaller systems and supplement the product line to satisfy all needs.

   As the Bell System is broken up, NET will have the opportunity to switch suppliers which could result in big savings. Western Electric Comapany (WECO) was a profit center within AT&T so transfer prices were high. NET can now shop around for the best products at the best prices.

3) Confrontation Strategy—The onslaught of many new competitors in the PBX market has forced NET to launch massive advertising and promotional programs which most competitors can't afford. More advertising of the total Bell system as 'The Knowledge Business' is being broadcast. With deregulation in 1983, the telephone company will be free to change the pricing structure. Bell's emerging business policy dictates that the Dimension PBX will be competitively priced across the product line.[28] Predatory pricing and price wars may result.

   NET will establish a specially trained group to be the initial telephone contacts with PBX customers. They will call existing clients to check on consistent satisfaction and potential ones to set up sales appointments. Professionalism will be stressed at all levels of customer relations. Each employee acts as a marketeer and can make or lose a sale.

   The firm's management will stop transferring and promoting non-marketing people into sales/marketing positions. These openings will be filled by professionals with a proven track record. Job function experience is more important than product knowledge.

   NET is considering the employ of an outside advertising agency. Hiring a specialist may cost less than maintaining an in-house division. Many NET advertising managers lack the contemporary background of managers in a progressive advertising organization.

C. Expand market share.

NET depends upon its capability to provide maintenance service far superior to the competitors'. Many customers who switched from NET PBXs to a competitor, in search of a lower price, have learned that the NET higher rates support 24-hour, seven days per week emergency service.[29]

The PBX delivery and service interval must be improved because this is a strong selling point. Sales people must be able to quote shorter intervals and guarantee they will be met. NET must not continue to let its size reflect the time span between delivery and service.

Since the residential market has proven unprofitable due to high delivery, installation, and maintenance costs, NET will lean into the business customer market. Fortunately, the residence customer can be adequately served from the "Phone Center Stores."

If the company cannot increase market share, it can at least stop the competitive market share from growing due to NET losses. As the total market expands, all companies experience proportional growth. The market leader maintains its lion's share and capitalizes on competitive mistakes and shortcomings. One major marketing campaign aimed at doing just this is the Migration project.[30]

THE MIGRATION PROJECT

As of May 1982, NET held 62.4% of the PBX market in the Northeast region. (See Exhibit 3.) This divides into four categories as shown on the graph: 1. New technology—stored program control, 2. Old technology—Electronic control, 3. Older technology—Electro-mechanical control, 4. Very old techonolgy—Mechanical control.

The potential market volatility is not in the total expansion of the market, which is important and relatively assured, but in the potential loss of customers using old technology. The old machines can be characterized by a lack of modern capabilities, a high rate of maintenance failures, and the difficulty and expense of maintaining parts inventories.

The old technology customers, though pleased with the lower rates, are bombarded by proposals from competitive vendors. They then become potential emigrants, customers who switch to a competitive product. NET's Migration program consists of the following plan: Substantially raise the monthly payment of the old technology users to a point higher than present new technology rates. Simultaneously, hit the user with an aggressive marketing proposal for a new technology PBX. This will force a customer decision.

The least likely alternative is retention of the old PBX which still results in NET revenue. The second likelier alternative is to emigrate to a competitive PBX. This could still result in a net gain for NET because some of the old technology PBXs have become maintenance nightmares. The third alternative is to buy a new technology PBX from NET. This would result in increased revenue and decreased maintenance for NET.

It is a risky plan. Many customers could take offense at the raising of their rates and retaliate by going to the competition. However, the strategy has been implemented in the Northeast and theoretically should maintain the NET market share and prevent additional competitive market growth due to unforeseen NET losses.

Thus far, the Migration policy has been successful. It gives NET the opportunity to direct-sell new technology before a competitor can lure the client away. Odds are that NET will retain market leader's share of the migrated customers and save money by phasing out the high maintenance old technology.[31]

FUTURE PREDICTIONS FOR THE PBX MARKET

What will we witness in the next ten years? Of the 225,000 PBXs now installed, AT&T controls 67%. The market is expected to grow to 325,000 systems. Shipments will increase from 23,600 in 1981 to 44,000 in 1985, to 64,100 in 1990. New technology PBX installations will rise from 11% in 1982 to 99% in 1990. And the installed PBX life will decline by one third to 6.9 years, increasing turnover. The value of shipments will accelerate from $2.1 billion in 1981 to $6.3 billion by 1990.[32] NET can prosper in this expanding market. In the PBX business, NET can remain the market leader, not just by doing any one thing very well, but by doing everything right.

REFERENCES

(1) "Telecommunications—Everybody's favorite growth business, the battle for a piece of the action," *Businessweek*, October 11, 1982, p. 60.

(2) Ray Blain, "Telephones join Computers to Usher in Office of the Future," *Telephony, The Journal of Telecommunications since 1901*, July 5, 1982, p. 46.

(3) "AT&T Anti-trust Settlement," *The Wall Street Journal*, January 11, 1982, p. 4, column 1.

(4) Joe Rose of XYZ Planning Institute. New York, New York.

(5) John F. Malone, *Telephone Engineering and Management*, (Harcourt Brace Jovanovich Publications, Geneva, Ill.) , March 15, 1982.

(6) "AT&T Anti-trust settlement," *The Wall Street Journal*, January 11, 1982, p. 1.

(7) "Telecommunications . . . ," *Businessweek*, p. 60.

(8) "Heads of AT&T Local Units Tell Congress Divestiture Won't Hurt Their Operations," *The Wall Street Journal*, February 24, 1982, p. 52.

(9) Sanford B. Jacobs, "Business Phone Rates Likely to Soar after AT&T Breakup," *The Wall Street Journal*, January 21, 1982, p. 41.

(10) "Telecommunications . . . ," *Businessweek*, p. 62.

(11) Ibid.

(12) *The Wall Street Journal*, February 24, 1982. p. 52.

(13) Ibid.

(14) "New [East] Telephone Seeks Boost," *The Wall Street Journal*, May 18, 1982, p. 42.

(15) "New [East] Telephone Applies for Rate Increase in Maine," *The Wall Street Journal*, May 5, 1982, p. 21.

(16) "Telecommunications . . . ," *Businessweek*, p. 70.

(17) John Paul Jon, "Semiannual PBX Product Performance Report," June 1982 edition.

(18) John F. Malone.

(19) John Paul Jon.

(20) Ibid.

(21) Philip Kotler, *Marketing Management: Fourth Edition* (Prentice-Hall, Inc., Englewood Cliffs, New Jersey, 1980) , p. 652–657.

(22) Product Manager—PBX Marketing, New [East] Telephone.

(23) "AT&T adds Features to its Dimension Line of PBX Equipment," *The Wall Street Journal,* May 19, 1982, p. 40.

(24) Elizabeth Ellison and Robert Nersesian, "The Small Business Branch,"*Focus,* August, 1982, p. 6.

(25) "New Perspectives in Productivity," *Focus,* June, 1982, p. 3.

(26) Ibid., p. 5.

(27) Ibid., p. 6.

(28) Ibid,. p. 4.

(29) John Paul Jon.

(30) Ibid.

(31) Product Manager—PBX Marketing.

(32) John F. Malone.

SOURCES

"AT&T Anti-trust Settlement." *The Wall Street Journal.* January 11, 1982. p. 1.

BLAIN, RAY.   "Telephones join Computers to Usher in Office of the Future." *Telephony, The Journal of Telecommunications since 1901.* July 5, 1982.

ELLISON, ELIZABETH and ROBERT NERSESIAN,   "The Small Business Branch." *Focus.* August 1982.

"Heads of AT&T Local Units Tell Congress Divestiture Won't Hurt Their Operations." *The Wall Street Journal.* February 24, 1982. p. 52.

JACOBS, SANFORD B.   "Business Phone Rates Likely to Soar after AT&T Breakup." *The Wall Street Journal.* January 21, 1982, p. 41.

JON, JOHN PAUL.   "Semiannual PBX Product Performance Report," June 1982 edition.

KOTLER, PHILIP.   *Marketing Management: Fourth Edition.* Prentice-Hall Inc. Englewood Cliffs, New Jersey. 1980.

MALONE, JOHN F.   *Telephone Engineering and Management.* Harcourt Brace Jovanovich Publications: Geneva, Ill. March 15, 1982.

"New [East] Telephone Applies for Rate Increase in Maine." *The Wall Street Journal.* May 5, 1982. p. 21.

"New [East] Telephone Seeks Boost." *The Wall Street Journal.* May 18, 1982. p. 42

"New Perspectives in Productivity." *Focus.* June, 1982.

Product Manager-PBX Marketing, New [East] Telephone.

ROSE, JOE.   Of XYZ Planning Institute. New York, New York.

"Telecommunications-Everbody's Favorite Growth Business, The Battle for a Piece of the Action." *Businessweek.* October 11, 1982, p. 60.

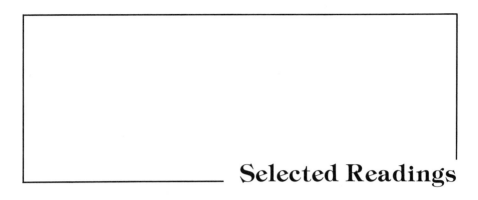

Selected Readings

GENERAL SOURCES ON BUSINESS AND PROFESSIONAL WRITING STYLE, EDITING, AND REVISION

BATES, JEFFERSON D. *Writing with Precision.* Washington DC: Acropolis Books, 1978.

DIGAETANI, JOHN L., JANE DIGAETANI, and EARL HARBERT. *Writing Out Loud: A Self-Help Guide to Clear Business Writing.* Homewood, IL: Dow Jones-Irwin, 1983.

ELBOW, PETER. *Writing with Power.* New York: Oxford University Press, 1981.

LANHAM, RICHARD A. *Revising Business Prose.* New York: Scribners, 1981.

MUNTER, MARY. *Guide to Managerial Communication.* Englewood Cliffs, NJ: Prentice-Hall, 1982.

O'HAYRE, JOHN. *Gobbledygook Has Gotta Go.* Washington, DC: General Printing Office, 1980.

ROMAN, KENNETH, and JOEL RAPHEAELSON. *Writing That Works.* New York: Harper and Row, 1981.

Simply Stated (monthly). Washington, DC: Document Design Center.

STRUNK, WILLIAM, JR., and E. B. WHITE. *The Elements of Style,* 3rd ed, New York: Macmillan, 1979.

VAN BUREN, ROBERT, and MARY FRAN BUELER. *The Levels of Edit,* 2nd ed. Pasadena, CA: Jet Propulsion Laboratory, 1980.

ZINSSER, WILLIAM. *On Writing Well: An Informal Guide to Non-Fiction,* 2nd ed. New York: Harper and Row, 1980.

TEXTS, WORKSHOPS, AND GUIDES FOR PRACTICAL WRITING

BRILL, LAURA. *Business Writing Quick and Easy.* New York: AMACOM, 1981.

DEMARE, GEORGE. *Communicating at the Top.* New York: Wiley, 1979.

Ewing, David. *Writing for Results.* 2nd ed. New York: Wiley, 1979.

Houp, Kenneth, and Thomas Pearsall. *Reporting Technical Information.* 4th ed. New York: Macmillan, 1980.

Internal Revenue Service, *Effective Writing: A Workshop Course.* Washington, DC: IRS, 1975.

Environmental Protection Agency, *Be a Better Writer.* Washington, DC: EPA, 1982.

General Accounting Office. *From Auditing To Editing.* Washington, DC: Government Printing Office, 1976.

Janis, J. Harold. *Writing and Communication in Business.* 3rd ed. New York: Macmillan, 1978.

Lesikar, Raymond U. *Basic Business Communication.* Homewood, IL: Irwin, 1979.

Lord, William and Jessaman Dave. *Functional Business Communication.* 3rd ed. Englewood Cliffs, NJ: Prentice-Hall, 1983.

Malickson, David L. and John W. Nabon. *Advertising-How to Write the Kind that Works,* Rev. Ed. New York: Scribner's, 1982.

Mills, Gordon H., and John A. Walter. *Technical Writing.* 4th ed. New York: Holt, Rinehart, and Winston, 1978.

Oliu, Walter, Charles Brusaw, and Gerard Alred. *Writing That Works.* New York: St. Martins, 1980.

Sigband, Norman, and David N. Bateman. *Communicating in Business.* Glenview, IL: Scott, Foresman, 1981.

Shulman, Joel J. *How to Get Published in Business/Professional Journals.* New York: AMACOM, 1980.

United States Air Force Effective Writing Course. Washington, DC: Department of Defense, 1980.

Weis, Allen. *Write What You Mean.* New York: AMACOM, 1977.

CORRESPONDENCE AND REPORTS

American National Standard for Writing Abstracts. New York: American National Standards Institute, 1979.

Cohen, Randy. *Modest Proposals.* New York: St. Martins, 1981 (satirical letters).

Correspondence Management (Records Management Handbook Series). Washington, DC: GSA, 1972.

Form and Guide Letters. Washington, DC: GSA, 1975.

Gallagher, William. *Writing the Business and Technical Report.* Boston: CBI, 1981.

Linton, Calvin D. *Effective Revenue Writing 2.* Washington, DC: IRS, 1976.

Lesikar, Raymond. *Report Writing for Business.* Homewood, IL: Irwin, 1973.

Lesley, Philip. *Public Relations Handbook.* Englewood Cliffs, NJ: Prentice-Hall, 1978.

Nygren, William. *Business Forms Management.* New York: AMACOM, 1980.

Plain Letters. Washington, DC: GSA, 1973.

Rogers, Raymond A. *How to Report Research and Development Findings to Management.* New York: Pilot Books, 1973.

Swenson, Dan. *Business Reporting: A Management Tool.* Chicago: SRA, 1983.

U.S. Civil Service Bureau of Training. *Writing Effective Letters.* Washington DC; GPO, nd.

WILKINSON, C. W., PETER B. CLARKE, and DOROTHY C. WILKINSON. *Communicating through Letters and Reports*. 8th ed. Homewood, IL: Irwin, 1983.

PROPOSALS

CRAWFORD, JACK, and KATHY KIELSMEIR. *Proposal Writing*. Corvallis, OR: Oregon State Press, 1971.

HALL, MARY. *Developing Skills in Proposal Writing*. Corvallis, OR: Orgegon State Press, 1971.

HILLMAN, HOWARD. *The Art of Winning Corporate Grants*. New York: Vanguard Press, Inc., 1980.

————. *The Art of Winning Foundation Grants*. New York: Vanguard Press, 1975.

————. *The Art of Winning Government Grants*. New York: Vanguard Press, 1977.

HOLTZ, HERMAN. *Government Contracts: Proposalmanship and Winning Strategies*. New York: Plenum Press, 1979.

————. *The $100 Billion Market: How to Do Business with the U.S. Government*. New York: AMACOM, 1980.

———— and TERRY SCHMIDT, *The Winning Proposal: How to Write It*. New York: McGraw-Hill, 1981.

IEEE Transactions on Professional Communication. June 1983 (special proposal issue).

JORDAN, STELLA. *Handbook of Technical Writing Practice*. New York: Wiley, 1971. (Section devoted to proposals.)

LEFFERTS, ROBERT. *Getting a Grant: How to Write Successful Grant Proposals*. Englewood Cliffs, NJ: Prentice-Hall, 1978.

Proposals and Their Preparation, Anthology Series. Washington, DC: Society for Technical Communication, 1973.

WHALEN, TIM. *Preparing Contract-Winning Proposals and Feasibility Studies*. New York: Pilot Books, 1982.

GRAPHICS, ORAL PRESENTATION, OFFICE AUTOMATION

Graphics

American National Standard for the Preparation of Scientific Papers for Written or Oral Presentation. New York: American National Standards Institute, 1979.

Illustrations for Publication and Projection. New York: The American Society of Mechanical Engineers, 1979.

International Paper Company. *Pocket Pal*. New York: International Paper.

LEE, MARSHALL. *Bookmaking*. New York: Harcourt Brace Jovanovich, 1980.

MAGON, GEORGE, *Using Technical Art: An Industry Guide*. New York: Wiley, 1970.

The Mead Paper Company. *Short Course in the Graphic Arts*. Dayton, OH: Mead Paper Company.

S. D. WARREN COMPANY *Paper Surface Comparison*. Boston: Scott Paper Company, nd.

Oral Presentations

Don't Talk . . . Communicate. Washington, DC: Department of the Navy, 1980.

DYER, FREDERICK. *Executives' Guide to Effective Speaking and Writing.* Englewood Cliffs NJ: Prentice-Hall, 1962.

HAAS, KENNETH and H. Q. PACKER. *Presentations and Use of Audio-Visual Aids.* Englewood Cliffs, NJ: Prentice Hall.

HOWELL, W. S., and E. G. Bergmann, *Presentational Speaking for Business and Professions.* New York: Harper and Row, 1971.

MEUSE, LEONARD F. *Mastering the Business and Technical Presentation.* Boston: CBI, 1980.

STARR, DOUGLAS. *How to Handle Speech Writing Assignments.* New York: Pilot Books, 1978.

Office Automation

IBM Displaywriter System General Information Manual. New York: IBM, 1981.

MULLINS, CAROLYN, J., and THOMAS W. WEST. *The Office Automation Primer: Harnessing Information Technologies for Greater Productivity.* Englewood Cliffs, NJ: Prentice-Hall, 1982.

OA, a special quarterly supplement to *Computerworld.*

Wang Office Systems Executive Introduction. Lowell, MA.: Wang Laboratories, 1979.

RÉSUMÉS, MERCHANDISING YOUR SKILLS, INTERVIEWS

BOLLES, RICHARD, N. *What Color Is Your Parachute? A Practical Manual for Job-Hunters and Career Changers.* Berkelley, CA: Ten Speed Press (frequently revised).

College Placement Annual. Bethlehem, PA: College Placement Council, Inc. (revisions for the Fall).

GERBERG, ROBERT JAMESON. *The Professional Job Changing System.* Parsippany, NJ: Performance Dynamics (updated annually).

NEW YORK STATE DEPARTMENT OF LABOR. *Guide to Preparing a Résumé.* Albany, NY: New York State Office of Public Information (updated frequently).

U.S. Department of Labor. *Merchandising Your Job Talents.* Washington, DC: GPO, 1980.

HANDBOOKS OF GRAMMAR AND STYLE, REFERENCE GUIDES

ALRED, GERARD. *Business and Technical Writing* (Bibliography). Metuchen, NJ: Scarecrow Press, 1981.

BROOKS, CLEANTH, and ROBERT PENN WARREN *Modern Rhetoric,* 3rd ed. New York: Harcourt Brace Jovanovich, 1970.

The Chicago Manual of Style, 13th ed. Chicago: University of Chicago Press, 1982.

CORBETT, EDWARD. *Classical Rhetoric for the Modern Student,* 2nd ed. New York: Oxford University Press, 1971.

DANIELLS, LORNA, M. *Business Reference Sources.* Boston: Harvard Business School, 1979.

EMORY, C. WILLIAM. *Business Research Methods,* rev. ed. Homewood, IL: Irwin, 1980.

Government Printing Office Style Manual. Washington, DC.

IRMSCHER, WILLIAM E. *The Holt Guide to English.* New York: Holt, Rinehart, and Winston, 1972.

KRESS, GEORGE. *The Business Research Process.* Ft. Collins, Colorado: Kandid Publications, 1974.

MLA Handbook for Writing Research Papers, Theses, and Dissertations. New York: Modern Language Association, 1977.

The New York Times Style Manual. New York: The New York Times Co.

ONG, WALTER J. *Rhetoric, Romance, and Technology: Studies in the Interaction of Expression and Culture.* Ithaca, NY: Cornell University Press: 1971.

WHITE, JANE F. and PATTY G. CAMPBELL. *Abstracts of Studies in Business Communication 1900–1970.* Urbana, IL: ABCA, nd.

Index

Abstraction 36–42
  dangers of 40–42
  levels of 37–39
Abstracts 211
*Abstracts of Studies in Business
  Communication* 212
Active voice 26–29
  advantages of 28–29
  letters and 97
  memos and 79
Advertising Age 214
Ambiguity 34–36
American Mathematical Society 296
*American National Standards, Bibliog-
  raphic References* 234
Analogy 62, 204
Android, John 211
And/or, discussion of usage 18–19
Application letters 302–07; 316, 318,
  323–25
**Applied Science and Technology
  Index** 212
Argument 154
  developing ideas and 206–10
  patterns of 209–10
Arrangement 134–39
  analysis and 137–38
  comparison and 137
  contrast and 138
  dramatic 135
  importance and position 139
  reason and 138–39
  space and 136–37
  time and 135–36

Asset list, résumés and 304–06
Audience analysis 1–2, 4–6, 16
  awareness of 5
  gathering information for 6
  letters and 94–96
  meeting readers' needs and 15
  memos and 79, 82
  placement of information 139
  proposals and 194–96
  questions for 5
  readers' needs and expectations 4–5
  short reports and 145–46
Automated equipment 290–300
  advanced functions of 291–93
  advantages for the writer 293–96
  basic functions of 290–91
  communication and 292
  dictionary and 292
  drawbacks of 296–98
  formatting 295–96
  memory and 290, 294–95
  security and 290
  technical problems 298–99
Ayer Directory of Publications 213
Asian Wall Street Journal 214

Baker, Sheridan 205
Barriers to automated writing 297–
  98
Barron's 214
Bibliographies 211
"Blind" ads for employment 306–07,
  318

Boiler-plate   294
Bolles, Richard N   323
Books in Print   211
Brainstorming   1–2, 7–13, 120, 144,
    148–50, 327
  determining subject and   148–50
  letters and   93
  storyboarding and   150, 176–79
  types of,   148–49
Brooks, Cleanth, and Robert Penn
    Warren   xiv, 125, 156, 158
Bullet letters   303
Business and Economic Books and Se-
    rials in Print   211
Business letters   91–119
  boiler-plate   104 (*see also* Chapter 12)
  bullet   97
  form   103–04
  hand-written   101–02
  negative   104–06
  parts of   106–111
  routine   102–03
  unique   98–101
Business Periodicals Index   212
Business Reference Sources   211
Business Week   214

Camera-ready document   293
Captioning graphics   249, 253
Captions   253
Cause and effect   62
Checklist(s)
  editing   225, Table 10–1
  get it written: a strategy   201
  meeting preparation   272
  memo situations   74
  when to write, phone, meet   68
Citation, difference between note and
    bibliography   235
Citing information from uncommon
    sources   233, Table 10–2
Classification   62
Classifying   133
Coherence   157
Colons   329
Commas   329
Commerce Business Daily   171
Common ground   121
Communication, barriers to   44–45
Communications models   43–44
Comparison   62
Computerworld   214, 286
Conclusions   147, 156, 158–59, 161

Conciseness   29
Concreteness in memos   78
Concrete language   36–42
Connotation   16–19, 23–24
Consultants and Consulting Organiza-
    tions Directory   213
Contents, table of   184–86
Context, and writing process   2
Contrast   62
Cooling off   1–2, 82, 143, 146
Coordination   57–58
Correspondence, selected reading   353–
    354 (*see also* business letters)
Courtesy   52–3, (*see also* Tone)
Cover letter   303–10
  attitude and   303
  instead of a résumé   303–04
Curbstone analysis and interviews   321–
    22
Cut-and-paste   293

Dangerous method, brainstorming   7
Daniells, Lorna M.   211
Deadlining, brainstorming technique   1,
    148
Debriefings   195–96
Deductive reasoning   208
Definition   19–23, 62, 120–24
  analysis of   21
  classical formula for   21
  connotation and denotation and   23
  context and   21–22
  extended   122
  formula for   21, 120
  glossary   20
  jargon   124
  LIFO defined   122–23
  placement of   20–21, 123–24
  types of words that require   20
Denotation   16–19, 23–24
Description   126–32, 154–55
  format of   128
  requirements of   128
  three kinds of seeing   127–28
Dialoging as brainstorming technique   1,
    149
Dickens, Charles   286
Diction (*see* word choice)
Dictionary, text processing and   292
Dictionary for Accountants   213
Dictionary of Business and
    Economics   213
Dictionaries   213

Directions   132–33
Directories   213–14
Directory of Business and Financial Services   213
Dissertation Abstracts   212
Documentation   226, 232–36
  bibliography   235–36
  citing information from uncommon sources   233, Table 10–2
  notes   226, 232–35
Dow Jones News/Retrieval   213
Dr. Fox Hypothesis   32
Dunn and Bradstreet Reference Book of Corporate Management   215
Dunn's Review   214
Dunn's Census of American Business   214

Economic indicators   216
Editing and computer   295
Editing and revising   1–2, 223–32
  cut-and-paste   224
  flagged constructions   224, Table 10–1
  Peter Elbow's advice   224–25
  putting yourself in reader's shoes   223
  sample revision, Figures 10–5—10–7   226–31
Editing, selected readings   352
Effective Writing, IRS Workshop   23
Einstein, Albert   36
Elbow, Peter   12n, 224
Electronic data bases   212–13
Electronic file cabinet   295
Electronic mail   292
Elements of Style   16
Eliot, T. S.   4, 300
Emerson, Ralph Waldo   36
Emphasis   157–58
  flat statement   157
  position   158
  proportion   158
Employment opportunities, sources of   304
Encyclopedia of Associations   214
Encyclopedia of Business Information Sources   211
Energyline   213
Enhancement in graphics   249
ERIC   213
Etiquette of writing   296
Evaluation criteria   189–94
Evaluating memos   76–83

Exposition   153–54
Excluding irrelevant information   76

F & S Index   212
Fallacies   209
Financial Times   214
First things first   76–78, 81, 153
Flags for Rewrite Checklist   225, Table 10–1
Flipchart   281–82
Flying Pyramid   204–05, Figure 10–2
Focus   203
Focus in graphics   249
Forbes   214
Forbes, Malcolm   91
Forced Associations, demonstration of brainstorming   10–11
Form letter, computer generated   294
Fortune   214
Four-S Formula for business letters   92
Foundation Directory   214
Free writing   1, 8–10
  demonstration of   8–10
  forget-remembering and   10
Friedenreich, Ken   xvi, 59
Functional analysis   203, 327

Gatten, Tom   xvi
Getting to the point   1
Government Contracts   179
Government documents and publications   215–16
GPO Style Manual   234
Graphics   155, 249–62
  captioning   249, 253
  computers and   295–96
  enhancement   245–62, 284–85
  examples of   256–62
  focus and   249
  placement of   251–52
  positive qualities of   247–49
  types of   249–51 Table 11–1
  selected readings   354–55
Guide to the Preparation of a Résumé   304, 321
Guide to Reference Books   211
Guide to U.S. Government Publications   211
The Gulf Publishing Company Dictionary of Business and Science   213
Gupta, Maresh   xvi

Handbooks of grammar, selected
readings   355–56
Handouts   281–82
Hawthorne, Nathaniel   19
Headings   222–23, 240, 255
Hemingway, Ernest   147
Holtz, Herman   179

Illustration, qualities of   251–52
Impact of message   51
*Index of Economic Articles in Journals
and Collective Volumes*   212
Indexes   211–12
Indexing   292
Inductive reasoning   134, 154, 208
Industry Week   214
Inflation, language   31–36 (*see also*
Language Inflation)
INFORM   213
Information, accumulating   210–23
primary sources of   217–23
standard research materials   210–16
Instructions   132–33
Interviews   218–19, 312, 314–22
"curbstone analysis"   321–22
employment   312–22
follow-up   322
gathering information through   218–19
listening and   319
meeting face to face   316
selected readings   355
Interviewers, types of   317–19
*International Dictionary of
Management*   213
Interpretation   134
Interrogation   218–22
interviews and   218–19
letters and   218
questionnaires   219–22
Introductions   160–162
traditional openings   160–61
unusual openings   161–62
Invention   7, 12 (*see also* Brainstorming)
Inverted pyramid   81–82, 204, Figure
5–1, Figure 10–1
Irmscher, William   56
"Is" sickness   25–26

Jane's Major Companies of Europe   214
Jargon   124–26
positive uses of   125–26
Journal   326 (*see also* Brainstorming)

Joyce, James   22
Juillard, Louis   196

Keyhole, Figure 10–3   205–06
Kilcup, Arthur   10
Korzybski, Alexander   37

Language inflation   3–4, 31–36
Language Inflator   32–34
Lanham, Richard   24
Letters
application   302–07, 316, 318,
323–25
examples of   13–14, 94–96, 98–99,
100–03, 105–07, 113–19, 122–23
interrogation through   218
parts of   106–11
Letter of transmittal   174
LEXIS   213
LIFO   122–123
Limiting a topic   6–7
Logic, developing ideas and   206–10

Main idea "up-front"   152–53, 157–58
representation of, Figure 8–1
emphasis and   157–58
Management Contents   213
Management style and purpose   4
Manual of Style (Chicago)   234
MEDLARS   213
Meetings   68
Memos   69–90, 292
checklist   74
electronic mail and   292
evaluation of   76–83
hand-written   73
posterior coverage and   72
tone in   73–81
when to use   70–73
who receives   72–73
Memory, computers and   290–91
Million Dollar Directory   213
Misleading graphic, example of   248–49
Misspelled words   332–34
letters and   303
Mistakes, admitting in letters   96–97
MLA Handbook for Writers   226,
234–35
Modern Rhetoric   xiv 125–56, 158
Modifying Structures   60–61
*Monthly Catalog of Government
Publications*   216

Narration   154
National Directory of Newsletters and Reporting Services   214
Naturalness   53
The New York Times Information Bank   213
The New York Times Index   211–13
Notes   226, 232–35 (*see also* Documentation and Definition)
NTIS (National Technical Information Service)   212

OA (Office Automation)   286
Obituary résumé   310
"Official Style"   24
Office automation   286–301
  selected readings   355
Office information, types of   286–87
Ong, Walter   8
Optical character reader (OCR)   288
Oral presentation   262–85
  approaches to   269–70
  checklist for   272
  closing   278
  evaluation   279–86
  the hook and   271
  questions and   278–80
  rating an   280
  review of   280
  selected readings   355
  storyboards and   271–76
  types of   262–63, 267, 269
  visual aids and   281–84
Organization   134–39, 152–56 (*see also* Arrangement)
  reports and   204–08
Orwell, George   18, 39, 124
  on jargon,   124
Outlines   172–74, 202

P.A.I.S. (Public Affairs Information Service)   212
Paperless office   299–300
"Paperless" writing   299–301
Paragraphing, structure and   61
Passive voice   26–29
Performance and presentations   277–78
Periods   329
Personnel Management Abstracts   212
Persuasion (*see* Argument)
Photocomposition   292

Placement of information,   139, 152–53
  figure 8–1   153 (*See also* Arrangement, Argument, Emphasis, Organization)
Plain letters   92
Poe, Edgar Allen   19, 126
Positive expression   49–51
  in memos   78
Posterior coverage   72
"Politics and the English Language"   18, 37, 124
"Power of the Printed Word"   91
Plurals   330
Practical Stylist   205–206
The Presence of the Word   8
Prewriting (*See* Brainstorming)
Process of writing, essential steps   1, 82
Professional format for résumés   311–12
Proofreading   236–38
Proposals   170–99
  audience analysis and   194
  debriefing   195
  evaluation of   189–93, 196–98
  introduction   174
  loser   196–97
  parts of   174–75
  response index   183
  responsiveness   171–75, 179, 182
  sections of   184, 187–88
  selected readings   354
  solicited   176
  storyboards and   176–79
  unsolicited   176
  winner   197–98
Psychological Abstracts   212–13
PTS (Predicasts Terminal Systems)   213
Public Affairs Information Service Bulletin (P.A.I.S.)   212
Public Relations Society of America   216
Publication Manual   234
Publisher's Weekly   214
Punctuation   329–30
Purpose   1–2, 4, 13, 16, 79, 82, 93–95, 145, 203–04
  business letters and   93–95, 145
  management styles and   4
  memos and   79, 82
  questions to determine   13
  reports and   203–204

Pyramid
  flying   204–205, Figure 10–2
  inverted,   81–82, 204 Figure 5–1,
    Figure 10–1

Qualities of effective writing   15
Quantitative language   36–42 (*See also*
    Concrete language)
Questionnaires   219–22
  check response   221
  multiple choice   220
  short comment   221
  scaled response   221
  yes/no   220
Questions and interviews   319–22
Quotation marks   330

Rand McNalley Commercial Atlas and
    Marketing Guide   214
Reader's Guide to Periodical
    Literature   212
Recommendations   147, 156, 158–59,
    161
Reference guides, selected
    readings   355–56
RFP (Request for Proposal)   171–174
  extract from   180–82
  outine   172
Reports   150–60, 200–45
  budgeting time and   200–202
  evaluating   238–239
  long   200–43
  organizational patterns   204–08
  outlines   202
  research example Appendix B
    341–51
  selected readings   353–54
  short   150–60
  strategy for writing   200–01
Report Organizational Flowchart, Figure
    10–4   205–07
Request for Proposal   171–74
Response index   174
  example of   183
Résumé   302, 307–18
  education and   304, 308–09
  employment history and   304, 308
  examples of   313–18
  formats   310–312
  personal attributes and   305–306
  screeners and   307–308
  selected readings   355

Résumé, *(cont'd)*
  writing   308–312
Revision   159–60, 223–231 Table 10–1
Rewrite Flags, Table 10–1 Checklist
Rhetoric   xiv
Rogers, Tom   xvi

Sample reports   341–51
Script and presentations   272 (*See also*
    Storyboards)
Security, computer   290
Selected readings   352–55
Semicolons   329
Securities and Exchange Commission
    10K and 10Q reports   216
Sentence patterns   56–58
Sentence structure   56–58
Sheehy, Eugene. *Guide to Reference*
    Books   211
Short report   150–69
  audience analysis and   151
  ending a   158–59
  examples for discussion   164–69,
    Appendix B
  goals   150
  organizing   152–56
  planning   152–56
  scope of   151
Smith, Karen E.   341
Social Science Index   212
Sociological Abstracts   212–13
SOW(Statement of Work)   171
Spelling   330–40
  exercise   331–32
  misspelled words   332–34
  sound-alikes   334–40
  suggestions for improvement   330–31
Spicer, George   xvi
Standard Rate and Data Service   214
Standard Periodical Directory   214
Standard and Poor's Register of Corpo-
    rate Directors and Executives   214
Statement of Work (SOW)   171
Statistical Abstract of the United
    States   216
Storyboards   1, 176–79, 271–76
  as a brainstorming technique   1
  example   177, Figure 9–1, 273
  oral presentations and   271–76
Strategy for writing   93–98, 144–48
  business letters   93–98
  conclusions, recommendations   147
  cooling off   147

discipline   146
managing time   143–45
Strunk, William, Jr.   16, 36
Styles, selected readings   352, 355–56
Subheadings   222–23, 240, 255
Subordination   57–58
Suffixes   330
Summary   160–61
Syllogism   208

Tact   52–53 (*See also* Positive expression)
Tautology   17–18
Tebo, Bill   xvi, 174
Telephone   68
Text-editing   286–88, 290, 293
Thomas Register of American Manufacturers   214
Thompson, John   xvi, 7–8, 147
Time management and writing   149–50
Tone   45–49 (*See also* Audience analysis)
   attitude and   45
   business letters and   94–97
   professional ability and   48–49
   purpose and   46–48
Transformational grammar   56
Transitions   58–60
Tuer, David   213

United States Government Manual   214
Unity   156–157
Unwritten messages in writing   94–97 (*See also* Audience analysis, Tone, Positive expression)
"User-friendly" computers   287–88
Users of automated writing equipment   288

Variety (*See* Sentence patterns and structure)
VDT (Video Display Terminal)   288–89, 291
Verbs, active   24–26
   strong writing and   224
Viewgraphs   272–81
Visual aids and presentations   281–283, Table 11–2
The Visual Arts as Human Experience   127–128

Warren, Robert Penn   xiv, 125, 156, 158
Wasserman, Paul   211
Weismann, Donald   127–28
What Color Is Your Parachute?   323
White, E. B.   16, 36
White space as graphic aid   222–23, 240, 255
Who Owns Whom: North American Edition   214
Who's Who in America   215
Who's Who in Finance and Industry   215
Word choice   16–19
Words that sound alike   334–340
Writer's Digest   296
Writing. process of   1, 143 (*See also* Arrangement, Audience analysis, Brainstorming, Cooling off, Editing, Organization, Purpose, Revision)
   strategy for putting ideas to paper   143–47
   ten steps of   1, 143, 146
   qualities of effective   15–16
Writing that Works   306
Writing with Power   12n, 224